STRUCTURES IN SCIENCE

Heuristic Patterns Based on Cognitive Structures

An Advanced Textbook in Neo-Classical Philosophy of Science

SYNTHESE LIBRARY

STUDIES IN EPISTEMOLOGY,

LOGIC, METHODOLOGY, AND PHILOSOPHY OF SCIENCE

VOLUME 301

STRUCTURES IN SCIENCE

Heuristic Patterns Based on Cognitive Structures

An Advanced Textbook in Neo-Classical Philosophy of Science

by

THEO A.F. KUIPERS
University of Groningen, The Netherlands

KLUWER ACADEMIC PUBLISHERS
DORDRECHT/BOSTON/LONDON

A C.I.P. Catalogue record for this book is available from the Library of Congress.

ISBN 0–7923–7117–8

Published by Kluwer Academic Publishers,
P.O. Box 17, 3300 AA Dordrecht, The Netherlands.

Sold and distributed in North, Central and South America
by Kluwer Academic Publishers,
101 Philip Drive, Norewell, MA 02018, U.S.A.

In all other countries, sold and distributed
by Kluwer Academic Publishers, Distribution Center,
P.O. Box 322, 3300 AH Dordrecht, The Netherlands

Printed on acid-free paper

Printed and bound in Great Britain by Antony Rowe Limited.

TABLE OF CONTENTS

FOREWORD

Although there is an abundance of highly specialized monographs, learned collections and general introductions to the philosophy of science, only a few synthetic monographs and advanced textbooks have appeared in the last 25 years. The philosophy of science seems to have lost its self-confidence. The main reason for such a loss is that the traditional analytical, logical-empiricist approaches to the philosophy of science had to make a number of concessions, especially in response to the work of Popper, Kuhn and Lakatos. With *Structures in Science* I intend to present both a synthetic monograph and an advanced textbook that accommodates and integrates the insight of these philosophers, in what I like to call a neo-classical approach.

The resulting monograph elaborates several important topics from one or more perspectives, by distinguishing various kinds of research programs, and various ways of explaining and reducing laws and concepts, and by summarizing an integrated explication (presented in *From Instrumentalism to Constructive Realism*, ICR) of the notions of confirmation, empirical progress and truth approximation. Finally, special attention is paid to design research programs, computational philosophy of science, the structuralist approach to theories, and research ethics. As this book is also an advanced textbook, it provides a detailed discussion of a number of the main topics in the philosophy of science aimed at advanced students in philosophy and in the natural, technological, social, behavioral, cognitive and neuro-sciences. To receive maximum benefit from the present work, familiarity with one introductory work in which the above indicated 'two parties' are given some serious attention is advisable, though not strictly necessary. Moreover, some elementary knowledge of logic and (naive) set theory is presupposed.

In view of the many introductions, monographs, and journals, present-day philosophy of science seems to have various appearances, ranging from superficial to deeply philosophical, from abstract to concrete, from verbal to formal, from descriptive to normative, from descriptive to explanatory, from general to specific, from expository to polemic. With *Structures in Science* I intend to steer a neutral middle course through all these often contentious viewpoints.

However, there are two other, related factors about which the present book will formulate a firm position. First, the philosophy of science, which generally bears only slight resemblance to an empirical science, should in my opinion aspire to become a genuine empirical science, or, more precisely, an empirical *meta-science*, with its own meta-theories. Second, philosophers of science usually are rather skeptical about the use-value of their results, despite the fact that these results are frequently presented in a rather prescriptive style. In my view the philosophy of science should explicitly aim at one or more use-values, not by strengthening its prescriptive claims, but by aiming at various kinds of *heuristic use-value*.

In *Structures in Science* I have tried to show that the philosophy of science may become a meta-science with a heuristic use-value. More specifically, on the basis of an analogy with the so-called social studies of science, one may conceive this book as the product of various cognitive studies of science, looking for cognitive structures that underlie scientific inquiry and that generate multi-purpose heuristic patterns. I will now first characterize and illustrate such cognitive studies, and then their possible use-value.

To begin with, it is important to note that 'cognitive' in 'cognitive studies/structures/patterns' does not refer to the notion of 'cognitive' used in mainstream cognitive science, where it primarily refers to the 'internal/psychological processes' of knowledge formation and processing. Of course, science is a kind of cognition, and cognitive science needs to study scientific knowledge formation, representation and processing as well. Therefore, it seems apt to call the relevant studies 'cognitive studies of science' and the findings 'cognitive structures or patterns'.

Cognitive studies of science may be characterized more specifically as follows: quasi-empirical studies of cognitive aspects of scientific knowledge, including its methods and its development. Empirical studies may of course be of a descriptive and explanatory nature. The qualification 'quasi-' (empirical) is used to leave room for normative problems and aims. In consequence, the research is frequently guided by one or more heuristic-normative points of view. For instance, the intention to apply patterns of successful cases to areas of research that so far don't show much progress.

The core of the program of cognitive studies of science is the idea that there is system underlying knowledge and knowledge production, and hence that theory formation about it is possible in principle. It is also plausible that such theorizing concerns part and aspect systems and hence different patterns, which may or may not be easily interrelated and harmonized.

In this book several cognitive structures will receive extensive attention; the most important ones are mentioned in the following survey of the book:

Part I, entitled 'Units of scientific knowledge and its acquisition', discusses:

- Four ideal types of research program, with some common features in structure, development, and strategies.
- The distinction between observational laws and proper theories based on an explication of the nature of theory-ladenness of terms, and the further consequences of this distinction for the hierarchy of knowledge and for the different relations between theory and observation, and giving rise to the main epistemological positions with respect to proper theories.[1]

Part II, entitled 'Patterns of explanation and description', presents:

- A five-step decomposition model of the explanation and reduction of laws by theories.
- A decomposition model of the common structure and diverging meaning of explanation and description by intentional, functional, and causal specification.

Part III, entitled 'Structures in interlevel and interfield research', focusses on:

- The decomposition of several forms of concept reduction and correlation, leading to a characterization of reductionistic and holistic research strategies.
- A matrix for the combination of ontological levels and the epistemological styles of causal, functional and intentional explanation and description, a matrix for the resulting four kinds of interfield research, and a model for interstyle and interlevel mind-body research.

Part IV, entitled 'Confirmation and empirical progress', is essentially based on ICR. After a coherent sketch of the nature of, and the relations between, qualitative (deductive) and quantitative (structural and inductive) confirmation, the main methodological lessons from ICR have been included in the present volume, notably:

- 'The evaluation report' of a theory in terms of general successes and (individual) counter-examples, and a systematic survey of the factors complicating theory testing and evaluation.
- The nature of comparative theory evaluation, and the resulting 'rule of success', which prescribes to select the theory, if any, that has so far proven to be the most successful. This enables the characterization of the notion of 'empirical progress', which turns out to be the glue between confirmation and truth approximation, and of the distinction between science and pseudoscience.

Part V, entitled 'Truth, product, and concept approximation', deals with:

- The nature of truth, product and concept approximation, and the relation of truth approximation to 'empirical progress'.
- Design research programs, and the similarities and differences between them and truth approximation (i.e. descriptive or explanatory) programs.

Part VI, entitled 'Capita selecta', provides:

- An overview of computational models of theory generation, evaluation, and revision, and a critical analysis of one of them.
- A review of the refined structure of theories, and their relations.
- A discussion of Merton's norms conceived as 'default-norms'.

More or less concrete and realistic patterns have been found about all these subjects on the basis of case-studies or formal analysis. These findings are partly restatements of views found elsewhere in the literature, partly they involve elaborations of lines of thought instigated by others, and partly they are the result of my own research. The title of Part VI, 'Capita selecta', makes it clear that I make no claim to have exhausted all the possible subjects in this area of study; on the contrary, several very interesting and important subjects are missing. Some of these are indicated in the introduction to Part VI.

Turning to the use-value of cognitive structures, they always involve informative patterns, which seem useful in one way or another. It is instructive to distinguish at least the following five kinds of possible use-value:

(a) cognitive structures may provide the 'null hypothesis of ideal courses of events', which can play a guiding role in the social studies of science; they may at least raise the question of whether certain patterns are favored over others in various psychological, cultural, sociological or economic circumstances, but stronger kinds of social or external influences are also conceivable;

(b) they may clarify or even solve classical problems belonging to abstract philosophy of science; e.g. ICR based explications of the correspondence theory of truth and of dialectical concepts on its truth approximation account;

(c) they may be useful as didactic instruments for writing advanced textbooks, leading to better understanding and remembrance; for forceful general pleas for use, and illustrations of using philosophy of science insights in science teaching, see Duschl (1990) and Matthews (1994);

(d) they may play a heuristic role in research policy and even in science policy, e.g. by stimulating applications of successful patterns of research in stagnating areas of research;

(e) last but not least, they may play a heuristic role in actual research, not only of the standard nature, but also in the increasing field of computational research; in the last respect cognitive structures at least provide a basis for generating new means for a further development of computational philosophy of science.

The last use-value, the heuristic role for research, providing the subtitle of this study, will be worked out a little further. Traditional philosophy of science has frequently been accused of studying ready-made science, i.e. the end products of science, rather than their production. Although it is certainly true that this can give very distorted pictures of science in the making, it is also true that structures recognizable in end products, which were apparently successful, may be used as heuristic means for new but, in one way or another, similar research: they may provide models of what one may be looking for. Or to use the phrase of one of the main forerunners of cognitive psychology and cognitive science, Otto Selz (1924), these structures may provide the 'schematic anticipation' ('schematische Antizipation') of the solution to a scientific problem. Of course, there is no guarantee that solutions to new problems should resemble solutions to old ones. For that reason it is important to stress that cognitive structures cannot provide prescriptive models for new research. At most, they can play the role of heuristic means, which may be rather much. In this respect, I would like to mention the paradigm example of cognitive studies *avant la lettre*, viz. systems theory, restricted to its philosophically modest forms. As is well-known, the main concepts of elementary systems theory can and do play a very useful structuring role in many areas of present-day research.

The example of systems theory also illustrates that heuristic means should not be applied scholastically. Hence, even if the tone of presentation in this book is sometimes rather prescriptive, the reader should keep in mind that a book like this one is only taken seriously when it is used as a free guide for research and other activities.

The title of this book, *Structures in Science*, and its subtitle, *Heuristic patterns based on cognitive structures: An advanced textbook in neo-classical philosophy of science*, have been explained in the foregoing. The specific version of the title alludes to Nagel's *The structure of science* (1961). That book is in some respects outdated, though much less than is generally assumed, as will become clear from the explicit and implicit use I will make of it. To my standards it is not sufficiently analytical. However, its richness of subjects, and its balanced and detailed treatment of them, makes that book in these respects still an ideal that I could only hope to emulate to some extent.

Structures in Science is heavily based on my previous research and publications. Material of the following earlier publications has been used with the kind permission of the publishers:

Onderzoeksprogramma's gebaseerd op een idee, inaugural address, Van Gorcum, Assen, 1989. (Chapter 1).

'Empirische wetten en theorieën' (with H. Zandvoort), *Kennis en Methode*, IX.1, 1985, 49–63. (Chapter 2).

'Reduction of laws and concepts', *Idealization I: General Problems*, J. Brzezinski, F. Coniglione, T. Kuipers and L. Nowak (eds), *Poznan Studies*, Vol. 16, 1990, pp. 241–276. (Chapters 3 and 5).

'Explanation by intentional, functional, and causal specification', in: *Epistemology and History: Humanities as a Philosophical Problem and Jerzy Kmita's Approach to it*, Anna Zeidler-Janiszewska (ed.), *Poznan Studies*, Vol. 47, Rodopi, Amsterdam/Atlanta, 1996, pp. 209–236. (Chapter 4).

'Explicating the falsificationist and the instrumentalist methodology by decomposing the hypothetico-deductive method', in: *Cognitive patterns in science and common sense*, T. Kuipers and A.R. Ruth Mackor (eds.), *Poznan Studies*, Vol. 45, Rodopi, Amsterdam-Atlanta, 1995. pp. 165–186. (Chapter 7).

'The logic of progress in nomological, design and explicative research', in *JFAK: Essays Dedicated to Johan van Benthem on the Occasion of his 50th Birthday*, J. Gerbrandy, M. Marx, M. de Rijke, and Y. Venema (eds.), CD-ROM, Amsterdam University Press, Series Vossiuspers, Amsterdam, ISBN 90 5629 104 1, 1999. (Chapter 9).

'Design research programs and the logic of their development' (with R. Vos and H. Sie), *Erkenntnis*, 37(1), 1992, 37–63. (Chapter 10).

'Computationele Wetenschapsfilosofie', *Algemeen Nederlands Tijdschrift voor Wijsbegeerte*, 85(4), 1993, 346–361. (Section 11.1.).

'On the architecture of computational theory selection', in: R. Casati and G. White (eds.), *Philosophy and the cognitive sciences*, Austrian Ludwig Wittgenstein Society, Kirchberg, 1993, pp. 271–278. (Section 11.2.).

'The refined structure of theories', in: *Idealization VII: Structuralism, Idealization, Approximation*, M. Kuokkanen (ed.), *Poznan Studies*, Vol. 42, Rodopi, Amsterdam, 1994, pp. 3–24. (Chapter 12).

'De integriteit van de wetenschapper', in: *Drieluijk: godsdienst, samenleving, bedrijfsethiek*, E. Kimman, A. Schilder and F. Jacobs) (ed.), *Liber Amicorum voor Henk van Luijk*, Thela-Thesis, Amsterdam, 1999, pp. 99–109.

Besides being based on earlier publications, *Structures in Science* is based at least as much on teaching experience at the University of Groningen, in which provisional versions of the material were applied and revised. The courses were not only intended for students in philosophy (of science) but at least as much for students in the cognitive, natural, and social sciences. Moreover, I have used the material for several years as the basis for a course for all PhD-students in the Groningen graduate school in Behavioral and Cognitive Neurosciences (BCN).

My optimism about the use-values of the presented cognitive structures is mainly based on the results of all kinds of student examinations: written, take-home, paper tasks, and masters theses. I was frequently struck by the fact that students delivered very plausible reconstructions of scientific findings based on one or more of the structures presented in this book. The optimism is also based on some of the work of my former PhD-students, mainly educated in philosophy *and* some empirical discipline. Those who have initiated, co-developed and applied structures in a critical-constructive way include Alexander van den Bosch, Roberto Festa, Lex Guichard, Bert Hamminga, Hinne Hettema, Maarten Janssen, Rick Looijen, Anne Ruth Mackor, Rein Vos, Henk Zandvoort, and Sjoerd Zwart. In almost all cases their work concerns specific (groups of) disciplines. Their contributions, and the related work of others, have already been presented in the 1995 collection of essays *Cognitive patterns in science and common sense: Groningen studies in philosophy of science, logic and epistemology*. However, their work will also frequently play an illustrative role in the present book.

Due to this background and my intentions the book is rich with examples from the natural, cognitive and social sciences, a feature that may make it useful for both established and novice scientists. Given my heuristic intentions for scientists the book is not particularly a book for philosophers, although it is nevertheless worthwhile to place it in a philosophical context. Albeit the book is not historically or naturalistically oriented in the popular sense, it claims to contribute to the meta-science of science, and hence to do justice to empirical science. It presupposes an introduction to the philosophy of science, but it is certainly not an advanced 'debate-book' in the philosophy of science. In the realism-empiricism spectrum it reflects a moderate, non-essentialist, form of realism. Finally, it is semi-formal: it uses some formal means, but no more than necessary.

Besides Nagel, a number of other authors should be mentioned as my main intellectual heros: Carnap, Hempel, Hintikka, Popper, Lakatos, Suppes, and Sneed. In this book I want to show that their work can be reconciled, or more precisely, that a synthesis of some of their best insights is possible. In making this attempt I have profited a lot by freely using the work of the already mentioned former PhD-students and other scholars, notably, Balzer, Bechtel, Burton, Causey, Darden, Kim, Millikan, Niiniluoto, Nowak, Panhuysen, and Thagard.

My resulting perspective on the present-day philosophy of science as a discipline will become quite clear in this book. However, I like to mention Ilkka Niiniluoto's *Critical Scientific Realism* (1999) as, as far as I know, the most learned recent exposition of some of the main themes in the philosophy of science in the form of an advanced debate-book, that is, a critical exposition and assessment of the recent literature, including his own major contribution, viz. *Truthlikeness* of 1987. Despite our major differences regarding the topic of truth approximation, I like to express my affinity to, in particular, his rare type of constructive-critical attitude in the philosophy of science.

Many people have contributed over the years to my writing and teaching. Again my former PhD-students should be mentioned. Their work and their comments have been of incalculable value to me, together with the comments of some other former and present members of the Promotion Club Cognitive Patterns (PCCP), viz. Atocha Aliseda, Renée Dalitz, Michel ter Hark, Ruurik Holm, Alfons Keupink, Erik Krabbe, Jan-Albert van Laar, Jeanne Peijnenburg, Anna Pilatova, Jan-Willem Romeyn, Barbara de Ruyter, Hauke Sie, Gerben Stavenga, Esther Stiekema, Yao-Hua Tan.

I gratefully acknowledge the permission of the co-authors Anne Ruth Mackor, Hauke Sie, Rein Vos and Henk Zandvoort to use parts of earlier joint publications.

Moreover, I would also like to thank many others who have critically commented on the underlying research and previous versions of various chapters: David Atkinson, Wolfgang Balzer, Jeroen Bartels, K. Bärwinkel, Anton van den Beld, Johan van Benthem, Henk Bodewitz, Theo de Boer, Anne Boomsma, A. Borghouts, Roger Cooke, Domenico Costantini, Trudy Dehue, Anton Derksen, Carl Doerbecker, Igor Douven, Douwe Draaisma, Job van Eck, Arthur Fine, Henk Flap, Francisco Flores, Kenneth Gemes, Wouter Gomperts, Carl Hempel, Johannes Heidema, Jaakko Hintikka, M. Hoekstra, Richard Jeffrey, M. Jeuken, Otto Kardaun, Peter Kirschenmann, David Miller, Graham Oddie, Jasper Oosten, Geert Panhuysen, David Pearce, Karl Popper, Theo Potts, Ivo Molenaar, Thomas Mormann, Hans Mooij, Ulises Moulines, Wim Nieuwpoort, Ilkka Niiniluoto, Hans Radder, Hans Reddingius, Menno Rol, Hans Rott, Willem Schaafsma, Heinz-Jürgen Schmidt, Gerhard Schurz, Abner Shimony, Brian Skyrms, Wolfgang Spohn, Sjaak Swart, Paul Thagard, Peter Urbach, Rineke Verbrugge, Philip Vergragt, Gerard Vreeswijk, Andrzej Wisniewski, Arno Wouters, Sandy Zabell, Gerhard Zoubek, and Jan Zytkow.

The opportunity to complete this book was made possible by three important factors. First, I got a sabbatical year from my home university, the University of Groningen, in the Netherlands. I am very grateful to all those who took over part of my courses, notably, Alexander van den Bosch, Michel ter Hark, Erik Krabbe, Jasper Oosten, Gerben Stavenga and Esther Stiekema. Second, I was, after fourteen years, again invited as a fellow for a year at the Dutch work-paradise for scholars, the Netherlands Institute for Advanced Study (NIAS, a Royal Academy Institute in Wassenaar, near Leiden). The support of the staff was in various respects very pleasant and efficient. I am especially grateful to Anne Simpson and Jane Colling for editing my English. Third, and finally, a number of people were willing to read the whole or parts of the manuscript: Wolfgang Balzer, Alexander van den Bosch, Roberto Festa, Lex Guichard, Anne Ruth Mackor, Eefke Meijer, Rick Looijen, Ilkka Niiniluoto, Hans Radder, Andrzej Wisniewski, and Arno Wouters. Of course, the responsibility for any shortcomings must be mine.

Theo A.F. Kuipers, April 2001, Groningen, The Netherlands

PART I

UNITS OF SCIENTIFIC KNOWLEDGE AND KNOWLEDGE ACQUISITION

INTRODUCTION

In the first three parts of the book our primary aim is to show in some detail the products that scientists are after, and the means by which they try to obtain them. Although theories will play a major role in these parts, questions of testing and comparing theories, and prospects for truth approximation are put aside and will be treated in Part IV (they are more extensively discussed in my *From Instrumentalism to Constructive Realism*, henceforth ICR).

In Part I the emphasis will be on patterns in scientific research programs, observational laws, and theories. Part II focusses on standard and non-standard patterns of explanation and description. In Part III we will explore patterns in the reduction of concepts and then we will apply almost all patterns distinguished so far to mind-body research.

In Chapter 1 we will present the more or less generally accepted view, introduced by Kuhn and Lakatos, that the development of scientific research takes place by means of encompassing cognitive units, called research programs. We will distinguish four kinds of programs: descriptive, explanatory, design, and explicative. Explanatory programs will be given the main attention, followed by descriptive programs. In Chapter 10 we will analyze design programs in detail. Finally, this book illustrates a number of explicative programs, while ICR deals with the truth approximation program. After describing the main structural and developmental features of programs and using Dalton's atomic theory program to illustrate them, we will address the strategic lessons that may be drawn. They involve the value of programmatic research as such and the interaction between programs as a result of competition or co-operation. Finally, we will address some strategies for the internal development of programs, in particular, idealization and concretization.

Nagel has stressed the distinction between experimental laws and proper theories, where the latter aim to explain the former by introducing theoretical terms. This 'law-distinction' is one of the main dynamic factors in the development of explanatory programs, and the interaction between descriptive and explanatory programs. It will be dealt with in Chapter 2. Since there does not seem to be anything like theory-free or theory-neutral observation terms, the law-distinction is explicated on the basis of a theory-relative explication of theoretical and observation terms. The ideal gas law will turn out to be an instructive example, viz. *prima facie* a proper theory according to the law-distinction, but on closer inspection (presented in an appendix) an observational law. The analysis suggests a disentanglement of the so-called theory-ladenness of observations. In particular, an observation may not only be laden by a theory, but even if unladen by it, observation may nevertheless be relevant to a theory, and even guided by it. An appendix will show how a similar explication of the main points can be obtained by starting from the so-called empirical basis; this possibility makes it even more surprising that Popper did not pay attention to the law-distinction. After indicating some structural features of proper theories, we will close with a brief presentation of epistemological positions involved in observational and theoretical knowledge claims of increasing strength.

CHAPTER 1

RESEARCH PROGRAMS AND RESEARCH STRATEGIES

INTRODUCTION

One of the most important insights of the philosophy of science since about 1960 was the awareness that the development of science should not be described in terms of the development of specific hypotheses and theories, but in more encompassing terms. The two main proponents of this insight are Kuhn and Lakatos. Kuhn first preferred the term 'paradigm' and later 'disciplinary matrix' (Kuhn, 1962/1969). Lakatos (1970, 1978) introduced the notion of a 'research program'. Half a dozen other terms are used to denote roughly the same concept, although their details may differ.

We prefer Lakatos's term 'research program' for its literal meaning: program of research. Although our favorite conception of research program will be somewhat weaker than that of Lakatos, we will use the same term. However, it should be stressed in advance that nobody means program in the detailed sense of a well-ordered sequence of things to do. At most a program of research in some global sense is meant. In this work, 'program' always means a research program.

Section 1.1 presents a necessarily incomplete summary of what one should know about modern philosophy of science as far as insights in global cognitive matters are concerned. Moreover, it distinguishes four types of research programs. In Section 1.2 a number of strategic lessons are suggested, which have been freely derived from both the global insights and the abovementioned distinction.

1.1 RESEARCH PROGRAMS

Introduction

After distinguishing four ideal types of research programs, we will discuss the structure of programs mainly in terms of five degrees of strength. Then we will discuss the development of programs in terms of an internal and external phase. The section concludes with a brief presentation of the atomic theory as a developing research program.

1.1.1 *Types of Research Programs*

The four types of programs to be distinguished are the following: descriptive, explanatory, design and explicative programs. They form ideal types. In consequence, mixtures are the rule, rather than the exception. However, it is often possible to describe a mixed program as a co-operative enterprise of two or more ideal program types, with one of them being in some way dominant. To put it still more cautiously, the first characterizations may well be read as descriptions of four types of research, which are, in practice, complex undertakings. However, in these four types of research it is then often possible, at least analytically, to identify the underlying ideal types of programs mentioned above.

Programs of the first three types are usually considered to belong to the empirical sciences. Programs of the fourth type, explicative programs, are not only characteristic for constructive analytic philosophy, but also occur elsewhere, viz. in mathematics and the empirical sciences.

In both this chapter and throughout the rest of the book, we will argue that these four types of research and research programs reflect the core of two divisions of labor, one figuratively between the cognitive products of scientific inquiry themselves, viz. descriptions, explanations, products, concepts, and the other literally between their producers, roughly speaking, experimentists, theoreticians, engineers, and philosophers/ mathematicians. Moreover, it will be claimed that these related divisions of labor on the level of products and producers, and their interaction, play a crucial role in the dynamics of science, which can be even more fully exploited by understanding their nature in more detail.

Descriptive programs are meant to describe a certain domain of phenomena, primarily in terms of individual facts (individual programs) or primarily in terms of general observable facts (general or inductive programs). Descriptive programs form a certain kind of observation program and may be fundamentally based on experiments, in which case it is plausible to speak of experimental programs. A famous example is Boyle's search for a relation between the pressure and volume of a gas, followed by Charles, Gay-Lussac and others with their quest for the relation with temperature. To mention just one other historical example for the moment, the famous investigation by Durkheim of what he called the social facts about suicide was typically a descriptive program.

Descriptive research takes place by more or less selective (experimentation and successive) observation, and the resulting facts are couched in so-called observation terms. These observation terms are not given by the natural world, but form the specific glasses through which the researcher in that program is looking. At the start of a descriptive program there usually is only some core vocabulary. For the rest it is not altogether clear which further observation terms are to be considered as relevant and precisely how certain observation terms are to be interpreted. Additional terms are only selected and shaped in the course of the development of the program. It should also be stressed that, at least as a rule, observation and hence observation terms are, and remain, laden by theoretical presuppositions which are considered to belong to the so-called unproblematic background knowledge.

Explanatory programs have another aim. Individual and general explanatory programs are directed at the explanation and further prediction of the observable individual and general facts in a certain domain of phenomena. Hence, an explanatory program has a (quasi-) deductive nature and is always built on an underlying descriptive program. For this reason explanatory programs are frequently developed along with underlying descriptive programs, in which case the two types of program can be distinguished only analytically. Paradigm cases of explanatory programs built on the previously mentioned descriptive programs are provided by the kinetic theory of gases on the one hand and the anomy theory of Durkheim on the other. The primary objective of the kinetic program was the explanation and detailed prediction of the precise relation between pressure, volume and temperature. To illustrate this fact, we confine ourselves to one representative of the many researchers conducting this type of research: Van der Waals. Similarly, Durkheim tried to explain the social facts about suicide with his anomy theory.

Other examples of explanatory programs are Newtonian mechanics, the transformational generative grammar of Chomsky, and the theory of rational choice or general utility theory, the latter providing the foundation of, among other things, neo-classical economics and so-called explanatory sociology.

It is important to be aware of the fact that several explanatory programs may arise on the basis of the same descriptive program. They may be competitive, but need not be.

The most important tools used by explanatory programs are so-called theoretical terms, denoting fundamentally new concepts, which have not yet been firmly established as observation terms, neither inside nor outside the program. Of course, the terms as such may have been used before to refer to a related concept. Examples of theoretical concepts are the concept of force in Newtonian mechanics, Chomsky's concept of deep structure, and the concept of utility in utility theory.

The new terms may refer to theoretical properties, relations and functions, as suggested by the examples, but also to newly postulated entities, such as atoms and genes. If an explanatory program introduces theoretical terms, it may also be called a theoretical program. If it does not, which certainly is possible, it belongs to the explanatory subtype of observational programs, to be distinguished from the descriptive subtype.

For most of the empirical sciences the above characterization of descriptive and explanatory programs makes sense and is useful. Although analogous programs occur in the historical sciences, especially programs about individual facts, the characterizations above of descriptive or explanatory programs are not suitable for historical research. In particular, general historical programs are rare, probably due to the fact that general historical facts, i.e. empirical laws and theories, are rare. Unfortunately, it seems that neither type of program in either their individual or general forms have yet been elaborated for the historical sciences.

There remain two further points to make about descriptive and explanatory programs of the general kind. The first may also be called inductive and the second deductive, but moreover, descriptive programs are always observational, whereas explanatory programs may or may not be theoretical.

In the current philosophy of the empirical sciences the main attention is paid to description, explanation and prediction. However, an important part of the empirical sciences is not primarily concerned with any of these three tasks. *Design* or constructive research programs involve the design and actual construction of certain products. Some examples are: programs directed at the production of new medical drugs, the improvement of breeding methods of plants, the design of training programs for certain types of handicaps, the design of so-called expert systems, and the construction of new materials. As the examples illustrate, the products of design programs need not be products in a strict sense but may also be processes, or their improvement. The product targeted by a design program has to satisfy certain previously chosen demands; these demands are of course derived from the intended use of the product being developed.

The examples also illustrate that design programs do not only occur in what are traditionally called the technical or technological sciences but also in other areas of scientific research. This is the reason for not choosing the term 'techn(olog)ical research programs', for that might be too narrowly interpreted.

Since design programs often use knowledge obtained in descriptive and explanatory programs, the design process will only be considered to belong to scientific research if it is not fully based on existing knowledge and techniques. That is, new theories have to be developed or new experiments have to be performed in order for a design program to be scientific in nature.

For philosophy and mathematics the fourth type of program, the *explicative* research program, is the most important type. Such programs are directed at concept explication, i.e. the construction of a simple, precise and useful concept which is, in addition, similar to a given informal concept (cf. Carnap, 1963,[2] pp. 1–18). For example, the concepts of 'logical consequence' and 'probability' have given rise to very successful explicative programs in the borderland between philosophy and mathematics. One of the main explicative programs dealt with in ICR is intended to explicate the intuitive idea of 'truthlikeness'. Although several analyses in the present book could have been explicitly presented as examples of concept explication, we have made this identifation in only a few chapters, and not even very rigorously at that, e.g. in Chapter 2 where we explicate the intuitive conceptual distinction between observational laws and proper theories, and in Chapter 4 where the intuitive concepts of intentional, functional and causal explanation are analyzed. The systematic presentation of as many topics as possible in concept explication form would take too much space.

The strategy of concept explication is the following. From the intuitive concept to be explicated one tries to derive conditions of adequacy that the explicated concept will have to satisfy, and evident examples and counter-examples that the explicated concept has to include or exclude.

Explication may go further than the explication of intuitive concepts, it may also aim at the explication of intuitive judgments, i.e. intuitions, including their justification, demystification or even undermining. A main example in ICR concerns the intuition about the functionality of choosing empirically more successful theories in order to enhance truth approximation. The strategy of 'intuition explication' is a plausible extension of that involving concept explication.

Concept explication has a kind of empirical counterpart called (concept) operationalization. This is concerned with the empirical specification of the application conditions leading to the so-called 'concept as determined' being as much as possible in agreement with the 'concept as intended' (De Groot, 1961/1969). However, it does not seem to make much sense to count concept operationalization as a separate type of program, nor to introduce a broader concept covering concept explication and operationalization. The reason is that concept operationalization usually forms a tightly integrated part of descriptive and explanatory programs.

1.1.2 *Similarities and Differences*

Although the four types of programs distinguished are different in many respects, they also have an important similarity. In all cases we can identify an internal goal, the true description, the true theory, the intended product and the intended concept.

The hypothetical totality of facts observable by the glasses of a descriptive program, assuming fully correct observation and registration, is called the true description of the domain. All other descriptions of the domain in terms of the program are either incomplete or (partially) false. This true description constitutes the internal goal of the descriptive program. It is important to note that the true description not only depends on reality but also on the program in which the choice of the observation terms delimiting its viewpoint, co-determines what will and what will not be observed. Hence, the true description is a program relative but nonetheless informative characterization of reality.[1] If the program concentrates on individual facts, i.e. object, place and time specific facts, we will speak of the true individual description, if it concentrates on general facts in some sense, we will speak of the true general description. In the latter case, the true description corresponds to the true theory within the (observational) vocabulary of the program.

In the case of a (general) explanatory program there is, usually, supposed to be a unique theory, in terms of the observation and eventual theoretical terms of the program, which not only explains, and predicts as far as relevant, all observable facts of the domain, but also uses only those theoretical terms in a substantial way that refer to something in reality. This theory, in fact the strongest true hypothesis, will be called the true theory of the domain, constituting the internal goal of the explanatory program. Like the true description, the true theory is determined by the specific combination of program and reality, hence it is program relative. If the vocabulary is observational it corresponds to the true general description.

The internal goal of design programs, the analogue of the sought-after true description or true theory, is of course formed by the intended product, i.e. a product that satisfies the demands put forward. Finally, in the case of explicative programs the analogue of the internal goal of the previously considered programs is constituted by the intended concept, i.e. a concept satisfying the conditions of adequacy.

Despite the fact that, similar to descriptive and explanatory programs, design programs always have internal goals, such goals differ greatly from those involved in description and explanation. In descriptive and explanatory programs internal goals are only indirectly characterized, and all the efforts are directed at the explicit characterization of the true description or the true theory. In design and explicative programs, the internal goal, the intended product and the intended concept are all explicitly characterized from the beginning, at least to a certain extent.

Another, related difference is the degree of uniqueness of the internal goal. As mentioned above, the true description is in principle uniquely determined jointly by the program and reality. Hence, it cannot change in the course of the program without either changing the program more or less radically or straining reality. The same holds *mutatis mutandis* for the true theory. In the case of design programs, on the contrary, the intended product need not be determined uniquely at all, for there is, as a rule, the possibility of functional equivalents, i.e. different products serving the same purpose, in which case it is also said that the purpose is 'multiply realizable'. Moreover, the desired product has to be determined in more detail in the course of its development, in which the strategic considerations of feasibility, affordability and salability play an important role. As an aside, it should be remarked that the intended product can also be over-determined by the sum of all demands.

Most of the differences mentioned also apply *mutatis mutandis* to explicative programs. This is no accident, for they form a kind of abstract design program, viz. of concepts.[2]

1.1.3 *The Structure of Research Programs*

So far the descriptions of the four types of programs may well be used to indicate just four types of goal-directed research. However, when we start to discuss the structure of programs it will become clear how programs acquire more identity than defined by their internal goals. We will discern an ordered set of five possible components of a program, viz. domain, problem, idea, heuristic, model. Since each component is supposed to presuppose the foregoing ones, this leads to five qualitative degrees of strength of programs.

A possibly disenchanting reformulation of the similarity between the four types of programs is that they are all directed at the solution of a certain problem, viz. to attain the internal goal of the program. This orientation seems the least one may expect of scientific research, i.e. that it is directed at the solution of a certain problem. Programs satisfying this minimal requirement might be called *programs with a problem*. In the practice of research policy, however, one even speaks of research programs when there is only a more or less well-defined domain of research, without a clear problem, for lack of an internal goal, in which case one might speak of *programs with (only) a domain*.

From the descriptions of the four types of programs it now follows that a research program is minimally conceived as a program with a problem. However, the prototypical meaning we want to advocate for the term research program is that of a *program based on an idea*, i.e. a program with not only a domain and a problem but also a fundamental idea governing the search for the solution to the problem. It could be called Popper's requirement, because more than anyone else he has stressed the equal importance of problems and ideas in scientific research. Of course, such a fundamental, leading idea is usually a complex idea, i.e. a set of coherent ideas. It will at least include the choice of a core *vocabulary*, and usually it includes one or more *principles* using that vocabulary. The idea should be strong so that it can provide secure footing for a research venture that should be able to withstand some critical blows. In other words, it should be possible to protect the fundamental idea somewhat dogmatically against falsification or similar threats. The standard way to do this is by trying to blame auxiliary hypotheses, but there are several other defence strategies.

According to Lakatos, the leading idea constitutes the hard core of a program. However, the notion of hard core has a double face. Lakatos's primary meaning is that a program is only one and the same program as long as the hard core remains the same. However, it frequently occurs that one feels obliged sooner or later to adjust the fundamental idea of a program, in which case one should strictly speak of a new program. But it seems more adequate to leave room for a semi-hard core of the program, a core that may be adjusted, when no other escape seems possible. We would like to stress another meaning component of the notion of a hard core. An idea, before or after a possible change, may be hard in the sense that it is supposed to be valid for the whole domain. It does not leave room for exceptions. This quality may be indicated by speaking of a 'core idea'. Incidentally, one way to retain the goal of no exceptions, in the face of persistent threats is to adjust the core idea, another is to adjust the domain. In sum, we conceive the fundamental idea on which a program can be based as a core idea of a semi-hard character.

To be sure, Lakatos only speaks of a genuine research program when there is, in addition to a hard core, also a so-called positive heuristic providing suggestions for protecting auxiliary hypotheses and their adjustment. Hence, a program, in Lakatos's sense, is a *program with a hard core and a positive heuristic*, that is, a program governed by two ideas, the first one directly bearing on the solution of the problem, and the second one concerning the way in which the first idea can be defended against attacks.[3]

Zandvoort (1984) has convincingly shown that the impressive examples of Lakatosian programs frequently are programs in which the positive heuristic is provided by an analogy or model, where he refers in particular to the notion of analogy as discussed by Nagel (1961). Such *programs with a hard core and a model as a positive heuristic* are maximally equipped to provide internal guidelines for research.

Research programs with a core idea and a stable positive heuristic, whether or not in the form of a genuine hard core and a model frequently occur in all forms of empirical science, not only in the natural sciences, but also in the social sciences and the humanities. However, many other programs have a semi-hard core idea about the way in which the goal has to be attained, without having a strong idea about the way in which that fundamental idea has to be protected. In other words, although they have a core idea, they don't have a stable positive heuristic. To put it differently, the historical claim in the beginning of this chapter can be stated more precisely as follows: the global history of science can best be described in terms of rising, winning and falling programs based on a core idea. For this reason we will henceforth assume the prototypical meaning of the term '(research) program' to be a program based on a semi-hard core idea, but not necessarily equipped with a stable positive heuristic.[4]

1.1.4 *Examples of (the Core Ideas of) Research Programs*

We now present a list of examples of ideas forming the core of many well-known programs, starting with explanatory programs. These are the idea in kinetic theory that gases consist of molecules which move and collide according to Newton's law; Mendel's idea that heredity is not a matter of transmission of one genetic factor from one of the parents, but of two factors, one from each parent; the idea in general utility theory that choices are governed by maximizing expected utility; Chomsky's original idea that the grammatical sentences of languages can be generated by the application of a limited number of transformation rules on an equally limited number of deep structures; and, finally, the central idea in classical computationalism, or symbolism (Newell and Simon), according to which human behavior should be (described and) explained in terms of problem solving.

The last example is a nice borderline case between explanatory and descriptive programs. An example of a purely descriptive program is network analysis, which is based on the idea that schemas using connecting arrows can result in very informative descriptions. Another example is fractal geometry, initiated by Mandelbrot, based on the idea that shapes in nature on different scales may nevertheless be congruent. A coastline of 10 m looks like a coastline of 10 km. So-called discourse-analysis provides a similar example. Behaviorism can also be viewed as a broad, descriptive program, with the core idea that one should restrict the scientific attention to the description of (patterns in) observable behavior. Finally, the goal of the Human Genome Project, more or less completed in 2000, was the true description of the (almost unique) composition of the 23 human chromosomes as pairwise sequences of the four bases C, T, A, and G, that is, the typical vocabulary of DNA. To be sure, although the sequences are almost unique, the individual variations provide an almost perfect means of identification.

The core idea of descriptive programs is frequently formed by a methodological searchlight principle, for instance the principle of causality, functionality or intentionality (see Chapter 4) or by a description or representation principle, as in the cases of network analysis and fractal geometry. Such leading principles usually open, for different domains, the possibility of a specific program directed at that domain. Such a representation principle may be guided, or is at least made available, by accepting a theory as an observation theory, that is, a theory that has become accepted as (approximately) true. This phenomenon will be explained in more detail in the next chapter. A typical example is the functional (descriptive) genomics program, building upon the 'structural' descriptive (Human) Genome Project, identifying the function of fragments of the chromosomes. That is, each fragment may or may not play a crucial role in ('code for') the generation of certain characteristics of organisms. This research is guided by the principle of functionality, rooted in the theory of evolution, and according to which features of organisms have functions. Even more than on the macro-level, there are many exceptions to this principle on the present micro-genetic level. As a matter of fact, most DNA-fragments seem without function.

In the case of a design program the leading idea is frequently called the lead. It is the core idea about the way in which the intended product should be construed and possibly with what material. Some examples are the following. To begin with an example of a purely technical, non-scientific, nature, the idea of a bicycle chain was developed at the end of the nineteenth century, and enabled the, still time consuming, design of a riding bike with two wheels of equal size, i.e. the bicycle, which was a very attractive feature of the walkbike of much earlier date (cf. Bijker, 1995). The idea of nuclear fission resulted in the development of nuclear power stations, and, to be honest, atomic bombs. The development of power stations based on nuclear fusion is still one of the main challenges of applied physics. One of the main starting points of computer science, the development of the standard Von Neumann architecture of digital computers, began with Turing's idea of a universal computing machine, the so-called Universal Turing Machine. Within the Von Neumann architecture the idea of so-called production systems, containing and generating complex 'if, do then' (production) rules, has turned out to be very successful, in particular for creative computing tasks. According to the central idea of the technological version of connectionism, learning mechanisms can be produced by connections between knots that are strengthened or weakened according to whether the previous response was or was not adequate. One of the leading ideas in cancer research, due to Judah Folkman (see T. Boehm et al., 1997), is to try to stop the formation of blood vessels leading to the tumor or, alternatively, to try to block their functioning. To mention an example of quite a different nature, the idea of conversation groups made up of people with similar personal problems was only recently introduced in all kinds of therapeutic contexts, with varying but increasing degrees of success. Their development was partly due to a systematic search for the best specific conditions in which to conduct therapy. For example, certain group therapies for breast cancer patients, an idea of David Spiegel (1989, 1993), seem to improve their immune system. Finally, in drug research (Vos, 1991, p. 62), the lead can be identified with the set of wished-for properties, the wished-for profile, and some idea about how to realize it. More specifically, the 'lead compound' comprises a chemical compound with certain operational characteristics, the operational profile, together with the wished-for profile. Only if there is already known to be an interesting overlap between the two profiles, it is a serious lead compound, and the challenge is to reduce the differences. In Chapter 10, design research programs will be analyzed along these lines, with drug research as a guiding example.

A nice example of an explicative program is the famous idea of Rawls, according to which the determination of the concept of a just society can best be undertaken by way of a thought experiment in which the future members of the just society to be construed do not know the place they are going to occupy in that society; that information is hidden behind 'the veil of ignorance'. Another example is the core idea of logical model-theory that the intuitive concept of 'logical consequence' should be explicated in terms of the models of the relevant language and according to the principle: the conclusion should be true in at least all models in which the premises are true.

In the last few decades several research programs have been described in detail and, of course, in the terminology preferred by the respective authors. To give one example, von Eckardt (1995), whose aim is to characterize cognitive science, describes the 'research framework', as a combination of domain-specifying assumptions (domain), basic research questions (problem(s)), and substantive assumptions (core idea(s)). In these terms she has given a lucid description of the framework for cognitive science, to be precise, as far as it is focused on adult, normal, typical cognition. In her approach, symbolism and connectionism appear as two different specifications of the 'computational system (substantive) assumption'.

1.1.5 *Additional Considerations*

A significant problem arises when any attempt is made to identify a respectable or even strong core idea. A plausible procedure involves the discerning of a research program based on an idea when, in the blindly refereed international literature, there appear to be several publications, from one or more authors, in which the idea is exposed, discussed and elaborated. In principle, the existence of international publications coalescing around one idea should be an adequate criterion since science is an international activity in the sense that national borders should not play an important role, in particular when the distribution of strong research ideas is concerned. However, presence in the international literature is certainly not an infallible criterion; it is neither a necessary nor a sufficient one. Referees and journals are necessarily selective, nor are they immune to trends and fashions. Hence, occasionally it may happen that bad ideas are promoted and that good ideas repressed. Moreover, there may well be strong ideas for which there is not very much interest in the discipline itself, but outside that discipline there may be considerable interest from other disciplines or scientific externals, i.e. from society and technology. The last case may particularly apply to ideas in design research, in which case it is not plausible to expect international scientific publications, as they may be prevented by the need for secrecy. Again, strong external interest is not a safe criterion, but the combination of international publications and lasting external interest is the best criterion we can think of for the identification of valuable research programs.

Although our concept of research program resembles Lakatos's concept of research program the most – it is a weakened version – this does not mean that ideas about structure and development of research programs can only be derived from the writings of Lakatos. As already mentioned, other authors have distinguished related cognitive units, and described their structural and dynamic features. Kuhn (1962/1969) speaks first about 'paradigms' and later about 'disciplinary matrices', Fleck (1935/1979) introduced the notion of 'styles of thought', and Laudan (1977) deals with 'research traditions'. We have also seen that von Eckardt (1995) has more recently dealt with the notion of 'research frameworks'. Additionally, 'theory nets' are distinguished in the so-called structuralist approach to scientific theories (Balzer, Moulines and Sneed, 1987), that will be discussed in Chapter 12. To conclude this incomplete list, Hamminga (1983) also uses the term 'research program', but gives it a detailed meaning tailored to economic research programs. We will make use of these and other variants whenever it seems appropriate.

In the next subsection we will deal with the main dynamic features of research programs, with emphasis on explanatory programs. In the rest of this book several structural and dynamic features of programs will be discussed in detail.

We conclude this subsection by mentioning one other structural feature, derived from the structuralist approach. The domain of a research program can frequently be divided into a number of subdomains. In such cases it is possible to make a distinction between the core idea associated with the core vocabulary, that is, the general or generic principles which are supposed to be valid for the whole domain, and special concepts and principles which are only supposed to be at stake for a subdomain. Think of Newton's general laws of motion and the special force laws. In many such cases the division into subdomains is such that it makes a lot of sense to speak of sub-programs, as the crucial idea for a special principle pertaining to a particular subdomain may well constitute a genuine research program in itself.

1.1.6 *Development of Research Programs*

As already mentioned in passing, our treatment of the dynamics of programs will mainly concentrate on explanatory programs, and close with a few remarks about the validity of the findings for other types of programs. One might prefer to read first the next section, dealing with the development of the atomic theory, and then return to this section.

We begin by elaborating a previously mentioned relativization of the term 'program'. A program is never fully mapped out in advance. At each moment only a few principal features are established. They enable researchers to look forward no more than a little bit, and depending on the results of their efforts, the program is adjusted and mapped out a bit further. This is the main reason why responsible bureaucratic middle- and long-term planning of research is impossible.

A program can pass through several phases. In cases of successful programs it is frequently possible to make a global distinction between an internal and an external phase.

In the *internal phase* the elaboration and evaluation of the core idea are central. When a program persists for some period of time it is usually possible to divide the internal phase into two subphases, viz. a heuristic and a test or, as we prefer to call it, evaluation phase. In the *heuristic phase* the new idea breaks through and the first auxiliary strategies are invented to protect the idea. This phase may or may not take place against the background of a so-called Kuhnian crisis of another program, for which seemingly unsolvable problems, called anomalies, have accumulated.

Gradually there comes a transition to the *evaluation phase*. The idea is elaborated for a small number of contexts or subdomains into specific theories, and these are evaluated. The core idea now constitutes the so-called core theory or generic theory, common to all specific theories. Evaluating a specific theory implies as a rule, that for the particular subcontext, a sequence of specific theories is developed, each containing auxiliary hypotheses, each resulting ideally in increasing success, and each including a decreasing number of (types of) counter-examples. Usually such a sequence satisfies the pattern of idealization and concretization: the consecutive theories take into account factors neglected by the foregoing ones.[5]

If this way of branched evaluation is not overall successful, the program is not necessarily deemed useless and made to disappear into the really wonderful museum of abandoned research programs. A failing program cannot only continue to inspire new research questions, it may also be the case that in a later stage someone succeeds in giving a successful turn to the program.

When the evaluation proceeds successfully, this usually leads to the more or less general acceptance of the core theory of the program and it has become clear for which domain and in what sense and to what extent the core theory can be assumed to be true. It should be stressed that many, if not most, programs in the empirical sciences, not to mention philosophy, do not reach this point. But if this stage is attained, the researchers in that program are left with two options. The first possibility is to look for another program presenting a new challenge. The second possibility is to try to direct the program for the benefit of questions that are *prima facie* independent of the program. The program then enters the *external or application phase*. The so-called Starnberg school (Schäfer, 1983) calls this finalization, and means by it in particular the application of the core theory to technological or social problems. This is seldom a matter of simple application of the theory. It usually requires highly specialized theory development, and may even lead to the start of a new scientific discipline, e.g. aerodynamics, in the case of the technological goal of aeroplanes and the like. We will nevertheless simply speak of application, in this case more in particular about the external application of science. Another form of application arises from the fact that an accepted theory may be usable as observation (or measurement) theory.

Zandvoort (1986, 1988, 1995) has convincingly established that research programs in the natural sciences, which have successfully passed the internal phase, are not always directly applied to problems external to science. They are at least as frequently applied to science internal problems. Hence, the terms 'internal' and 'external phase' of a program should be strictly interpreted as program relative:

internal and external to the program. As a rule, the science internal application of a successfully established program means that the program is directed at the solution of specific problems generated by other programs, possibly but not necessarily design programs. It may also be used for observations relevant to other programs, requiring the acceptance of the core theory as an observation theory. Zandvoort has shown that this type of co-operation between programs has unjustly been neglected by Popper, Kuhn, Lakatos and others: it constitutes the main part of successful interdisciplinary research within the natural sciences. Among others, Zandvoort's findings make it clear why it is not only difficult to show the practical, i.e. science external, relevance of natural science research programs that are still in the internal phase, even the practical relevance of programs in the application phase may well be only indirectly so. In the next section we will return to this and other types of co-operation between programs.

So far we have not paid any attention to the question of what is precisely meant when a program is successful or makes progress. To be sure, there are many sorts of success, not all of them need to be sufficient or even relevant for real progress. Success criteria for progress should of course be derived from what scientists themselves count as progress, as for instance expressed by Nobel prizes and other prestigious scientific recognitions. According to Popper and Lakatos the factual criterion for progress used for scientific recognition is not just increasing explanatory success,[6] predictive success is also required. That is, it is not enough that a program succeeds in explaining new facts, from time to time it should predict, and hence also explain, new facts. To put it differently, there should not only be postdictive but also predictive explanatory success. Although this criterion turns out to be logically too strong as an indicator of truth approximation,[7] a point which is also argued in ICR, it should be conceded that explanatory *and* predictive success is in practice the employed criterion, at least as far as the internal phase of (explanatory) programs is concerned.

For the external phase predictive success does not appear to be necessary, although this does not mean that only explanatory success is sufficient for progress. For the external phase another supplementary criterion to explanatory success is obvious: external success. From time to time the program should successfully solve external problems to which it is directed. Indeed, Nobel prize motivations frequently report, in addition to (new) explanatory success, either predictive or external success.

An important case-study undertaken by Zandvoort (1986) concerns the main theory transitions in the nuclear magnetic resonance (NMR-)program, which originates from nuclear physics and is based on quantum mechanics. He had to conclude that in almost all cases theory transition concerned theory accommodation on the basis of newly discovered facts. On closer inspection it also became clear that nobody doubted the possibility that the NMR-program could explain the new facts by some further articulation and hence that such doubts could not be the reason why the program was prolonged. In fact, the program was continued because it simultaneously solved important problems which were very relevant to other programs, in particular in chemistry and biology.

In this short overview of scientific progress we came across the following basic types of success in science: truth approximation, explanatory success, predictive success, external success. At least the following other types of success should also be mentioned: scientific recognition, textbook treatment, financial support, popular-scientific publicity, and institutional power. As is frequently claimed, and regretted, only the first two types are highly correlated with the basic types.

Although the foregoing is generally applicable to all explanatory programs, some distinctions may nevertheless only be made analytically. For instance, the application phase may well be started when the internal phase has not yet come to an end. However this may be, the proposed distinctions may also be applied in an adapted form to other types of programs. For instance, for descriptive, design and explicative programs similar phases can be distinguished, at least analytically. Moreover, the exposition about progress also applies for example to explicative programs. For such programs the analogue of (intended) increasing explanatory success, is intended increasing explicative success, i.e. succeeding in explicating the informal concept in a more satisfactory way, as determined by the conditions of adequacy, along with the evident examples and counter-examples that have been put forward. However, it is also considered to be very important that the proposed explication turns out to give rise to unintended explications, that is, to satisfactory explications of related concepts and intuitions. This type of success is the analogue of the extra, i.e. predictive or external, success of explanatory programs. Again the question is whether this form of success is formally defensible as a necessary condition for progress, but the fact remains that in practice this type of explicative success plays an important role. For descriptive and design programs it is less clear whether there are similar criteria of 'more than explicitly intended success'. In Chapter 10 we will elaborate the plausible claim that a transition in a design program is successful when a modified prototype satisfies more of the desired properties than the foregoing prototype, which evidently is the design analogue of explanatory success. Finally, in Chapter 7 we will introduce some refined ideas concerning the successes *and problems* of theories which are particularly relevant for explanatory programs. Moreover, Chapter 7 of ICR introduces refined notions of successes and problems of descriptions; they are relevant for explanatory as well as descriptive programs.

1.1.7 *The Atomic Theory as a Developing Explanatory Program*

The atomic theory and its development may well serve as an example of a successful research program. The following portrait is a simplification of the reconstructions given by Holton (1973) and Zandvoort (1989).

Dalton (1766–1844) introduced the theory of the atom in order to explain certain laws of chemical reactions and possibly to predict possibly some other ones. Hence, it is an explanatory program. Its *domain* consists of reactions between chemical substances. Along with the development of that program the distinction between pure substances and mixed substances (mixtures) and the division of pure substances into elements and compounds emerged as observational categories. The following exposition will presuppose that these terms were unproblematically available to Dalton. The development of the program can be described in three phases.

Phase 1: Dalton's primary *problem* was to explain certain relatively well established (observational) laws of reaction:

> LR1 (Lavoisier): the total weight of the substances before and after a reaction remains the same.
> LR2 (Proust: the law of definite proportions): compounds always decompose into components with constant weight ratios.

The core ideas of the program initiated by Dalton can be summarized into four principles, with the notions of (types of) atoms and molecules as the program specific theoretical vocabulary. The first two are *internal principles*, only dealing with postulated, hence theoretical micro-entities; the remaining two are *bridge principles*, in fact identity postulates, relating the theoretical terms to the observation terms.

> I1: *atoms* are indivisible, unchangeable, hence indestructible, small material particles of a certain type.
> I2: atoms are grouped into *molecules* of a certain type, and they may regroup into other types of molecules.
> B1: – pure substances consist of one type of molecule;
> in the case of elements these molecules consist of one type of atom;
> in the case of compounds they consist of more than one type of atom.
> – mixed substances consist of more than one type of molecule.
> B2: chemical reactions amount to systematic regrouping of the molecules of a substance.

Let us indicate the core idea consisting of I1&I2&B1&B2 by C, and the specific theory at stage i by Ti, consisting of C + Hi, where Hi indicates the auxiliary hypotheses at stage i. Hence, T1 = C, as there are not any substantial auxiliary hypotheses at the start. It is not difficult to check that T1 explains the two target observational laws RL1 and RL2.

In agreement with both Popper's and Lakatos's views, Dalton felt obliged to obtain a more impressive result and predicted a third law, viz.

> RL3 (the law of multiple proportions): when two different elements unite into two different compounds, the different proportions bear a simple relation to one another.

One successful test tuple of compounds consists of carbonic oxide and carbonic acid, both composed of carbon and oxygen. In the first case the proportion, in terms of weights, of oxygen to carbon is about 4:3, in the second case about 8:3. Hence, the ratio of these proportions is 1:2. For this prediction Dalton needed a strong rule of simplicity concerning the possible composition of molecules of the same type of atoms:

> A-s (simplicity assumption): if a certain type of molecule exists then all the conceivable more simple types of molecules composed of the same type of atoms exist as well.

In combination with the assumption that the number of existing compounds of two elements is rather limited (an auxiliary hypothesis that will be neglected in the remainder of this section), RL3 can easily be derived. Although A-s is certainly false according to our present knowledge, the derived predictions came true and RL3 became accepted. Hence, according to the progress standards proposed by Popper and Lakatos, the transition from T1 to T2 = C + A-s is progressive: intended (or postdictive) explanatory success is supplemented with predictive explanatory success.

Phase 2: However, in the meantime a severe anomaly was arising: the law of combining volumes, independently established by Gay-Lussac. First an example: two liters of hydrogen gas and one liter of oxygen gas result into two liters of water vapor. In general:

> RL4: pure gases combine with simple integer numbers of volume units, into an integer number of volume units of the compound gas, not necessarily equal to the sum total of volume units of the component gases.

It is easy to see that it is not possible to derive RL4 from T2, nor its negation, for the simple reason that T2 does not say anything about volumes of atoms and molecules. However, Dalton and some of his followers actually favored an auxiliary assumption A-g about the nature of gases:

> A-g (gas assumption): gases consist of non-moving, contiguous gas particles, in their turn consisting of a molecule and a caloric mantle.

It is quite obvious that the resulting T3 = C + A-s + A-g, constituting a specific theory for the subdomain of gases, predicts the negation of RL4. Indeed, Dalton was confronted with a very big explanatory problem. However, he himself was inclined not to blame his theory, but to put the truth of RL4 into question. To be sure, the transition from T2 to T3 cannot be called progressive.

Avogadro (1776–1856) took the explanatory problem more seriously. He not only proposed a modified version of the rule of simplicity, A-s*, but also a totally different and, at first sight, very surprising auxiliary hypothesis about gases:

A-g*: (gas assumption proposed by Avogadro: Avogadro's hypothesis): equal volumes of different gases, at equal pressure and temperature, contain equal numbers of molecules.

Note first that A-g* typically is an extra bridge principle. Even without specifying A-s*, it is plausible that Avogadro could show that the resulting specific theory, T4 = C + A-s* + A-g*, is indeed able to explain, in addition to RL1, 2, and 3, Gay-Lussac's law of combining volumes, RL4. Moreover, it enabled him to produce, using examples of weight and volume ratios in the line of RL3 and 4, molecular composition formulas and molecular reaction equations. However impressive these results were considered to be, they did not lead to the general acceptance of Avogadro's specific version of the atomic theory. Why? Referring once again to an insight developed by both Popper and Lakatos, the crucial point in this case is that, although T4 is explanatorily more successful than T2 and T3, it has, at this point, not yet achieved (specific) predictive success.

Phase 3: It took about half a century before Cannizaro (1826–1910) was able to derive a new prediction, using an additional auxiliary hypothesis concerning dissociation:

A-d (dissociation assumption): large molecules of a certain composition will fall apart in a gaseous state.

From the resulting specific theory T5 = C + A-s* + A-g* + A-d, using Avogadro's findings on substances that consist of large molecules, the following prediction could be derived:

RL5: substances so and so will dissociate in a gaseous state.

These predictions turned out to be largely correct, hence the transition from T2, or T3, or T4 to T5 is progressive in the sense of predictive success.

As a matter of fact, largely due to Cannizaro's results, Avogadro's hypothesis and the whole specific theory became generally accepted, i.e. its internal phase came to an end. It nicely illustrates that a research program can be very successful without a stable overall positive heuristic, which is difficult to discern in the case of the atomic program. At most there is something like a positive heuristic for partial use only, viz. to modify the rule of simplicity when problems arise involving molecular formulas. To be sure, C seems to have been really a hard core, in the sense of Lakatos. However, for the sub-program restricted to gases, the transition from A-g to A-g* may well be construed as a fundamental change of the core idea.

As is well-known, the atomic theory has been very successfully applied in other areas of scientific research, and has turned out to be indirectly of great practical use in chemical technology. Moreover, by accepting it, determination criteria for atoms and (the composition of) molecules were also established, and it not only became a background theory, but an observation theory.

This concludes our treatment of the example for the moment, but we will come back to it a number of times.

1.1.8 *A Pictorial Summary*

We conclude this section with a pictorial summary of its main content.

The concept of research tradition developed by Laudan (1977) can be interpreted as an even more global conceptual unit than that of a research program. As a matter of fact, a research tradition can be seen as the metaphysical and methodological core of a number of research programs. Behaviorism is a good example; it generated several research programs in psychology and biology. If we include this in a picture we get a double branched description of the state of affairs at a certain moment in a certain tradition, bringing together ideas from Kuhn, Lakatos, Sneed, Laudan, the Starnbergers and Zandvoort. In Figure 1.1 RP1/2/3 denote research programs belonging to a certain research tradition, CT2 the core theory of RP2, T2.i.j a specific theory of RP2, to be precise, the *i*-th attempt for the *j*-th subdomain. Moreover, RP* denotes a research program which may or, more likely, may not belong to the same research tradition, and DP denotes a design program for some science external product. A white arrow denotes 'gives rise to', a black arrow denotes a transition with explanatory and predictive-or-external success, and, finally, a thin arrow indicates 'tries to contribute to the solution of a problem of' a certain (research or design) program.

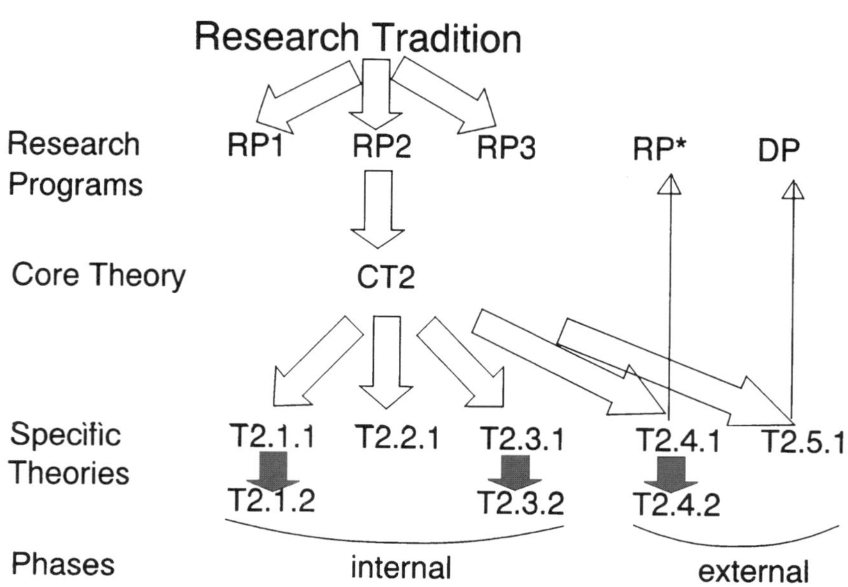

Figure 1.1 Research programs and their development.

1.2 RESEARCH STRATEGIES

Introduction

The above exposition of the structure and development of research programs raises at least four questions:

- In what way can research programs fruitfully interact?
- What is the detailed structure of explanatory arguments, in particular when different ontological and/or epistemological levels are involved?
- What are the kinds of successes and failures specific theories of a program can make, what are the choice criteria for specific theories, that is, how do we establish empirical progress, and how is empirical progress related to the idea of truth approximation?
- How are new specific theories developed, starting from the available ones?

The second and the third question bring us to the heart of the nature of scientific end products and of the nature of rationality in the production process. The second question will receive the main attention in the rest of this part and Part II, and the third in Part III (and in ICR).

The first and the fourth question essentially concern research strategies for the (further) development of research programs: on the one hand strategies directed at interaction between programs for the benefit of at least one of them, on the other hand program-bound strategies, that is, strategies within a single program aiming at improving the last specific theory. Both types of strategies will be discussed in this section. As far as the interactive strategies are concerned no detailed reconstructions of such strategies have been developed, and it is doubtful whether they ever will be. However, knowledge of their global nature is essentially sufficient for application in new contexts. The same is true for the program-bound strategies, but it is possible to analyze in detail the sense in which such programs are useful for truth approximation (see ICR, Chapter 7).

We start with a discussion of the importance of working within research programs, the program strategy itself. Here, as well as with respect to the interaction of programs, we expose the global lessons to be drawn from the work of Popper, Kuhn, Lakatos, Sneed, Hamminga, the Starnbergers, and Zandvoort, along with insights taken from Nowak (1980), Krajewski (1977), Darden and Maull (1977), and Bechtel (1988a/b). It is more or less a strategy oriented synthesis of their insights as far as they are compatible. The reader should understand that the assertive tone concerning strategies should not be taken too seriously. As with all heuristics, you may consider a strategy, you may even try it out, but you cannot blame it.

1.2.1 *The Program Strategy*

From the long-term development of the sciences it has become clear that scientific research can be aptly characterized in terms of research programs. This does not alter the fact that the historiography of science can frequently be accused of concentrating too much on the success stories, on the successful research programs. Historiography should also pay much attention to programs that have lost the competition. The results of these programs are of course not added to the body of knowledge, but they rightly deserve a decent place in the museum of knowledge. Unfortunately, the only programs arriving in the museum of knowledge are those that were winning programs until superseded by newer programs. But deposition in the museum of knowledge should for instance also take place when two competing programs started more or less at the same time, with one of them having to give up sooner or later in favor of the other. Hence, the claim is, more precisely, that the main lines of the history of science can well be described in terms of rising, winning and falling research programs. If that history is written with any attention being given to mutual interaction, it will not only become apparent that programs frequently compete,[8] but also that they may fruitfully co-operate.

The lessons from these observations seem to be the following. Anyone who wants to undertake frontier research, as suggested by my home university, the University of Groningen, by the slogan 'to work at the frontiers of knowing', will in general also aim at getting the results of his research sooner or later incorporated in the international knowledge base, or at least in the museum of knowledge. Taking all things into consideration, and whether one likes it or not, in order to achieve this goal participation in one or more of the internationally recognized research programs is virtually unavoidable.

Many university researchers consider this point to be obvious. However, there are nevertheless quite a few researchers who think otherwise. One seldom meets them in the natural sciences, but regularly in the human sciences, and, for sure, in philosophy. As a consequence, the feature of having in a certain domain of inquiry only a few (interacting) research programs is badly developed in the human sciences. In the natural sciences this fruitful characteristic certainly has been partly instigated by the high costs of experimental research.

A frequent objection to program participation is the claim that it inhibits creativity. But the converse seems to be the case. Given the enormous potential of competitors one needs stronger creative talents to deliver a substantial contribution. Moreover, some critics are of the opinion that it should be possible to develop a new program. Hence, in the human sciences many researchers start their own shop, complete with a new publication medium. It is strange that such initiatives are usually taken rather seriously. It is interesting to compare this with a researcher who announces that he is going to make an important new discovery and who will be regarded rather skeptically. But the invention of new ideas that can lay the foundation of new research programs is as rare and difficult as making pioneering empirical discoveries. In fact it concerns pioneering theoretical discoveries. For both types of scientific achievements Nobel prizes are awarded. Moreover, the inventors of ideas for new programs are frequently steered by a fresh look at the severe problems they and others met when conducting research in existing programs.

In the social sciences and the humanities there even seems to be an abundance of research ideas. One therapy against this condition is the foregoing type of plea for program-bound research, the other is a plea, which is to follow, for stimulating interaction between research programs, by co-operation and/or competition.

1.2.2 *Interaction Between Programs as a Research Strategy*

It is plausible to distinguish two main types of interaction between research programs, viz. interaction by competition and interaction by co-operation. Of course, it is also possible that after a period of interaction of one kind, the interaction turns into one of the other kind.

We will first concentrate on *competition*. When two programs are directed at the same domain and problem, and both are still in the internal phase, competition will concern the adequacy of the core ideas. When both programs are already in the external phase, competition concerns the question of which program is best suited to solving problems external to science or problems raised by a third program. When one program is still in the internal phase and the other in the external phase, competition usually takes the form of a challenge by the first to the supposed domain of validity or degree of accuracy of the second. A well-known example of the last kind is Einstein's questioning of Newton's theory.

The three indicated types of competition can all be very stimulating. However, when competition occurs, it is seldom seen as an explicit research strategy. Interestingly enough, one is not even always aware of being steered by a competing program, or one is at least not willing to admit that this is the case. These facts explain why the question of whether a further articulation of a competing program may lead to even more stimulating interaction is not always raised.

Population genetics provides a nice example of competition between two programs in the internal phase (Dolman and Gramsbergen, 1981). Concerning the problem of the origin and dynamics of variations in populations two programs can be distinguished, viz. the classical and the equilibrium program. The development of both programs cannot be described without bringing the stimulating interaction between them into the picture. Moreover, they gradually show a remarkable convergence, with the consequence that the competition increasingly transforms into co-operation in such a way that a fruitful synthesis has emerged. The same development seems to take place in the interaction between symbolism and connectionism in cognitive science (Smolensky, 1988; Stiekema, to appear).

Now let us consider *co-operation*. As in the case of competition, the forms of co-operation can be divided according to the three combinations of phases in which the two programs are situated. We have already seen that a program in the external phase can offer its services to another program, in the internal or external phase, that is confronted with a problem which the program itself is unable to solve. In Zandvoort's appealing terminology (1986, 1988, 1995) the latter program then functions as the *guide program* and the former as *supply program*. The core theory of the supply program may either be specialized (finalized) to the domain of the guide program, or it may be used as observation theory providing relevant observations for the guide program. For the particular problem the co-operation is of a fundamentally asymmetric nature. This character does not exclude the fact that the roles can be interchanged in dealing with another problem, in particular when not only one but both programs are in the external phase.

A typical form of the type of asymmetric interaction is provided by design research programs in the internal phase. They frequently function as guide programs, with descriptive and explanatory programs in the role of supply programs.

Besides the foregoing type of co-operation, in which at least one program is in the application phase, co-operation is possible between two programs in the internal phase, in which case they frequently stimulate each other in rotation, alternating in the role of guide and supply program. In this case, as in the case of two programs in the external phase, the co-operation is (although with respect to specific problems asymmetric) on the whole symmetric: the programs co-evolve (Bechtel, 1988a).

For instance, at least on the basis of accounts given in physics textbooks, one easily gets the impression that the interaction between phenomenological thermodynamics and statistical mechanics is a classical example of this type of co-operative co-evolution. However, it is well-known that the intentions of the researchers concerned were of a much more competitive nature. Apparently, this does not exclude the fact that the result of competitive interaction can make it plausible that the intention to co-operate could have been at least as productive.

The example shows that researchers themselves may be inclined to perceive the interaction between two explanatory programs as competitive rather than as co-operative. However, when one program evidently is of a descriptive nature and the other of an explanatory nature directed at the first, the interaction between them can unproblematically be conceived by the researchers as co-operative: it is a paradigmatic kind of co-evolution.

In the foregoing the basic aim of co-operation between programs was the solution of a problem encountered in one of the programs by the other. Of course, other goals of co-operation occur. For example, programs may jointly strive to articulate an overarching theory, or a synthesis of theories. Recall that the latter was the case in the example taken from population genetics. Still one other important form of co-operation involves the causal concatenation of two theories, requiring a third so-called interfield theory (Darden and Maull, 1977; Bechtel, 1988a). This type of theory will be discussed in the context of interlevel research (Chapter 6).

For all mentioned kinds of interaction the programs may be empirical programs of the same or different type. Moreover, interaction may also involve an empirical program of a certain type and an explicative program of a philosophical or mathematical nature. Current interactive research within cognitive science and between cognitive and neuroscience is of the latter nature. Some of the current philosophical mind-brain-body theories and theories of representation are not only of an explicative nature, they play at least some interactive role with some empirical programs (Bechtel, 1988a/b).

1.2.3 *Interdisciplinary Research*

It is worthwhile to consider Zandvoort's model of co-operation in tandem with his model for successful interdisciplinary research (Zandvoort, 1986, 1988, 1995).

IR-model: interdisciplinary research consists of some research programs, belonging to one or more disciplines, co-operating according to the following rules of the game:

– one program is the guide program which raises problems of a theoretical or experimental nature in the others,
– the other programs are supply programs, which have successfully passed their evaluation phase and hence can try to solve the problems provided by the guide program.

Compared with the popular ideas about interdisciplinary research, the above model has three fundamental differences. First, interdisciplinary research is not so much a matter of global co-operation between disciplines but, more specifically, co-operation between research programs. Second, it is a matter of asymmetric co-operation: one program poses the problems, the others try to supply solutions, and if successful they have the last word. Third, effective supply programs typically are in the external phase. Note that, if the guide program also has already passed its evaluation phase and is not a design program, then it will usually be directed at a science external problem of a technological or societal nature.

The IR-model suggests that the failure to start successful interdisciplinary research may well be due to the lack of relevant supply programs in the external phase. Moreover, it may be due to the collision of cognitive and social factors; in addition to the necessity of 'cognitive asymmetry' there is an inclination to as much 'social symmetry' as possible: all participants are supposed to deliver contributions of equal importance.

The stress on asymmetric co-operation between programs needs a counter-balance on the level of disciplines. It is conceivable that all interdisciplinary research directed at some science external problem area (e.g. health, environment, education) develops into a state in which one discipline provides all the guide programs, whereas the other participating disciplines provide only supply programs (the hierarchical model). Alternatively, it is also possible that there arises on the level of disciplines a more symmetric situation (the symmetric model). On the level of science and research policy, when setting up long-term strategic interdisciplinary research in some science external problem area, it seems very important to start with the symmetric model. The reason is that it is easy to imagine that starting from the symmetric situation purely scientific reasons may gradually lead to a hierarchical situation, whereas it will be much more difficult to reach a symmetric situation starting from a hierarchical one, let alone to reverse that hierarchy.

For reasons of completeness we conclude by noting that the IR-model does not seem to be appropriate as a point of departure for the investigation of a science external problem area when one wants to obtain short-term practical results. In planning that type of research ad hoc considerations seem to be unavoidable.

1.2.4 *Program Pluralism as an Education and Research Strategy*

Program-bound research has one main disadvantage. One can readily become very indoctrinated with a program. The postgraduate schooling in program-bound research may well degenerate into the delivery of program-bound researchers. As a counterbalance to the importance of program-bound research, a program pluriform education and subsequent research career seems equally important.

Based on the work of Kuhn and Lakatos it may be inferred that mature science frequently consists of dogmatic research, i.e. research sticking to the hard core of ideas, executed by dogmatically inclined researchers. In Chapter 8 we will see how such practices can be rational. However, it is also our conviction that science would profit still more if dogmatic research is performed by non-dogmatic researchers.

The way to learn to do program-bound research, without becoming a prisoner in one program, is to get research experience in at least two programs. They need not be competing programs. According to Kuhn, such program pluralism is almost impossible, due to the Gestalt-switch that it is claimed to require. Although this thesis, together with the so-called incommensurability thesis, has been criticized severely (Franklin, 1986; Hintikka, 1988, to mention a few), it is important to stress that it may be at least as instructive to undertake research in two programs that might co-operate.

The short-term effects are very positive. To begin with, when one alternatingly does research in two programs, the period engaged in one program may function at the same time as a form of productive breathing space for the work on the other. It may even occur, to say it in popular terms, that the right hemisphere is further stimulated to do its work, for, as suggested by the literature on creativity and serendipity (van Andel 1994), successful, or even unsuccessful, attempts at problem solving in one program may be transformed into successful solutions for the problems of the other.

This short-term favorable effect concerns the stimulation of largely unconscious processes, the second to be mentioned results from the conscious stimulation of interaction. If two programs have something to do with each other, knowledge of both leads in a natural way to questions of co-operation or competition. For instance, one may ask whether one program may be of help for the other. Can it solve a problem the other is unable to solve? Additionally, competition questions may arise. If on closer inspection both programs essentially claim to be able to solve the same problem, and if the one has already succeeded in doing so, the other cannot remain behind.

At first sight one may find the second favorable short-term effect of doing research in two programs unimportant, for competition and co-operation questions also may arise in the research team, or in the study of the literature, as specific research questions may be transmitted from one individual researcher to the other.

In theory such exchanges should occur, but in practice they are limited, because it is understandable and even productive that one primarily works in program-bound research groups. Moreover, as far as competition questions are concerned, there is the additional fact that it is only attractive to raise such questions when the researcher has acquired affinity with both programs.[9] Hence, it seems plausible that the suggested interaction questions are best stimulated by promoting research training and further practice in more than one program.

The plea for pluralistic research experience should not, however, be misunderstood as a plea for unlimited and diffuse eclecticism. On the contrary, it is a plea for experience within and interaction between a limited number of (analytically) well-distinguished research programs.

The favorable long-term effects of this plural-program-bound research experience also seem advantageous. To join new developments in international research it is of great importance that one has learned about methodological multiplicity and the various perspectives from which one can investigate the world, not only in theory but also in practice. Plural-program-bound research experience stimulates the flexibility of the individual scientist in his further research career. The desirability of this flexibility for scientific research in general and interdisciplinary research in particular is obvious.

1.2.5 *Interaction of Holistic and Reductionistic Research Programs*

One of the most exciting forms of competition and co-operation occurs between reductionistic and holistic research programs involving essentially the same domain. In most cases the interaction can be described in Zandvoort's terms of a guide program and one or more supply programs. Moreover, in one sense or another the guide program is reduced, i.e. there is a reduction of concepts, or laws, or both. In this case, the guide program is called holistic and the supply program reductionistic, terms which are of course relative qualifications. The reduction may conform to one of the three kinds of reduction distinguished in the pluralistic model for the reduction of laws that will receive extensive attention in Section 3.3.

None of these basic forms implies the elimination of the guide program. One of them only amounts to one or more corrections in the tentative laws that guided the research. The other two are variants of a non-eliminative reduction of a higher to a lower level. In these cases the laws and concepts of the higher level of the guide program are reduced to laws and concepts of lower levels, without explaining away the higher level.

Below we give a number of illustrative examples of studies of interacting research programs, where reductionistic and holistic perspectives play a major role. All studies mentioned are related to Groningen. Although they represent a type of 'cognitive studies' in philosophy of science which is typical for Groningen (see Kuipers and Mackor, 1995), numerous examples from elsewhere could, of course, be given. In a detailed study Janssen (1993) analyzes the so-called micro-economic foundation of macro-economic concepts and laws. The current micro-foundation is a non-reductionistic attempt to interaction between the descriptive guide program of macro-economics and the explanatory program of neo-classical micro-economics, in particular general equilibrium theory. According to Janssen the results are problematic because the supposed individualistic foundation of general equilibrium theory is doubtful. He sketches another way to explain macro- and micro-economic laws and theories in a strictly methodological-individualistic way. In this approach game-theoretic adaptation of utility theory plays a crucial role: it serves as a supply program.

Looijen (1995, 1998/2000) investigates the structure and dynamics of ecological communities. His working hypothesis is that the co-operation between three kinds of research programs could be improved: holistic guide programs describing the structure and dynamics of communities, radical reductionistic programs that try to explain these patterns in terms of the species composing the communities and their environmental needs, and moderately reductionistic (or moderately holistic, for that matter) programs that try to explain these patterns using theories about the interactions between the composing species, such as predation and competition.

Guichard (1995, 1997) starts by documenting that stress researchers, though striving at co-operation between psychological and biological research programs, were unsuccessful in the strategies employed to achieve this end. The reason for this systematic failure seems to have been that these strategies essentially presupposed a dualistic explication of the mind-body problem. Guichard argues that monistic explications, in particular of a materialistic-reductionistic nature, are more appropriate for such co-operation. His intervention uses a philosophical explication program to get new perspectives for co-operation between (relatively holistic) psychological guide programs and (relatively reductionistic) biological supply programs. More specifically, Guichard argues that the proper function theory of Ruth Millikan provides the ideal 'interfield theory' for this purpose.

Mackor (1995, 1997) argues along the same lines, but more generally, that a conceptual unification is possible not only between biology and psychology but that this can be extended to sociology by analyzing meaningful and rule-guided behavior in terms of Millikan's notion of proper functions. The result is a new, naturalistic, mildly reductionistic perspective on the spectrum of disciplines and their possibilities of co-operation. From this perspective, the most important boundary between the sciences runs between physics and chemistry on the one hand and biology on the other, although, successful co-operation between them is never excluded.

Festa (1993, 1995) initiates a confrontation between three research programs which were developing almost independently: inductive logic or inductive probability theory (Carnap), truth approximation (Popper) and Bayesian statistics. In the first place he shows that inductive logic, despite Popper's dismissive attitude about it, can be considered as a part of the truth approximation program, directed at the approximation of the true, objective chances. In the second place Festa elaborates the claim that, using De Finetti's representation theorem, (relatively holistic) systems of inductive probabilities can be reduced to special types of Bayesian statistics, which can hence be used as a supply program for further development of inductive logic. Last but not least, he shows that it is possible to define an optimum inductive system in terms of the available prior information about the domain to be investigated.

The general outline and the examples suggest at least three different research strategies for attacking a domain on the macro-level of some macro-/micro-level distinction. The *radical holistic* strategy is to try not only to describe but also to explain the phenomena at the macro-level in terms of that level, or even higher levels, and to refrain from lower level theories. The *radical reductionistic* strategy is to try not only to explain but also to describe the macro-phenomena in micro-terms as much as possible. The third strategy is a mixed one: according to the *mixed* strategy, one describes the macro-phenomena in macro-terms, and tries to explain them in micro-terms as far as possible, and hence in macro-terms as far as necessary. In Section 4.3. these strategies will be discussed in some more detail in relation to related philosophical positions.

1.2.6 *Program Development Guided by Idealization and Concretization*

We now come to sketch two strategies for the internal development of a research program, particularly for the succession of improving specific theories. To be sure, these strategies can also be used without assuming the boundaries of a research program, the only claim is that they are frequently used within a program. We start with idealization and concretization. Idealization is frequently applied in empirical scientific practice as an unavoidable step in theory formation. This is certainly true in the natural sciences; in the human sciences and also in philosophy the necessity of explicit idealization is not yet generally accepted.

Surprisingly enough, on closer inspection Marx developed his ideas in *Das Kapital* rather systematically according to the method of idealization and successive concretization. Nowak (1974, 1980) has pointed out that this procedure was used by Marx; in particular he shows how Part I and Part III in their succession can be seen as illustrations of what Marx used to call 'rising from the abstract to the concrete'. Another Polish philosopher, Krajewski (1977), freely following Nowak, has also contributed importantly to the growing awareness of the systematic role of what he calls 'idealization and factualization'.

Although idealization-and-concretization (I&C from now on) also occurs in qualitative theorizing, it is primarily explicated for quantitative theorizing, in particular the succession of specific theories within a research program. The general idea is that it is frequently possible to make an ordering in the degree of importance or relevance of all the factors that influence the value of a certain quantity G, which may even lead to a division of primary and secondary factors. Starting from such an ordering of factors $f0, f1, \ldots$ (fm), in the n-th stage of concretization factors $f0$ up to fn have been accounted for, while the remaining factors are still neglected, leading to the typical I&C-formulation of the n-th specific theory:

$$\text{if } f0 \neq 0, f1 \neq 0, \ldots, fn \neq 0 \text{ and } f(n+1) = 0, f(n+2) = 0, \ldots$$
$$\text{then } G = Gn(f0, f1, \ldots, fn)$$

In the 0-th stage there is maximal idealization and when all factors have been concretized, maximal concretization has been achieved. Note that, although any given functional representation of a factor is alloted the value 0 on a formally arbitrary basis, the neglect of a certain factor is empirically speaking usually not arbitrary, in which case the functional representation can be chosen in accordance with this.

The transition of the ideal gas law to the Law of Van der Waals is a paradigm case. This transition can be represented in a stepwise decomposition, of which the crucial formulas include:

(0) $P = RT/V$

(1) $P = RT/V - a/V^2$ (or, alternatively, $P = RT/(V - b)$

(2) $P = RT/(V - b) - a/V^2$ (or the standard form: $(P + a/V^2)(V - b) = RT$)

where P, V, T and R indicate pressure, volume, temperature and the ideal gas constant, respectively, and a and b refer respectively to specific gas constants related to mutual attraction between the molecules and the volume of the molecules.

The book series *Poznan Studies for the Philosophy of the Sciences and the Humanities* includes many examples of I&C taken, in particular, from biology, economics and sociology.

I&C can be used to structure theories in their research stage as well as in textbooks. Although it seems very plausible to do so, to say the least, it is very surprising that it seldom is explicitly done. However, in general expositions about what one has been doing or how one should do it, there is frequently reference to I&C.

A specific reason for the relative neglect of I&C in the social sciences may be the great social pressure to avoid very strong idealizations: fear of being accused of distorting reality too much seems to be very rampant.

The above mentioned paradigm example raises a very interesting question concerning explanations: is it possible to (re)construct the explanation of a concretized law as a concretization of the explanation of the (more) idealized law? Chapter 10 of ICR deals with this question in some detail.

Another, at least as important, question is whether and in what sense the I&C-strategy is functional for truth approximation in the empirical sciences. A detailed positive answer is given in ICR (Section 10.4).

As already mentioned, the I&C-heuristic can also be used in the qualitative theorizing that often occurs in mathematics and philosophy. The ordered textbook presentation of first propositional logic and then predicate logic provides a famous example. In this book we will occasionally apply the I&C-heuristic in the presentation and development of qualitative meta-theories, e.g. the HD-method for the separate evaluation of theories in Chapter 7, the set-theoretic approach to design research in Chapter 10 and the structuralist approach to theories in Chapter 12. On closer inspection it will turn out that the theory of truth approximation, developed in ICR, is an example of the variant of the I&C-heuristic being discussed in the next subsection.

1.2.7 *Program Development Guided by Interesting Theorems*

A related strategy of theory development in economics was made explicit by Hamminga (1983). The time when economists thought that economics could and should in principle be done along naive Popperian lines has passed, but the question remains how economists in fact do their job.

Hamminga studied the development of the theory of international trade in the period 1930–1970 and reached the following diagnosis. Economists direct their attention to theorems which they find interesting and they try to prove their validity for an increasing number of conceivable cases, probably with the following motive in the back of their mind: the desire to increase the plausibility that the theorem also holds in the actual world (or the nomic world, see Section 2.4.3). Apart from this motive, the world does not play a clear role: it is all and only mathematics, or so it seems. Nevertheless, or precisely because of this, one can find a large amount of systematics in the details of what theoretical economists do.

To begin with, it is possible to systematize the specific claims of the research program to an even greater extent than Lakatos could have imagined: under such and such conditions it is possible to prove the interesting theorem (*IT*), or in a formal schematization:

$$V_{lmn}\ (C_1 \ldots C_i; C_{i+1} \ldots C_j; C_{j+1} \ldots C_k \rightarrow IT)$$

The division of conditions here is as follows. V_{lmn} indicates the *field conditions* which describe the domain of the claim; in the example, l, m, and n indicate, respectively, the number of countries, goods and production factors taken into consideration. $C_i, \ldots, C_i$ indicate the generic or *basic principles*, i.e. the core ideas of the general research program with which the problem area is attacked. In the case of international trade this basic program is that of neo-classical economics, of which the core consists of utility theory. $C_{i+1}, \ldots, C_j$ indicate the special or *specific principles* for the particular subdomain of international trade, e.g. that the production functions are the same in all countries, while the endowments of production factors may vary greatly. Finally, $C_{j+1}, \ldots, C_k$ indicate the *technical conditions* of a mathematical nature, e.g. that the production functions are continuous. An example of an interesting theorem is that of factor-price-equalization: the price of a certain factor becomes equal in the dynamic equilibrium with international trade.

Theory development or, more precisely, results which are considered to be important theoretical achievements consist of both new specific claims and their proof. In the latter, field and/or technical conditions have been liberated in one or more of the following ways:

- field extension: increasing the number of countries, goods or factors (2 is for each the point of departure).
- weakening technical conditions.
- substituting more plausible conditions for technical conditions.
- introducing alternate technical conditions.

In ICR (Section 10.4) such developments are shown to be formally similar to concretization. They are therefore functional for truth approximation.

The picture that Hamminga draws seems representative of neo-classical economics; reconstructions of the theory of the market (Janssen and Kuipers, 1989) and of capital structure theory (ICR, Section 11.2, based on Cools, Hamminga, Kuipers, 1994) confirm this diagnosis. Of course, Hamminga does not provide a complete picture of the whole of the science of economics; in particular applied econometric models are overlooked. His views do however characterize an important part of the discipline, viz. so-called theoretical economics. Moreover, his work suggests the question of what the systematics is, if anything, in theory development in those areas of economics where the picture drawn seems inadequate.

The sketched diagnosis of the mathematical nature of economics may be an important underlying motive for the striking ambivalence of economists about the question of whether economics is an empirical social science or not. Moreover, the diagnosis illustrates that the cognitive aims of the social sciences in general and of economics in particular appear to be less evident than philosophers of science use to assume on the basis of an analogy to the natural sciences.

To be sure, the described strategy is certainly not restricted to economics. For instance, in population biology similar strategies are used with respect to the Law of Hardy–Weinberg (Lastowski, 1977). As already suggested, in mathematics and philosophy the 'theorem-strategy' is also frequently used, e.g. in ICR, Chapter 7–10, proving the 'success theorem', stating that a theory closer to the truth than another will also be more successful under more and more realistic conditions.

CONCLUDING REMARKS

In this opening chapter we have argued that the notion of a research program is very useful in globally describing scientific research, and in specifying global research strategies. In Chapter 2 and in Part II and III we will study in detail the main products of scientific research, usually resulting from isolated or co-operative explanatory programs, viz. observational laws and proper theories, and types of explanatory arguments. We will pay special attention to cases in which different ontological and/or epistemological levels are involved. Part IV deals with general methodological issues, such as the detailed structure of the hypothetico-deductive (HD-)method, the truth testing of hypotheses, and, in particular, the test implications of theories. Moreover, it analyzes the separate and comparative evaluation of theories in order to explain empirical progress within and between research programs, and to distinguish science from pseudoscience, and, more specifically, scientific from pseudoscientific dogmatic research. In ICR it is argued, following the I&C-strategy, that empirical progress is functional for truth approximation. For that purpose, the main task is to explicate the idea that one theory is closer to the truth than another, which will lead to a basic and a refined interpretation. Part V gives a self-contained version of the resulting approach and the related analysis of design research programs. Finally, the Capita Selecta Part VI deals with two specific research programs in the philosophy of science: computational approaches to theory discovery, evaluation and revision and the structuralist approach to scientific theories. Moreover, it deals with some ethical issues encountered in undertaking research.

CHAPTER 2

OBSERVATIONAL LAWS AND PROPER THEORIES

INTRODUCTION

In the empirical sciences the informal distinction between observational laws and proper theories plays a crucial role. Observational laws are supposed to describe observationally, usually experimentally, established regularities. Different names for roughly the same concept are: empirical, experimental or phenomenological laws, reproducible effects, inductive generalizations, general facts. Proper theories or systems of theoretical laws, on the other hand, are supposed to explain such laws and to predict new ones, by postulating underlying mechanisms. For easy reference, we will call this distinction the law-distinction. The law-distinction forms a crucial construction principle for the hierarchy of knowledge and therefore an important heuristic factor in the dynamics of knowledge development. However, it has occasioned philosophers of science much brain-racking to explicate the law-distinction in a defensible way. Without doubt, the distinction is strongly related to the distinction between observational (or empirical or experimental) and theoretical terms. Whereas proper theories introduce theoretical terms, observational laws do not. But how can one make sense of this term-distinction?

Although the law-distinction may seem at first sight to be parallel to the different internal goals of descriptive and explanatory research programs, it is actually not so. To be sure, proper theories cannot be the internal goal of descriptive programs. However, the true description, aimed at by a descriptive program, may concern the true general description, i.e. the true observational theory in the sense of the (relative to the program) complete set of observational laws of the domain, which is an improper theory. The complicating fact is that this true observational theory may also be the true theory at which an explanatory program is aiming, in order to explain and further predict certain observational laws, usually those restricted to some sub-vocabulary of the explanatory program. In terms of the term-distinction the failing parallel is even easier to formulate: whereas a descriptive program can not by definition introduce new theoretical terms, an explanatory program need not introduce such terms. When it does, it is frequently called a theoretical (explanatory) program.

The point of departure of the classical logical empiricists was a theory-free, hence theory-neutral, observational vocabulary. Starting from this postulate their explication of the distinction was obvious. Observational laws were by definition all those laws that could be formulated in this observational vocabulary. Proper theories on the other hand introduce new concepts not belonging to this observational vocabulary. Given their preference for the observational vocabulary the important question remained whether the new terms introduced by these theories could be reduced, in some way or other, to the observational vocabulary. However this may be, the existence of a theory-free observational vocabulary and the law-distinction were interwoven for the logical empiricists. Looking backwards, the standard examples of observational laws, such as Galilei's law of free fall or the Balmer series ordering the spectral lines of hydrogen, must have been dubious from the very beginning, for they are, at least *prima facie*, not couched in a pure observational vocabulary.

Gradually it became clear, even in empiricist circles, that the postulate of a neutral observational vocabulary was an unfortunate creation of the empiricist mind, a paradigm of wishful thinking not corresponding to anything in the empirical sciences. Non-empiricists were eager to embrace the doctrine that all observation was theory-laden. The most popular became the other extreme view, called meaning holism, which states that all terms occurring in a theory are laden with that theory, with the immediate consequence that an interesting distinction between a theory and the observational laws explained by it became impossible.

Empiricists, like Nagel (1961), Hempel (1966, 1970) and Sneed (1971), started to elaborate the idea that certain terms occurring in a theory may be laden with that theory, whereas other terms may not. These latter terms may, however, nevertheless be laden with observational laws and theoretical connotations. In the relevant literature, however, these 'theory-relative ideas' have been presented or at least understood as just a reinterpretation of the *two-level* distinction between an observational and a theoretical level. These two levels may enable accountability for part of the dynamics in science, the short-term dynamics, in particular the interaction between invention, evaluation and correction of observational laws and proper theories. However, the picture hides the long-term dynamics. When a proper theory is accepted as (approximately) true, it usually enables the establishment of criteria for the determination of its theoretical terms. In this way it becomes an observation theory, and the corresponding theoretical level transforms into a higher observational level, enabling new observations and hence the establishment of new observational laws, requiring new, 'deeper' theories to explain them. Moreover, the acceptance of a theory enables experimental or technological applications of the theory, that is, applications presupposing that it is true. In terms of the first chapter, in both cases we have entered the external phase of the program that generated the accepted theory.

In this chapter we will show that the theory-relative ideas essentially lead to the suggested *multi-level* picture of knowledge and knowledge development, such that a two-level picture may either concern just a fragment of it or must be the result of a pragmatic contraction of essentially different observational levels. Only from this multi-level picture does it become clear that the theory-relative move is a way of rejecting the idea of a neutral observational vocabulary that enables a new explication of the intuitive law-distinction such that it not only accounts for the short-term dynamics, but also for the indicated long-term dynamics.

However, in terms of the first chapter, it is the (relatively) short-term dynamic role of the law-distinction that relates to the development of research programs.

The first context in which the law-distinction is used in short-term dynamics is just one explanatory research program, more in particular a theoretical program. Here a proper theory is revised on the basis of its successes and failures in explaining and predicting observational laws. The second context is the interaction between a theoretical program and a descriptive program. Here the two programs develop hand in hand, each challenging the other with new results. On the one hand, there are potential observational laws that have been predicted by the theoretical program and which have to be tested within the descriptive program. On the other hand, there are observational laws that have been established independently, in particular inductively, within the descriptive program (which may be called to that extent an inductive program) and that have to be explained by the theoretical program. Finally, the third context is of course the competition between two theoretical programs claiming to be superior with respect to the explanation and prediction of observational laws for a certain domain. Of course, the latter laws may or may not hang together within a descriptive program. In all three contexts the relevant theoretical programs are in the internal phase, whereas the relevant descriptive programs may or may not already be in the external phase.

In our explication of the term- and law-distinction we do not try to do justice to the way in which some philosophers like to use the distinctions, but to the ways in which scientists use these distinctions. An impressive exposition of the far-reaching theory-laden character of what scientists call observations is given by Shapere (1982) under the revealing title 'The concept of observation in science and philosophy' in which he uses several examples taken from astrophysics. Except when stated otherwise, we will also follow the scientific practice of saying that an observational law is already explained by a theory when it can (approximately) be derived from the theory, whether or not one has good reasons to assume that the theory is true; it may even be known to be false.

After the presentation, in Section 2.1, of some clear examples of observational laws and related proper theories, and a preliminary inventory of the characteristic differences, as potential conditions of adequacy, we will introduce in Section 2.2 the theory-relative distinction between theoretical and non-theoretical terms and use this distinction of terms for an explication of the law-distinction. In Appendix 1 we show that the ideal gas law, being *prima facie* a proper theory according to the law-distinction, is on closer inspection an observational law, as one might hope. The explication of the law-distinction will then make the postulate of a multi-level hierarchy of knowledge in terms of observational laws and proper theories highly plausible. The law-distinction will function as the construction principle for this hierarchy.

In Appendix 2 we will pay attention to the surprising fact that Popper pays so little attention to the law-distinction. He was not only the first proponent of the view that all observation is theory-laden, but he did so without falling victim to the other extreme of meaning holism. It will be shown that it is also possible from his 'basis-relative' perspective to explicate the law-distinction. Given that this is not a difficult task and given Popper's evident interest in the internal mechanisms of the development of knowledge, his neglect of the distinction is indeed surprising. It will appear to be instructive to dwell upon the good and poor reasons that may have been responsible for that neglect.

In Section 2.3 it will also be shown that the developed perspective of Section 2.2 sheds new light on the idea of theory-laden observation. Three related notions can be clearly distinguished: theory-laden, theory-relevant and theory-guided observation.

In Section 2.4 we will first give an elementary account of the structure of theories, starting with the important distinction between epistemologically and ontologically stratified theories. In Chapter 12 we will present the sophisticated structuralist representation of (the structure of) theories. We will close the section by briefly characterizing, mainly in terms of aspects of theories, the leading epistemological positions: epistemological relativism, along with observational, referential, constructive and essentialistic realism.

2.1 EXAMPLES AND *PRIMA FACIE* CHARACTERISTICS

Introduction

We start the explication of the law-distinction by listing first a number of evident examples of both entities, and a number of *prima facie* characteristic differences that may serve as conditions of adequacy. Here, and later, we will speak of testing a (complex) claim when we are only interested in its truth-value, and of evaluation when we are interested in its merits and failures. The first is usually the case with potential observational laws and the second with proper theories.

2.1.1 *Examples of Proper Theories*

In this section we will use 'theory', except when otherwise stated, to refer to a proper theory, a concept which is exemplified by the following theories represented here by a brief statement of their core ideas:

(a) Newton's theory of gravitation, which states that all physical objects have a definite mass, that the sum of all forces exerted on an object is equal to the product of its mass and its acceleration, and that two objects exert an attractive force on each other proportional to their masses and inversely proportional to the square of their distance.
(b) The kinetic theory of gases. This theory postulates that gases consist of particles, called molecules, which exert forces on each other and which move in accordance with Newton's laws of motion.
(c) Dalton's theory of the atom (the example of Chapter 1). This theory claims that all chemical substances are composed of indivisible atoms. According to the theory these atoms can group together in certain ways to form molecules. The formation of molecules is associated with chemical reactions. Chemically pure substances are supposed to consist of one type of molecule.

(d) Bohr's theory of the internal structure of the atom. According to this theory atoms are particles consisting of a nucleus and one or more electrons which circulate around the nucleus in fixed orbits. However, the electrons can jump from one orbit to another, absorbing or emitting electromagnetic radiation at the same time. This example is dealt with in detail in ICR (Chapter 11) as an example of potential truth approximation.

(e1) Mendel's theory of genetics. According to Mendel the characteristics of organisms are inherited by means of genetic factors. For each feature there are two different factors in the game, each individual has a combination of two factors, the same or different ones, and each parent transmits one of them to its offspring. That this selection is by chance amounts to Mendel's first law, while the fact that the transmission of factors related to different types of characteristics is independent is the second law.

(e2) The theory of chromosomes. This theory states that the nucleus of each cell of an organism contains a number of so-called chromosomes, consisting of two separate threads, called the chromatids. Each parent transmits by chance one chromatid to its descendant. The link with Mendel's theory results of course from the fact that genetic factors are materialized in the chromosomes and that the factors of any pair are supposed to be located on the corresponding positions of the two chromatids of a chromosome.

(e3) The molecular theory of genetics. This theory tells that the characteristics of organisms are determined by the specific structure of the so-called DNA-molecules, which each individual carries in the nucleus of all of its cells. Moreover, this information is transformed by a special molecular mechanism to the offspring. The link with the previous theory results of course from the fact that molecular theory analyzes the chemical composition and working of the chromosomes.

(f) Festinger's theory of cognitive dissonance. According to this theory the presence of cognitive dissonance, being psychologically uncomfortable, gives rise to pressures to reduce the dissonance and to achieve consonance. The strength of the pressures is a function of the magnitude of the existing dissonance.

(g) Utility theory or rational-choice theory. According to this theory people choose out of a set of alternative actions the action from which they expect the highest utility.

2.1.2 *Examples of Observational Laws*

The mentioned theories are said to be able to explain the following observational laws:

(a*) Galilei's law of free fall stating that falling objects near the earth have constant acceleration.

(b*) The law that the velocity of sound is higher in gases with a lower density.
(c*) Proust's law (i.e. the law of definite proportions, RL2 in Chapter 1) according to which chemical compounds always decompose into component substances with constant weight ratios.
(d*) The Balmer series, which states that the wavelengths of light emitted by glowing hydrogen gas fit in a simple algebraic series.
(e*) Mendel's interbreeding law on the fact that inherited characteristics manifest themselves after two generations in a certain statistical pattern.
(f*) The (quasi-)law stating that when people have made a decision there is active seeking out of information which is consistent with the action taken.
(g*) The macro-economic consumption function, which claims that total national consumption increases with increasing (average, and hence) national income.

2.1.3 *Some Characteristic Differences*

Let us now mention a couple of the characteristics of observational laws and proper theories, features that can help to consolidate the intuitive distinction of these two types of statements. We begin with an unimportant difference. To call a statement an observational law means that it is well enough supported that it may be assumed to be (approximately) true. On the other hand, talking about a theory does not imply any veracity. Here we are essentially concerned with potential observational laws and theories apart from their truth-value, i.e. as hypotheses that may be true or false. Let us now turn our attention to relevant differences.

(i) Whereas an observational law is usually represented as one, possibly complex, statement, a theory is usually presented as a system, a coherent set, of statements. Of course, this does not exclude the possibility of an artificial representation of a theory as one conjunctive statement. With or without some extra definitions, even a reformulation in an elegant compact statement may be possible, in which case it is again tempting to speak of a law. The ideal gas law (see below) and the law of Archimedes (the upward force exerted on a solid body in a fluid is equal to the weight of the displaced fluid) are examples of this.
(ii) An observational law specifies what will happen under certain experimental conditions. Hence, it gives a partial characterization of what is, not only conceptually, but really possible in the context. The claim of a theory may be stronger: it may not only specify some necessary conditions for being really possible, it may claim to give a complete characterization of what is really possible in the context. But such a (relative) completeness claim is certainly not associated with every theory.

The first two differences do not only leave room for proper theories, but also for 'observational theories', i.e. coherent sets of (potential) observational laws for a certain context. Moreover, it may or may not be possible to summarize any theory in an elegant compact statement, and there may or may not be associated with it a completeness claim. Hence, the first two *prima facie* differences are not acceptable as strict conditions of adequacy.

(iii) Proper theories, however, not only use concepts which are used in the observational laws to be explained, but introduce also new concepts, called 'the theoretical terms' of the theory. For instance, Newton's notions of mass and force do not occur in Galilei's law; Dalton's concepts of atom and molecule do not occur in Proust's law; the notions of subjective utility and probability do not occur in the consumption function, etc. (Of course, it may be that old terms are used, but then their old meaning is replaced by a new meaning provided by the theory.) On the other hand, observational laws do not introduce such new terms; for all non-logico-mathematical terms occurring in them there are independent application criteria in the form of experimental and argumentative procedures.

(iv) If an observational law can be explained by a theory, it can nevertheless be tested independently from that theory. This is of particular importance when some potential (corrections of) observational laws are predicted by the theory (and hence can be explained by it), and have still to be tested.

(v) The same observational laws can in principle be explained by different theories. It is for example conceivable that there would have been developed a new theory explaining the same laws as explained by Dalton's theory, in which the notion of atom did not occur, although one or more rather different notions did occur. Hence, a theory can be rejected, without the consequence that the observational laws explained by the theory are also dragged down in its fall. When Bohr's theory of the structure of the atom was rejected, this did not imply that the Balmer formula lost its descriptive adequacy.

From the classical logical empiricist point of view it was plausible to think that the fundamental difference between observational laws and theories, responsible for the above mentioned *prima facie* differences is that observational laws are or at least can be expressed in pure observation terms, free from further assumptions. Although this assumption might be able to explain the differences, it should be stressed that nothing which has been said so far implies that observational laws express regularities which can be established by pure observation, i.e. observation not presupposing instruments or assumptions. On the contrary, it is not difficult to see that the testing of the observational laws mentioned presupposes all kinds of auxiliary assumptions.

Let us consider, as a tribute to Nagel, who used the same example, the testing of the innocently looking law (b*) about the velocity of sound in gases. To test this law we have to know how to produce and to register sound, and how to measure its velocity; further we should know how to distinguish gases from substances in other aggregation phases, such as a liquid and a solid state, and how to measure the density of gases. All these identification and measurement procedures presumably presuppose the truth of certain theories. The measurement of the (mass-) density for instance requires the measurement of volumes and masses: both presuppose at least some general assumptions of stability and the like, and the first presupposes in principle a (naive or sophisticated) theory of space geometry, the second a theory of mechanics. Moreover, replication-measurements seldom lead to exactly the same results: to arrive at unique values on the basis of the test results presupposes general principles of dealing with 'measurement mistakes'.

But if observational laws have no immediate relation to reality, what then is the fundamental difference between observational laws and proper theories? As suggested, the characteristic differences (iii)–(v) will serve as conditions of adequacy in the explication to follow.

2.2 THEORY-RELATIVE EXPLICATIONS

Introduction

We will start with the theory-relative explication of the distinction between theoretical and non-theoretical terms, after which the explication of the law-distinction will be possible. This will naturally lead to the epistemological hierarchy of knowledge.

2.2.1 *Theory-Relative Theoretical and Non-Theoretical Terms*

Let us consider in some detail the theory that is supposed to explain (b*), viz. the kinetic theory of gases (b). In the context of this theory, sound is associated with wave movements jointly performed by the gas particles under certain conditions, and the velocity of sound then is identified with the velocity of these waves. The mass density of the gas is in this theory identified with the product of the number of gas particles per unit volume (the number density) and the mass of one particle. Theory (b) together with the mentioned auxiliary assumptions explains the law (b*) in the sense that the law is derivable from it.

Let us now look at the (non-logico-mathematical) terms of the theory, i.e. 'gas', 'gas particle', 'sound', 'velocity of sound', 'wave movement performed by gas particles', etc. It is easy to see that some of these terms can be understood independently from the kinetic theory of gases, viz. 'gas', 'mass density of a gas', 'velocity of sound in a gas', etc. We know their meaning even if we do not yet know the kinetic theory. But also within the context of this theory these terms still have the same meaning: for example, it is still the case that we indicate with 'gas' a substance which is in certain respects different from liquid and solid substances. The same can be said about terms like 'sound' and 'velocity of sound': they have a clear meaning independent of the theory and they retain this meaning within the context of the theory. They are 'antecedently understood', to use Hempel's phrase.

Let us now turn our attention to terms like 'gas particle', 'wave movement performed by gas particles', etc. These terms are not antecedently understood. On the contrary, what we have to understand by a gas particle is specified, or implicitly defined by, the theory itself, because it is the theory that introduces the term. By consequence, the correct use of the term presupposes the truth of the theory.

We will use the last point as the basic criterion for a general distinction between two kinds of (non-logico-mathematical) terms in relation to a statement *S*. We say that term *t is S-laden* if the correct use of *t* presupposes the truth of *S*, at least to some extent, and we say that *t is S-free* if *t* is not *S*-laden. We assume that this criterion can be made precise in such a way that it can always be applied unambiguously. Note that we do not assume that *t* occurs in *S*; in this way we leave room for the case that *t* may be, in some way or other, indirectly laden with *S*. It is also plausible to define that statement *S*1 *is (un-)laden with S*2 if *S*1 does (not) contain terms that are laden with *S*2.

Applying this definition to a theory *X*, conceived as a complex conjunction of statements, *X*-laden terms are also called *theoretical terms with respect to X* or *X-theoretical* terms, and *X*-unladen terms are called antecedently understood or *non-theoretical terms of X* or *X-non-theoretical* terms.

It is important to notice that in this way we do not make an absolute distinction between two kinds of terms in scientific language in general, but a *theory-relative* distinction: a term like 'mass-density' is non-theoretical with respect to the kinetic theory of gases. But, as noted before, the correct use of this term, defined as mass per volume unit, presupposes the truth of general assumptions and other theories, concerning the (macroscopic) notions of volume and mass. 'Volume' is theoretical with respect to Euclidean geometry; 'mass' is laden with Newtonian mechanics. With respect to these theories the term 'mass-density' is not antecedently understood but theoretical.

The foregoing definition of *X*-non-theoretical terms may even be liberalized in two respects. First, it may well be that the theory leads to a meaning enrichment in the sense that the theory may provide new criteria of application of the term to the already existing criteria. A new way of determining the term may be the result. Second, it may even lead to a proper meaning change in the sense that the old criteria of application are changed, but then only in such a way that the new criteria of application, though suggested by the theory, do not invoke it. In the following the concept of *X*-non-theoretical terms is taken in this sophisticated sense.

In the context of a particular theory *X* the terminology can be simplified by just speaking of theoretical and non-theoretical or even observation terms when the theory-relative qualifications are meant. The qualifications 'non-theoretical' or 'observation(-al)' may then of course not be misunderstood as implying 'not laden with theories'.

2.2.2 *Observational Laws as Improper Theories*

With the distinction between *X*-theoretical and *X*-non-theoretical terms we are close to a general explication of the intuitive distinction between observational laws and proper theories. The following formulation seems adequate at first sight: a theory is only a proper theory when it has at least some theoretical terms of its own, i.e.

terms laden with the theory itself. An observational law, on the contrary, is an improper theory, a theory which has no theoretical terms of its own, i.e. no terms laden with the law itself. According to this characterization an observational law does not contain terms for which the correct use depends on the truth of the law.

It is easy to check that law (b*) satisfies this condition, and also that the other examples of observational laws satisfy it. However, there are also examples of laws which are not, according to the proposed definition, observational laws because they have theoretical terms of their own, whereas we are intuitively inclined to qualify them as observational law. A nice example is the ideal gas law $PV = RT$ (P: pressure; V: volume; T: empirical absolute temperature; R: the ideal gas constant). Everyone calls it an observational law, whereas it is at the same time generally known that T and R are laden with the law itself in one way or another, with the consequence that according to our definition the law has to be qualified as a proper theory.

Closer inspection (Kuipers, 1982) shows that the situation is as follows. Some observational laws can be formulated in the strict sense suggested above (hence, without R and T and also without other theoretical terms of their own) which are together sufficient to define R and T explicitly, with the surprising consequence that their conjunction is precisely equivalent to the, indeed very elegant, ideal gas law. Hence, although the terms R and T are, according to our theory-relative distinction between theoretical and non-theoretical terms, theoretical with respect to the ideal gas law, these terms can be explicitly defined on the basis of observational laws in the strict sense, and hence can be eliminated. The example is presented in some detail in Appendix 1.

Many scientific terms are introduced by explicit definition on the basis of observational laws in the strict sense. For instance, these laws provide the existence and uniqueness conditions enabling the explicit definition, which is also the case in the above example. Another example of this kind is the notion of weight when based on the observational laws of the slide-balance, to be treated in Chapter 12.

Given the frequency of such definitions it is worthwhile to take them into account in our final definition. In this definition we will use 'lawlike statement' as a primitive term indicating general statements that may be considered for the qualification 'observational law' or 'theoretical law'. Moreover, we will make a distinction between two kinds of inductive jumps that are made by accepting lawlike hypotheses as true. In case of *observational induction* or inductive generalization one essentially remains within the available observational vocabulary. In the case of *theoretical induction*, the acceptance of the relevant statements, based on the observational laws they can explain, implies the conclusion that the new terms refer.

Definition:
An *observational hypothesis in the strict sense* is a lawlike statement not containing theoretical terms of its own. It is called an *observational law in the strict sense* when it is accepted as (approximately) true, a condition which

requires *observational induction*. An *observational hypothesis* is a lawlike statement containing at most theoretical terms of its own which can be eliminated with the aid of an explicit definition based on observational laws in the strict sense. An observational hypothesis is called an *observational law* when it is accepted as (approximately) true, a condition which requires indirectly the observational inductions necessary for the defining observational laws in the strict sense and perhaps also directly some new observational inductions.

An *observational theory* is a coherent set of observational hypotheses. There may or may not be an associated (relative) completeness claim to the effect that the conjunction of the observational hypotheses is the strongest true observational law for a certain domain that can be formulated with a given set of terms not laden with this particular potential law in a non-eliminable way. If there is such a completeness claim we speak of a *strong* observational theory. An observational theory is called an *observational observation theory*[1] when it is accepted as (approximately) true, e.g. as the strongest (approximately) true observational law, and used for observation.

A *proper theory* is a coherent set of lawlike statements, called *theoretical hypotheses*, their conjunction containing at least one non-eliminable theoretical term of its own; there may or may not be associated a completeness claim with it. If there is such a completeness claim we speak of a *strong* proper theory. If the theory is accepted as (approximately) true, which requires a *theoretical induction*, the constitutive statements are called *theoretical laws*, and the theory itself, when used for observation, a *theoretical observation theory*.

Finally, a lawlike statement is an *observational law with respect to a theory* if it is an observational law which is not laden with that theory.

One might question whether the definition of a 'proper theory' cannot better be replaced by a corresponding definition of 'theoretical hypothesis'. This would however be an unfortunate move, because of the fact that the theoretical hypotheses constituting a theory are usually interwoven in such a way that an isolated evaluation of, for example, the eliminability of terms would be unjustified.

It is also important to note the following. It frequently occurs that certain statements are considered as observational laws (to be) explained by a certain theory *X*, whereas they are in fact clearly formulated in *X*-theoretical terms. In such a case, if it is right, a reformulation should be possible which avoids the use of *X*-theoretical terms. The resulting *X*-non-theoretical statements are, in that case, the genuine observational laws (to be) explained. For example, the results of Wilson chamber experiments, designed for the evaluation of elementary particle physics, are usually couched in terms of orbits described by the particles postulated by the theory under evaluation. But the relevant aspects of these evaluation results can be formulated in terms of traces of water-drops.

From their definition it is directly clear that observational laws and proper theories satisfy the intuitive characteristic difference (iii), which was proposed as condition of adequacy, and that they do not satisfy (i) and (ii), for an observational law may be a compact reformulation of an observational theory, and such observational laws as well as proper theories may or may not be relatively complete, in contrast to what (i) and (ii), respectively suggested. The defined distinction also satisfies the conditions of adequacy (iv) and (v). An observational law can be tested independently from the theory which is supposed to explain the law, and it can remain well supported even when that theory becomes falsified, in which case the question then is: by what other theory can it be explained? This is easy to check in our example of the law about the velocity of sound in gases and the kinetic theory. If, however, a statement contains *X*-theoretical terms, i.e. when it is *X*-theoretical, it cannot be tested independently from *X*. Take for instance the statement that gases consist of particles. To test this statement we will have to know what to understand by gas particles, but it is precisely the kinetic theory that specifies this. We will have to presuppose this theory, or at least a part of it, in order to test the statement in question.

It may be true that a potential observational law can be tested independently of the theory that is supposed to explain it, but this fact makes testing not just the unproblematic affair suggested by common scientific parlance. For there are always underlying observational laws and (observational or theoretical) observation theories in the game. In most cases, however, the underlying laws and theories, with which the potential observational law in question is laden, are not in dispute; to put it differently, in most cases the underlying laws and theories belong to the background knowledge, i.e. they are assumed to be true. Against this background, assuming it as 'underground', one wants to know whether the potential observational law itself is true and whether the theory proposed to explain it does indeed imply the law. This brings us to the hierarchy of knowledge.

2.2.3 *The Epistemological Hierarchy of Knowledge*

Given that one statement may or may not be laden with another there are three possible relationships between two statements. They may be *interwoven* in the sense that the one is laden with the other and vice versa. Two statements are *disconnected* when neither of them is laden with the other. Finally, *S*1 is an *underlying* statement of *S*2 when *S*2 is laden with *S*1, but not vice versa. Of course, being an underlying statement of another is an asymmetric relation and it is safe to assume that it is also transitive, although exceptions to such a case are not inconceivable.

Conceiving observational laws as well as proper theories as (complex conjunctions of) statements, the relation of being an underlying statement of another leads to interesting cases of observational laws and proper theories underlying other observational laws and proper theories.

It is instructive also to consider the relation of an observational law being explained by a proper theory. Such a situation is of course an asymmetric relation. We do not presuppose that explanation of observational laws is always deductive; in Chapter 3 we will include a corrective type of explanation in which case only a deductive explanation of some approximation (a corrected version) of the law to be explained can be given. In that chapter we will also see that explanations of laws by proper theories always need auxiliary hypotheses, although they will not often be mentioned in the following.

If an observational law is explained by a proper theory then it has to be of course an observational law with respect to that theory in the sense defined above and hence the law may not be laden with the theory. The converse, however, is not excluded: an observational law *L* explained by proper theory *X* may or may not be an underlying law of that theory, depending on whether or not *X* uses terms presupposing *L*. Moreover, if *L* is explained by *X* and if, in addition, *X** is an underlying proper theory of *L*, then *X** will also be an underlying theory of *X*. Note, moreover, that it is perfectly possible that two observational laws explained by the same proper theory share one or more underlying proper theories.

On the basis of the foregoing asymmetric relations there arises a hierarchy of crucial pieces of knowledge: a proper theory *X*, the observational laws explained by *X*, and their underlying proper theories and observational laws.

In Figure 2.1 we represent the suggested ordering, which we will call the epistemological hierarchy of the context.

The diagram indicates, for instance, that proper theory *X* explains observational law *L* and that *X*1 (or *L*1) is an underlying proper theory (or observational law) of *L*, and hence of *X*. Of course, Figure 2.1 presents only a connected fragment of knowledge which can in principle be extended to all sides. For example, if *X* has been accepted as (approximately) true, it can then be used as an observation theory

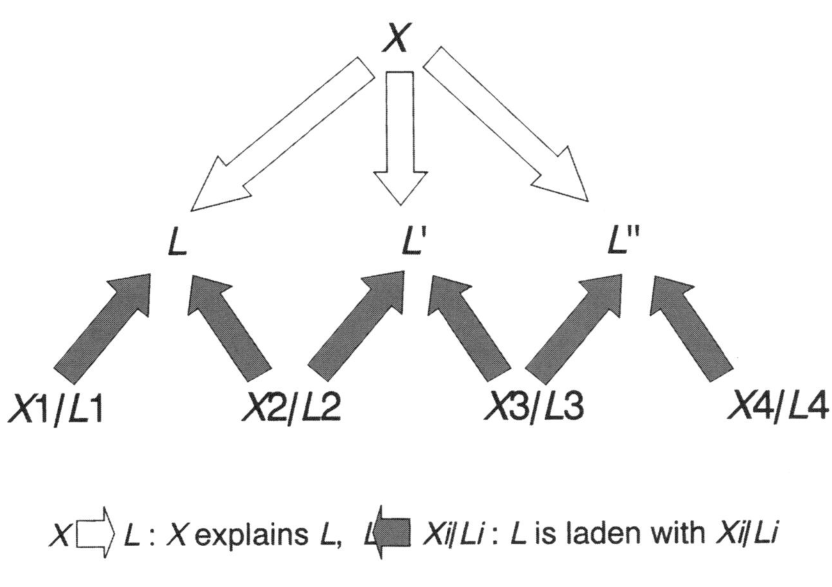

Figure 2.1 *Fragment from the hierarchy of knowledge.*

with corresponding (new) observation terms. This use of it may lead, in combination with other accepted observation terms, to new observational laws. Furthermore, Figure 2.1 neither forces us to assume that there are fundamental observational laws and proper theories without underlying proper theories, neither must we assume that they do not exist.

It is important to note that Figure 2.1 presents an epistemological order which may not be interpreted ontologically in the sense of the lower the fragment the deeper the concerned level of reality. On the contrary, in particle physics and Mendelian genetics for instance, the ontological whole-part relation will roughly correspond to the upward direction. However, in cosmology for instance, the upward direction will sometimes correspond to the part-whole relation, e.g. from theories about heavenly bodies to theories about galaxies. Hence, there is not supposed to be any standard correspondence between our epistemological hierarchy of knowledge and an intuitive ontological hierarchy of the corresponding objects of knowledge.

The crucial question is of course whether local epistemological hierarchies frequently occur in scientific practice. Inspection teaches us that the observational law (b*) about the velocity of sound, explained by the kinetic theory (b), does not indeed contain terms that are laden with the kinetic theory. Moreover, we have already noted that the law contains terms, such as 'mass density', that are laden with other theories, such as space geometry and mechanics. It is easy to see that these theories are laden neither with the law nor with the kinetic theory, hence they are indeed underlying theories of the law as well as of the kinetic theory.

Neither is it difficult to verify that the other examples of proper theories (a), (c)–(g) and observational laws (a*), (c*)–(g*) explained by them are in accordance with the hierarchy.

The picture of the hierarchy has been restricted to the most essential elements. The following three features have been left out. First, an explanation of an observational law by a proper theory always needs auxiliary hypotheses, including observational laws and proper theories. Second, domains and subdomains have been overlooked. Third and finally, besides observational laws explained by proper theories there are also other important forms of explanation, in particular, an observational theory explaining an observational law or theory and, last but not least, a proper theory explaining another proper theory.

We confine ourselves to some examples of the last kind: Bohr's theory of the atom (d) explains (a corrected version of) Dalton's theory (c). Mendel's theory (e1) is explained by the theory of chromosomes (e2), which is in its turn explained by the theory of molecular genetics (e3).

It is interesting to dwell further upon Mendel's theory itself. It does not only explain observational laws it also explains the core theory of population genetics, constituted by Hardy–Weinberg law, which states that, when there are no outside influences and no mutations, the gene ratio in a population remains constant over the generations. It is clear that this law is laden with Mendel's theory, so it certainly is no observational law with respect to Mendel's theory. On the other hand it predicts patterns of inheritance of outer characteristics that are, when approximately true, observational laws with respect to Mendel's theory as well as population genetics.

In terms of the first chapter the situation can be illuminated as follows. Even before Mendel's theory became accepted as (approximately) true; it became the object of explanation by deeper theories, which led to the theory of chromosomes and molecular genetics. When Mendel's theory became accepted, its terms became available as (relative) observation terms and one started to elaborate its consequences for large populations, not so much as a new way to evaluate it, but as a supply program for population genetics. The Hardy–Weinberg law was the resulting idealized point of departure, requiring concretization in both senses distinguished in the first chapter. On the one hand researchers started to explore the changing conditions under which the law could still exactly be derived from Mendel's theory. On the other hand they tried to account for all kinds of disturbing influences, such as mutations and genetic drift, by correcting the law (Lastowski, 1977).

Returning to our general discussion, the three elements not included in the above diagram would not alter the hierarchical nature of a more refined picture for the two relations involved: 'to be underlying' and 'to be explained by' remain asymmetric. We are therefore justified in using the simplified figure.

The underlying theories in the diagram essentially represent the proper theories and observational laws with which the terms occurring in the explicitly represented observational laws are laden. But in their turn, these proper theories and observational laws explain (other) observational laws, formulated in terms laden with other proper theories and observational laws, of still lower levels. By way of contraction we can collect together all the terms of all lower levels. Let us call this the contracted observational level, the observational level in short. Hence, the terms of the observational level presuppose the (approximate) truth of all their underlying proper theories and observational laws, in consequence, the truth of all the observational laws explained by them. The combination of these laws and theories is called the background knowledge.

The acceptance of a (general) observational law, however it was generated, logically requires an 'observational inductive jump' from a finite number of singular observation statements, i.e. the data. The acceptance of a proper theory also requires a 'theoretical inductive jump' from the observational laws explained by it. Hence, although the background knowledge may be rather strong, it is not at all a foundation deductively based on data. It is a hierarchically ordered set of observational and theoretical inductions. In ICR (Chapter 13) we speculate more about its lowest level, if that exists at all. We argue there that all terms are at least laden with some general assumptions, e.g. concerning durability and mutual relations, which guide their application. Apart from that, terms may be indirectly or directly applicable, that is, their application may or may not presuppose (observational or theoretical) inductions.

So far we have neglected a complication that arises due to the fact that one may accept a proper theory as true as far as its observational consequences are concerned, and that one may use it as an observation theory as far as its observation terms are concerned. Such occurrences may be called the *empiricistic* acceptance of a proper theory and the empiricistic use of a proper theory as an observation theory, respectively. Of course, the empiricistic use of a proper theory at least presupposes the empiricistic acceptance of it, and the inductive jump required for the empiricistic acceptance of a proper theory amounts to the observational induction or the theory's observational consequences. In contrast to the empiricistic acceptance and use of a proper theory we may call the above straightforward definitions of the acceptance and use of a proper theory *realistic*. That is, the realistic acceptance of a proper theory amounts to the theoretical induction of all consequences of the theory, including its referential claims. The realistic use of a proper theory as an observation theory amounts to its use for the application of all its terms, which of course presupposes its realistic acceptance.

The foregoing distinction between realist and empiricist attitudes is an initial indication of the different kinds of epistemological positions that are at present taken seriously. In the final section of this chapter we will present a hierarchical survey of such positions.

In this section we have discussed the relative distinction between the theoretical and the observational level of a specific theory, and developed a theory-relative distinction of levels. However, these distinctions did not prevent us form giving a theory independent characterization of observational laws. This approach to the level distinction is primarily derived from Sneed (1971), but it was also anticipated by Nagel (1961) and Hempel (1966). There is, however, another, completely compatible, approach the level distinction and to the possible law-distinction, one which was also anticipated by Nagel and Hempel, and which is in addition easy to connect with Popper's work. In Appendix 2 we briefly present this approach and then speculate about Popper's motives for neglecting the law-distinction.

2.3 THEORY-LADENNESS OF OBSERVATION

Introduction

The foregoing analysis of proper and improper theories can also be used to throw light on the so-called theory-ladenness of observation. Although it may be argued that all observation is in a sense theory-laden, we will see that this position does not imply that all observation is laden with any theory for which that observation is relevant, even if that observation was guided by that theory.

2.3.1 *Theory-Laden Observations*

The insight in modern philosophy of science that all statements about the world, however direct and unproblematic they may seem, rest on certain general assumptions, originates to a large extent with Karl Popper. For some other philosophers of science, such as Paul Feyerabend (1962, 1975) and, to a lesser degree, Thomas Kuhn (1962/1969, 1963), this insight has some rather negative implications for the possibility of theory evaluation. Such thinkers hold that all statements formulated in the context of a certain theory are laden with that theory. So it seems that Feyerabend's ideas about theory-ladenness of terms boil down to the claim that all non-logico-mathematical terms are laden with all theories in which they occur. But this would inevitably lead to the consequence that, for instance, evaluating the theory of Newton with Galilei's law of free fall cannot be more than a circular procedure. For under these circumstances the empirical establishment of the regularity expressed by Galilei's law would already be laden with the theoretical principles of mechanics or, as it is also stated, the meaning of the terms occurring in the formulation of the law (distance, time) is determined, among other things, by the principles of mechanics. Some of Kuhn's expositions seem to lead to the same conclusion.

The correct answer to this threatening impasse seems to be the following. We can concede to authors like Feyerabend and Kuhn that such statements as the distance covered by an object near the earth is proportional to the square of the elapsed time required are laden with theories; viz. theories concerning space and time measurement. But such statements therefore do not need to be laden with the specific principles of mechanics. In other words, the meaning of concepts such as position and time, needed for the formulation of such laws, may be fixed in and by theories existing independently from mechanics. In fact, it is in principle possible to reject mechanics, without changing our opinions about space and time measurement. For this reason, the proposed way of evaluating mechanics is not circular, although it is a conditional evaluation: assuming that the theories with which Galilei's law is laden are true, Newtonian mechanics is supported by this law, at least in the sense that it can explain it. As is well known in the history of science, the condition had to be abandoned: Newton's ideas about space and time had to be fundamentally revised, a need that was satisfied by Einstein's theory of relativity. However, this does not exclude the possibility that the described condition could have been fulfilled, and that the suggested positive result of the conditional evaluation is still defensible as a first approximation of actual history.

The above mentioned discussions are frequently put in terms of the theory-ladenness of observations or facts, instead of statements. We shall reformulate the purport of the foregoing in terms of observations. To begin with, it is plausible to make a sharp distinction between the unspecific statement that a certain observation is theory-laden and the specific statement that this observation is laden with theory *X* or, in short, is *X*-laden.

We call an observation (or fact) *X*-laden when it is, and has to be, formulated in an *X*-theoretical statement, that is, a statement essentially using an *X*-theoretical term in the sense of Subsection 2.2.1. *X*-laden observations can not of course be used to evaluate *X* without being confronted with the kind of circularity of which Feyerabend and Kuhn were thinking.

Fortunately, however, not all observations that might be relevant to *X* are laden with *X*: there are also *X*-unladen observations, i.e. observations that are or can be phrased in terms of *X*-non-theoretical statements. Such observations can of course be used to evaluate *X* without running into circularity problems. Though *X*-laden observations cannot be used to evaluate *X* itself, they can perhaps be used to evaluate another theory, in which case it is necessary to presuppose *X* as an unproblematic background theory.

Let us now also consider the unspecific statement: 'All observations are theory-laden'. The logical empiricists assumed in the beginning that there are observations which are not laden with general assumptions at all. According to them, the corresponding statements which would hence be testable without assuming any theory constituted the class of neutral, theory-free observational statements. The insight that this class is empty might be called Popper's insight: all observations are theory-laden, or at least laden with general assumptions. This, however, does not imply the view that all observations that we can make in order to evaluate a certain theory *X* are laden with *X* itself. The latter view was held by Feyerabend and, to a lesser degree, by Kuhn. It has always been severely criticized by Popper, roughly along the suggested lines: for every empirical theory *X* there are theory-laden but *X*-unladen observations, which can be used to evaluate *X*.

2.3.2 *Theory-Relevant and Theory-Guided Observations*

Now we want to discuss two questions which are usually also at stake in discussions about theory-laden observation. The first is that observations may or may not be relevant to or interesting for a certain theory. The second is that (relevant) observations may or may not have been governed or guided by a theory. In both cases we are primarily concerned with observations that are not laden with the theory in question. *X*-laden observations which are not relevant to *X* in one way or another are difficult to conceive, and they are by definition guided in a certain sense by *X*, although it is not necessary that one always realizes this point. *X*-unladen observations on the contrary may or may not be relevant to *X* (*X*-relevant), and if they are relevant to *X*, they may or may not be guided by *X* (*X*-guided). The foregoing is summarized in Figure 2.2, which gives a classification of observations in relation to theory *X*.

As far as relevance is concerned, the foregoing can also be formulated in terms of facts, which may be individual facts or general facts, i.e. observational laws. In this perspective we see that *X*-laden facts are always relevant to *X*, but *X*-unladen facts may or may not be relevant to *X*. For the case of a fact being guided by a theory it would only be possible to speak of 'facts which are or are not revealed by *X*-guided observation'.

The notion of relevance that is at stake in the present context is the idea that a certain (*X*-unladen) fact is relevant to theory *X* if *X* is not indifferent with respect to this fact, i.e. if *X* explains this fact or if it contradicts it, in both cases with or without some relatively unproblematic auxiliary hypotheses.

There is also, however, a second sense of relevance which may be even more important because there need not be any consensus about it among different scientists. Facts with respect to which a certain theory is indifferent can nevertheless be considered as relevant to that theory in the sense that one may

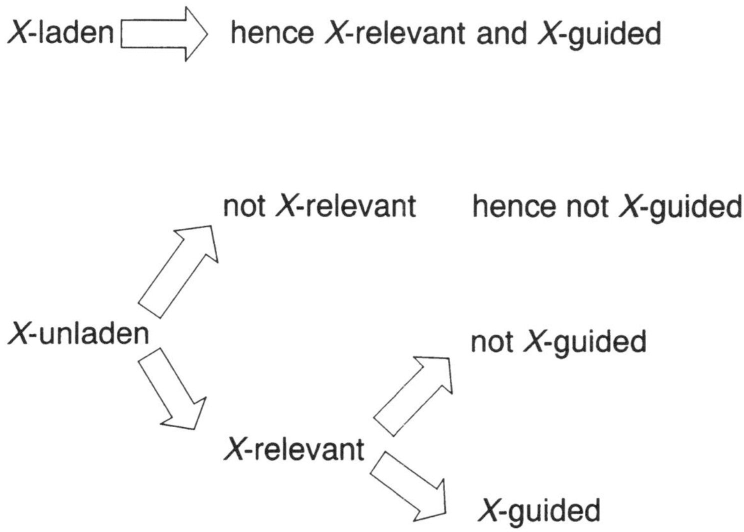

Figure 2.2 Classification of observations in relation to X.

think that the true theory in question should not be indifferent with respect to these facts. The proponent of a certain theory can of course be inclined to consider a fact as irrelevant when his theory is indifferent with respect to that fact.

Newton for instance did not consider it important that his theory neither explained nor contradicted Bode's law, which gives a simple mathematical relation between the radii of the planetary orbits; the indifference of his theory with respect to this law was, according to Newton, no objection to his theory. Kepler, on the contrary, insisted that the true theory about the solar system should be able to explain Bode's law, and Kepler did have an explanation, which he based on Pythagorean ideas about numerical harmony. Following Newton, present day astronomers also think that Bode's law is irrelevant, an accidental feature of actual orbits with a questionable status as a proper observational law. As a result, not only do we not have any explanation of this law but we also do not feel any need to have one. The phenomenon that a later theory does not give an explanation for a fact which had an explanation before but for which an explanation is no longer considered to be required is called *Kuhn-loss*.

If we want to evaluate a theory X we aim at X-unladen observations that are relevant to X in the first sense: we let ourselves then be guided by X in helping us to decide what to pay attention to. But also after the successful closure of the evaluation phase, when the theory has been accepted for the time being, much research is guided by the theory. The periodic table of Mendeleev (PTM) is not only a perfect example of a theory which was evaluated by PTM-unladen, but PTM-relevant and PTM-guided predictions of chemical elements. It later also became an important means for predicting the possibility of artificial production of new elements (see Chapter 12 for some more examples). All these cases concern observations which are guided by a theory but not laden with that theory.

Observations which are not guided by a theory are frequently called 'accidental observations (or discoveries)'. Accidental observations can of course be perfectly relevant to a theory. A nice example is the Balmer series which he discovered by trial and error on the basis of data provided by (the descriptive research program of) Angström. Hence, the discovery was not guided by a specific theory, but at most by some global Pythagorean ideas. Even so it was recognized as very relevant for the later developed theory of Bohr, for that theory was far from indifferent to the Balmer series: it could explain the series.

2.4 THE STRUCTURE OF PROPER THEORIES AND THE MAIN EPISTEMOLOGICAL POSITIONS

Introduction

In a first analysis of the structure of theories we will emphasize the distinction between two main types of stratification, viz. epistemological and ontological. We will also pay some attention to non-empirical theories. Finally, we will present the main epistemological positions with respect to proper (empirical) theories and theoretical terms.

2.4.1 *Epistemological and Ontological Stratification*

Let us start by summarizing some of the main points made in this chapter so far. A proper theory *X* has been defined as an epistemologically stratified theory in the sense that it contains terms, and hence statements, that are laden with one or more of its principles: *X*-theoretical terms. The other terms of *X* are called *X*-non-theoretical. In contrast to proper theories, observational hypotheses are defined as improper theories, containing no theoretical terms of their own. A set of connected observational hypotheses is called an observational theory. It should be noted that being *X*-non-theoretical is a theory-relative, to be precise, an *X*-relative qualification of a term or a statement: they may well be laden with underlying theories. However, when the theory is clear from the context, a point which we will assume from now on, we will simply speak of theoretical and non-theoretical terms and statements, respectively.

The main function of a proper theory is the explanation and prediction of observational laws relative to the theory, i.e. true general hypotheses containing no terms laden with themselves or the theory. For this function the distinction between observational laws and proper theories is of course crucial.

There is much more than this to say about the structure of proper and observational theories. Here we confine ourselves to some main points. In several chapters we will present theories in the so-called structuralist way. In Chapter 12 we will present this approach to theories in detail.

Besides epistemological stratification there is ontological stratification: they frequently go together, but are essentially independent. A theory is said to be ontologically stratified when there are two or more kinds of entities involved and when entities of one of these kinds are components of entities of the other kind. It is then plausible to speak of a lower, micro-level and a higher, macro-level. In this case some principles of the theory concern only the micro-entities, and their properties and relations, and are called *micro- or internal principles*, whereas others connect the different kinds of entities, and their properties and relations, and are called *bridge principles*. The example of the atomic theory dealt with in the first chapter provides a nice example of an ontologically as well as (along the same lines) epistemologically stratified theory. To be sure, auxiliary hypotheses may also have an internal or a bridge character. In the case of the atomic theory, the simplicity auxiliary hypotheses are of an internal nature, whereas the other ones, including Avogadro's hypothesis, are bridge principles.

Another feature of some theories, already met in the first chapter, is that the principles of a theory, whether ontologically and/or epistemologically stratified or not, can frequently be differentiated into core or generic principles, claimed to be true for the whole domain concerned, and special principles, only claimed to be true for a certain subdomain. Of course, a similar distinction can be made for auxiliary hypotheses.

In the case of an epistemologically stratified theory it is plausible to define three types of statements: non-theoretical, purely theoretical and mixed theoretical statements. The division of theoretical statements in purely and mixed ones, however, seems only useful when the epistemological stratification reflects an ontological stratification, in which case the purely theoretical principles, i.e. the internal principles, constitute a clearly separable theory dealing only with the theoretical level. Compare the atomic theory as opposed to the (ontologically unstratified) theory of gravitation. In the latter example the distinction between pure principles (e.g. action = – reaction, the third law) and mixed principles (e.g. '$f = ma$', the second law; and the special law of gravitation) plays no significant role.

In the case of epistemologically and/or ontologically stratified theories there is a natural distinction of two vocabularies: the complete vocabulary in which the theory is formulated, including theoretical and/or micro-terms, and the sub-vocabulary generated by the non-theoretical and/or macro-terms.

Of course, even if none of both stratifications apply, viz. when we are considering an ontologically unstratified observational theory, it may still be useful to make a distinction between the full theory and the corresponding vocabulary and a sub-theory and the corresponding sub-vocabulary. For an observational theory may be designed to explain an observational sub-theory.

Whatever kind of theory, our discussion more or less implicitly assumes that a theory can be formulated in terms of a finite number of principles. This feature can be conceived of as a very informal type of finite axiomatizability, which is a *conditio sine qua non* to talk about a theory at all. However, this condition should not be confused with the claim of finite axiomatizability in the sense of first or higher order logic. As a matter of fact, in Chapter 12 we will only illustrate the structuralist claim that it is possible and instructive for many theories to finitely axiomatize them in the set-theoretic sense of set-theoretic structures, defined by a finite number of axiom schemes, using as much mathematical language as necessary.

Explanation and prediction of observational laws have already frequently been mentioned as functions of theories. As additional functions, or at least additional forms of observational success, we should mention: unification, correction and enrichment. A theory may unify, by explanation, a number of *prima facie* rather heterogeneous observational laws. It may predict successfully a corrected version of an observational law, implying that the latter apparently was at most approximately true. Finally, it may predict observational laws concerning new observable phenomena. Of course, theories may also have theoretical success. One example is the conceptual unification of two previous theories into a new theory which is observationally equivalent to their conjunction. Another example is a theory providing a 'deeper' explanation of a proper theory.

2.4.2 *Conceptual Theories*

Theories are up to now understood as empirical theories. Following Popper, we say that a theory is an empirical theory in the strict sense if it is, in combination with certain special or auxiliary hypotheses, falsifiable, and it is an empirical theory if it is intended to become an empirical theory in the strict sense, i.e. if one aims at special or auxiliary hypotheses that make the theory falsifiable. For instance, a generic theory, like Newton's general theory of motion, may well be unfalsifiable as it stands, but become falsifiable together with appropriate special principles, such as the law of gravitation.

However, it also makes sense to leave room for theories that are not intended to be made falsifiable. In Chapter 12 we will distinguish a number of kinds. Here we will restrict our attention to conceptual theories. A conceptual theory is intended to provide a perspective, a way of looking, at a certain domain without making a general empirical claim.

The claims which are associated or made with a conceptual theory are either logico-analytic or restricted to individual intended applications. A typical logico-analytic claim is a theorem stating that the instances (models) satisfying the theory can be proven to have a certain explicitly defined property. A typical specific claim states that a certain intended application is (or is not) an instance (model) of the special theory. The very distinction advocated between observational laws and proper theories is an example of a conceptual (meta-)theory for the domain of lawlike statements. This example makes clear at the same time that a conceptual theory may well be the result of concept explication. Of course, the claim that the result of concept explication, a conceptual meta-theory, roughly captures an intuitive concept or distinction is a (quasi-)empirical meta-claim. However, the main point is that, although it may always be possible to formulate a falsifiable general claim with a conceptual theory, the (meta-)claim that all theories are observational theories need not be intended.

As already suggested, generic theories may well be unfalsifiable as such. They cannot only be made falsifiable, they can also be used as purely conceptual theories. It is clear that conceptual theories may or may not be ontologically and/or epistemologically stratified.

2.4.3 *Epistemological Positions*

Returning to empirical theories, the core of the ongoing instrumentalism-realism debate concerns the nature of proper theories, or rather the attitude one should have towards them. Here we will briefly sketch the most important epistemological positions in that debate, viz. instrumentalism, constructive empiricism, referential realism and theory realism. In the introductory chapter of ICR, they are extensively introduced and ordered according to the ways in which they answer a number of leading questions, where every next question presupposes an affirmative answer to the foregoing one. Moreover, the questions are considered from four perspectives on theories. On the one hand, theories supposedly deal primarily with 'the actual world' or primarily with 'the nomic world', that is, with what is possible in the natural world. On the other hand, one may primarily be interested in whether theories are true or false, or whether they approach 'the truth', regarding the world of interest. It should be stressed that 'the truth' is always to be understood in a domain-and-vocabulary relative way. Hence, no language independent metaphysical or essentialist notion of 'THE TRUTH' is assumed.

ICR is a study of confirmation, empirical progress and truth approximation, and their relations, with important implications for the way the epistemological positions are related. The survey of positions and the analysis in ICR are restricted to the investigation of the natural world and hence to the natural sciences. Several complications arise if one wants to take the social and cultural world into account. However, the survey of epistemological positions in the natural sciences may well function as a point of departure for discussing epistemological positions in the social sciences and the humanities.[2]

As we have seen, proper theories arise from the two-level distinction between observation and theoretical terms, as opposed to observational laws and theories, which only use, by definition, observation terms. The resulting two-level distinction between observational laws and proper theories gives rise to a short-term dynamics in the development of scientific knowledge. The long-term dynamics is generated by the transformation of proper theories into observation theories, by accepting them as true, and giving rise to a multi-level distinction according to which proper theories may not only explain or predict a lower level observational law, but also be presupposed by a higher level one. This description of the long-term dynamics typically has a theory realist flavor. However, the other positions have their own way of describing such dynamics. In the following brief survey of questions and answers we restrict ourselves to (the ingredients for) the short-term dynamics as seen from the different positions.

Question 0: Does a natural world that is independent of human beings exist?
Question 1: Can we claim to possess true claims to knowledge about the natural world?
Question 2: Can we claim to possess true claims to knowledge about the natural world beyond what is observable?
Question 3: Can we claim to possess true claims to knowledge about the natural world beyond (what is observable and) reference claims concerning theoretical terms?
Question 4: Does there exist a correct or ideal conceptualization of the natural world?

In the following elucidation, we always presuppose an affirmative answer to the foregoing question. *Question 0*, about the existence of an independent natural world, is not an epistemological question, but a preliminary ontological question. The negative answer leads to *ontological idealism*, and the positive one to *ontological realism*. A negative answer to the first epistemological question, *Question 1*, about the possibility of true claims about the natural world, leads to the position of *epistemological relativism* or *skepticism*. It has two forms: *experiential skepticism*, that is, skepticism with respect to claims about sensory and introspective experiences, and *inductive skepticism*, that is, skepticism merely with respect to inductive extrapolations in the sense of inductive predictions and inductive generalizations. The positive answer to *Question 1* leads to epistemological objectivism or *epistemological realism*.

Question 2, about the possibility of more than observational knowledge, brings us to the heart of the distinction between observation and theoretical terms. A negative answer assumes that the notion of observability is relatively fixed. It indicates *observational realism* or just *empiricism*, of which there are two versions. According to *instrumentalism*, advocated for instance by Schlick (1938) and Toulmin (1953), talking about the reference of theoretical terms does not make sense, let alone talking about true or false (proper) theories. The only function of proper theories is to provide good derivation instruments; that is, they need to enable the derivation of as many true observational consequences as possible and as few false observational consequences as possible. Hence, the ultimate aim of the instrumentalist is the best derivation instrument, if any. According to the second type of empiricism, called *(constructive) empiricism* by its inventor and main proponent van Fraassen (1980, 1989), it may make sense in principle to say that theoretical terms have referential value and that proper theories can be true or false. The problem is that we will never know if such is the case beyond reasonable doubt. Hence, what counts is whether such theories are empirically adequate or inadequate or, to use our favorite terminology, whether they are observationally true or false.

A positive answer to *Question 2* amounts to so-called *scientific realism*, according to which proper theories, or at least theoretical terms, have to be taken seriously. Since the books by Hacking (1983) and Cartwright (1983), there is a weaker version of realism than the traditional one, which amounts to a negative answer to *Question 3* on the possibility of more than (observational and) referential knowledge. Primarily thinking of the referentiality of entity terms, they call their position *entity realism*. However, it seems highly plausible to extrapolate that position to attribute referentiality, in some plausible sense, to many types of terms, and speak of *referential realism*.

The positive answer to *Question 3* brings us to so-called theoretical or *theory realism*, in some version or another advocated by, for instance, Peirce (1934), Popper (1963), and Niiniluoto (1987a). Theory realism adds to referential realism that theories are claimed to be true and that we have from time to time good reasons to further assume that they are true, that is, to carry out a theoretical induction.

A positive answer to the last *Question 4*, about the existence of a correct or ideal conceptualization, brings us to a position that is not purely epistemologically built on the positive answer to the preliminary, ontological *Question 0* (i.e. ontological realism); viz. it amounts to an extreme kind of metaphysical realism which we like to call *essentialistic realism*. According to that view, for instance, there must be natural kinds, not only in some pragmatic sense, but in the sense of categories in which entities in the natural world *perfectly* fit. Philosophers of science like Boyd (1984) and Harré (1986) seem to come close to this point of view.

The negative answer to *Question 4* gives rise to what we call *constructive realism*. It combines theory realism with the view that vocabularies are constructed by a human mind guided by previous results. Of course, one set of terms may be more appropriate than another, in the sense that it produces, perhaps in co-operation with other related vocabularies, more and/or more interesting truths about the domain than the other set of terms does. The fruitfulness of alternative vocabularies will usually be comparable, at least in a practical sense, despite the possibility of fundamental incommensurability. There is however no reason to assume that the improvement of vocabularies will ever become impossible.

We summarize the preceding survey in Figure 2.3.

The four perspectives, indicated above, imply that all (non-relativistic) epistemological positions have an 'actual world version' and a 'nomic world version'. Moreover, they may be restricted to 'true-or-false' claims, or emphasize 'truth approximation claims'. In both cases it is plausible to distinguish between observational, referential, and theoretical claims and corresponding inductions. Instrumentalists, in parallel, speak of theories as 'reliable-or-unreliable' derivation instruments or as 'approaching the best derivation instrument'.

All four perspectives occur, in particular in their realist versions. Standard or traditional realism focusses on 'true/false' claims about the actual world. Giere (1985), who introduced the term 'constructive realism', focusses on the nomic world, but does not take truth approximation into account. Peirce, Popper and Niiniluoto, however, do take truth approximation into account. Moreover, whereas Peirce and Niiniluoto focus on the actual version, Popper and Giere seem to have

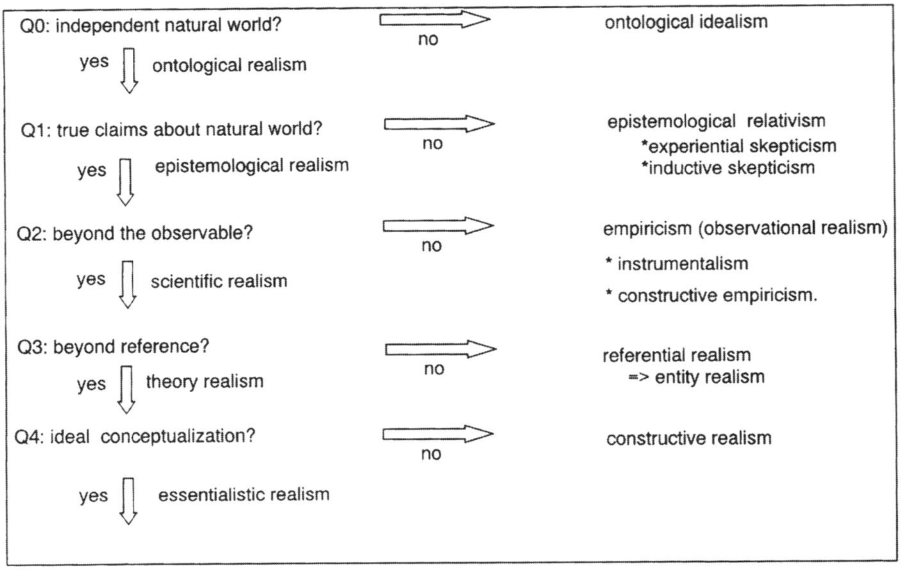

Figure 2.3 The main epistemological positions.

primarily the nomic version in mind, without excluding the actual version. In our view, the nomic version of constructive realism is best suited to scientific practice.

In our view, instrumentalism, constructive empiricism, referential realism, and constructive realism are the most interesting epistemological positions. In ICR they are, with the emphasis on their nomic interpretation, further characterized and compared in the light of the results of the analysis of confirmation, empirical progress and truth approximation in the rest of that book. The conclusions reached in that study are encapsulated in the following summary.

There are good reasons for the instrumentalist to become a constructive empiricist; in his turn, in order to give deeper explanations of success differences, the constructive empiricist is forced to become a referential realist; in his turn, there are good reasons for the referential realist to become a theory realist. The theory realist has good reasons to indulge in constructive realism, since there is no reason to assume that there are essences in the world. As a result, the way leads to constructive realism and amounts to a pragmatic argument for this position, where the good reasons mainly deal with the short-term and the long-term dynamics generated by the nature of, and the relations between, confirmation, empirical progress and truth approximation.

Besides these epistemological conclusions, there are some general methodological lessons to be drawn. There will appear to be good reasons for all positions not to use the falsificationist but the instrumentalist or 'evaluation(ist)' methodology. That is, the selection of theories should exclusively be guided by empirical success, even if the better theory has already been falsified. This common methodology, directed at the separate and comparative evaluation of theories, is extensively presented in Chapters 7 and 8.

According to the evaluation methodology, the role of falsifications has to be strongly relativized. This does not at all imply that we dispute Popper's claim that falsifiable theories are characteristic for empirical science; on the contrary, only falsifiable theories can obtain empirical success. Moreover, instead of denouncing the hypothetico-deductive method, the evaluation methodology amounts to a sophisticated application of that method. As suggested, the evaluation methodology may also be called the instrumentalist methodology, because the suggested methodology is usually associated with the instrumentalist epistemological position. The reason is, of course, that it is quite natural for instrumentalists not to consider a theory to be seriously disqualified by mere falsification. However, since we will argue that the instrumentalist methodology is also very useful for the other positions, we want to terminologically separate it from. From the instrumentalist epistemological position, by calling the former the evaluation methodology, and the latter 'instrumentalism'.

We close this section with a warning. The suggested hierarchy of the heuristics corresponding to the epistemological positions is, of course, not to be taken in any dogmatic sense. That is, when one is unable to successfully use the constructive realist heuristic, one should not stick to it, but try weaker heuristics: hence first the referential realist, then the empiricist, and finally the instrumentalist heuristic. For, as with other kinds of heuristics, although not everything goes always, *pace* (the suggestion of) Feyerabend's slogan 'anything goes', everything goes sometimes. Moreover, after using a weaker heuristic, a stronger heuristic may become applicable at a later stage: 'reculer pour mieux sauter'.

CONCLUDING REMARKS

It will not be possible to systematically hide our preference for constructive realism nor to use a terminology that is as liberal as possible and, for instance, acceptable to instrumentalists. In the foregoing we have already made one proposal in this direction. We agreed to speak of the testing of a (complex) claim when we are only interested in its truth-value, and of its evaluation when we are interested in its merits and failures. The first is usually the case with potential observational laws and the second with proper theories, in which case the merits and failures are usually supposed to be of an observational nature. This differentiation will not only lead us to explicate the standard ideas about testing hypotheses, in order to confirm or falsify them, but also to discuss the separate and comparative theory evaluation of theories (see Chapters 7 and 8).

On the other hand we have also accepted the need to follow the usage of scientists and to say that an observational law is already explained by a theory when it can (approximately) be derived from it, even if there is no good reason for assuming that it is true or even if it is known to be false. This possibility is not very attractive to empiricists, as they are not inclined to speak of explanation by proper theories at all, neither does it appeal to realists, because they are inclined to speak of explanation only if there are good reasons to assume that the theory is true. However, in Chapters 3 and 4, dealing with the detailed analysis of the argumentative steps for different kinds of explanation, we will explicitly take the (more general) empirical condition into account, which insists that there have to be good reasons for assuming the statements constituting the explanation to be true.

In Chapters 2–7 of ICR we frequently refer to the various epistemological positions, and at the end of Chapter 9 we reconsider them in the light of the findings about truth approximation. In Chapter 13 of ICR we complete the characterization of constructive realism, including its criteria for reference and truth.

Parts II and III of the present book are devoted to the further analysis of the various kinds of explanation, with special emphasis on the role of two epistemological and/or ontological levels. This leads to the explication of the idea of reduction of laws (Chapter 3) and of concepts (Chapter 5), and hence of the heuristics of the corresponding research strategies. Chapter 4 deals with explanation by specification, which may or may not be of an interlevel nature. We conclude Part III with a chapter (Chapter 6) on interlevel research, or more generally, 'interfield research' with emphasis on mind-body research.

APPENDIX 1: THE IDEAL GAS LAW

The main conclusion of (Kuipers, 1982) is that it is possible to disentangle the reduction of the ideal gas law to the kinetic theory of gases in such a way that the required reduction postulates can be conceived as ontological identities without leading to the reduction in the opposite direction. As a by-product it turned out that the ideal gas law is laden with eliminable terms, which provided a crucial example for the discussion in Section 2.2. The purpose of this appendix it to illustrate the latter point in detail. At the end we will also make some general remarks about the resulting reduction, and indicate the nature of the reduction postulates. They will also be occasionally referred to in the chapters to come.

As background we presuppose some macroscopic and molecular, but non-thermal notions:

G the set of all pure gases
$Z(g)$ the set of equilibrium states of an isolated mole g in G

The notion of an equilibrium state is here considered as a primitive one, leaving aside the problem of how to determine that a gas is in an equilibrium state. Of course, moles of the same (pure) gas are assumed to behave in the same way, which justifies our speaking of a mole g as if it were a representative mole.

The concept of mole does not presuppose any thermal theory. To be precise, the concepts of mole and pure gas are based on some macroscopic laws concerning masses of gas mixtures, some molecular hypotheses concerning types of molecules and their relative molecular-masses and the trivial, but crucial, reduction postulates that a sample of gas is a sample of molecules and that the mass of an amount of gas is equal to the sum-total of the absolute masses of the molecules. In principle it is now possible to determine whether an amount of gas is pure, i.e. is of one type of molecule. A mole amount of pure gas is then defined as an amount with mass in grams equal to the relative mass of the molecules of this gas. The molecular hypotheses, if true, assure that a mole of any pure gas contains the same number of molecules: Avogadro's number N. Of course, determination of the value of N is another story, but we will not assume to know this value.

The basic notion of any thermal theory is the empirical relation of thermal equilibrium between states. This relation happens to be an equivalence relation (0-th law of thermodynamics). The generated equivalence classes constitute:

S the set of thermal states

The following two notions remain implicit in textbook expositions, but they will turn out to be of fundamental importance.

Definition 1: g in G is an *Asymptotic*[3] *Boyle Gas* (g in *ABG*) iff it can be represented by $<Z(g), V, P, t>$ such that

$Z(g)$ indicates the set of equilibrium states of an isolated mole of g

V $Z(g) \to \mathbb{R}^+, V(z)$ being the measurable volume 'of state z'

P $Z(g) \to \mathbb{R}^+, P(z)$ being the (measurable) pressure of z

t $Z(g) \to S, t(z)$ being the (measurable, i.e. identifiable) thermal state of z, such that Range $(t/Z(g)) = S$

there is a one-one function $B_g: S \to \mathbb{R}^+$ such that $P(z)V(z) \to B_g(s)$ if $V(z) \to \infty$ and $t(z) = s$

Definition 2: g and g' in *ABG* are *Comparable* iff $B_g = B_{g'}$

With these notions we are able to formulate the three empirical laws which in conjunction appear to be equivalent to the (asymptotic) ideal gas law. The first and the second provide the required existence and uniqueness condition for the introduction of the notions of (empirical) absolute temperature and the (universal) gas constant. Addition of the third law completes the final claim.

E-EL	some *G* are *ABG*	(Existential E-law)
C-EL	all *ABG* are Comparable	(Comparability E-law)
U-EL	all *G* are *ABG*	(Universal E-law)

By EC-, EU-, UC-, and EUC-EL we indicate in an obvious way conjunctions of E-laws. Note that the addition of 'E' to 'U' gives a universal law existential import.

Definition 3: For arbitrary g in *ABG*
3.1: the (empirical) *absolute temperature* is the function $T: S \rightarrow \mathbb{R}^+$ defined by $T(s) = (B_g(s)/B_g(s_3)) \cdot T(s_3)$, with s_3 and $T(s_3)$ conventional; current convention: s_3 is the (thermal state of the) triplepoint of water and $T(s_3)$ is 273.16
3.2: the (universal) *gas constant* R is the quotient $B_g(s_3)/T(s_3)$ (and hence $RT(s) = B_g(s)$)
3.3: by natural extension we define T: $Z(g) \rightarrow \mathbb{R}^+$ by $T(z) = T(t(z))$ (implying $T(z) = T(z')$ iff $t(z) = t(z')$)

It is evident that E-EL provides the required existence condition for the definition of T and R and it is easy to check that C-EL provides in addition the uniqueness condition. To be precise, T is already unique on the basis of the law that all *ABG* are quasi-comparable, i.e. B_g and $B_{g'}$ have constant proportion. However, this law, together with E-EL, implies C-EL if R is also required to be unique. In conclusion, T and R presuppose E- and C-EL, i.e. EC-EL.

Supposing EC-EL we may now define

Definition 4: g in *ABG* is an *Asymptotic Ideal Gas* (g in *AIG*) iff $P(z)V(z) \rightarrow RT(z)$ if $V(z) \rightarrow \infty$

Some reflection shows that EC-EL is now equivalent to the law:

(i) some G are *ABG* and all *ABG* are *AIG*

If we use 'all G are *ABG*', i.e. U-EL, we can derive from (i) the law:

(ii) some G are *AIG* and all G are *AIG*

Using the fact that 'all *AIG* are *ABG*' and 'all *ABG* are *G*' are true by definition we see that (ii) implies not only (i), and hence EC-EL, but also UC-EL. Therefore (ii) is equivalent to EUC-EL. But (ii) amounts precisely to the ideal gas law if that law is considered as a universal one, restricted to asymptotic behavior and with existential import.

In sum, the asymptotic ideal gas law essentially is a conjunction of three empirical laws, enabling the explicit definition of T and R (Definition 3). One important consequence is that it suffices for the reduction of the asymptotic ideal gas law to reduce EUC-EL or, equivalently, its three component laws. In (Kuipers 1982) it is shown that the solution of this reduction problem also leads to the reduction of the ideal gas law if that law is extended (idealized) to non-asymptotic behavior. However, given the previous analysis this extension is in no way required for the notions of absolute temperature and the gas constant. Moreover, it turns out that the reduction postulates that have to be introduced for the asymptotic case are sufficient for the extended, non-asymptotic, case. These reduction postulates amount to:

RP1: identifying $P(z)$ with the kinetic pressure, i.e. the pressure we measure is, ontologically speaking, the kinetic pressure

RP2: thermal equilibrium, i.e. $t(z) = t(z')$, is identified with having equal mean kinetic energy, i.e. precisely this is the case, ontologically speaking, at thermal equilibrium

In textbooks one usually introduces the kinetic temperature relation KTR instead of RP2:

$$\text{KTR: } RT(s) = (2/3)\, N\bar{u}(s)$$

where $\bar{u}(s)$ indicates the unique mean kinetic energy that can be associated with s according to RP2. In contrast to RP2, KTR is, via T and R, evidently laden with the empirical laws constituting the asymptotic ideal gas law. The main problem with KTR, however, is that it not only enables the reduction of this law from the kinetic theory, but also the other way around. The cause is that KTR is also laden with the kinetic theory, which is not the case for RP2. Not surprisingly, KTR can be explained by EUC-EL, RP2 and the kinetic theory. For details, see (Kuipers 1982).

APPENDIX 2: THE EMPIRICAL BASIS

Introduction

As mentioned at the end of Section 2.2, besides the theory-relative approach, there is another, completely compatible, approach to the level distinction and to the law-distinction, which is easy to connect with Popper's work. We will first present this approach briefly and we will then speculate about Popper's motives for neglecting the law-distinction.

The Basis-Relative Approach

We start with some of Popper's core concepts. According to Popper it is possible to reach provisional agreement in every scientific context about what belongs to the level of observation or, to use Popper's favorite term, the (empirical) basis or the basic level. As already noted before, Popper has more than anyone else stressed the theory-laden, swampy character of this observation basis. It is however surprising that he did not make a number of plausible distinctions, let alone exploit them.

Let us call the (non-logico-mathematical) terms occurring at the basic level the *basic terms*. As a matter of fact Popper reserved the term 'basic statement' for a special type of statement that can be formulated in basic terms, viz. so-called singular existential statements, i.e. precisely those statements in basic terms about singular facts which can be in conflict with general statements.

The law-distinction can now be introduced as follows. Calling general statements which can be formulated completely in basic terms (general) observational hypotheses makes it also plausible to call such hypotheses observational laws when they are accepted for the time being after severe testing.

We are only concerned with a proper theory and hence with a theoretical level if it concerns a set of statements which, at least partly, breaks through the framework of basic terms. In other words, these statements should postulate new entities or attributes for which new terms have to be introduced that cannot be defined explicitly in terms of the available basic terms.

It is plausible to call the present approach to the level distinction the basis-relative approach. Notice that the corresponding characterization of observational laws is also basis-relative and hence not basis-independent, whereas in the theory-relative approach it was possible to give a theory-independent characterization of observational laws. It is nevertheless clear that the two approaches are essentially the same and that preferences will only depend on one's further purposes.

Why Neglect of the Law-Distinction?

Popper pays little attention to (the possibility of) the law-distinction, let alone to the importance of the distinction for the dynamics of science. We have to guess at Popper's reasons for his lack of interest, because he is not explicit about it. This guessing may, however, be instructive. The only good reason we can think of is the impressive fact (cf. Chapters 7 and 8) that strictly speaking the distinction is not necessary to characterize the logic of theory evaluation in the abstract terms of (singular) basic statements leading to falsification or confirmation (corroboration) of a theory. However, although this shortcut is possible, the distinction has to be introduced in a more realistic and sophisticated characterization of the evaluation of theories, and hence of the structure and development of science, as we will see many a time in this book.

There is also a reason which has to be respected: Popper does not show any interest worth mentioning in the didactic of scientific textbooks. For someone who does have this interest and thinks that the distinction can be made relatively sharply it is clear that the distinction is not yet sufficiently exploited in textbooks.

Poor reasons for Popper's lack of interest are also easy to conceive, however. In the first place, Popper was without doubt a victim of the misunderstanding that the law-distinction itself, or its importance, was related to the assumption of a theory-free observational vocabulary propagated by the classical logical empiricists. It is true that observational laws were considered by them as theory-free universal statements, or at least as reducible to such statements. In the light of the examples that were always mentioned as paradigms of observational laws, such as the laws of Galilei and Kepler, it gradually became clear that two interwoven, but distinguishable claims were involved. We have already shown above that the distinction can be unproblematically based on a relative distinction of levels, and that this was already done by Nagel in 1961. The (rightly made) objections to a theory-free observational vocabulary do not provide a good reason for avoiding the distinction.

A second poor reason has perhaps more of a psychological nature and is related to Popper's oversimplified fight against induction. Popper is of course completely right in claiming that induction does not play and cannot play a role in the invention of proper theories. It is however also perfectly clear that observational laws, as far as they are not found by way of prediction by a theory, are frequently found by observational induction; i.e. they are thought of by way of inductive extrapolation. That is also the reason why they are frequently called inductive generalizations. Computational philosophy of science (Chapter 11) exploits and elaborates the various methods of induction. Such methods do not alter the fact that inductively devised potential observational laws still have to be tested in the standard hypothetico-deductive way. A consequence of this is that the recognition of the importance of the law-distinction is almost impossible without simultaneously recognizing that as a matter of fact induction frequently guides the formulation of general observational hypotheses, but again does this not provide a good reason for avoiding the distinction.

The third and last poor reason concerns of course the under-estimation of the importance of the law-distinction for a realistic characterization of the structure and development of science. In general, one can say that Popper placed theories so central, that his attention to the role and nature of experiments became rather one-sided; they were seen from the point of view of theories. 'The neglect of experiment', the telling title of a book by Franklin (1986), was much severer in Popperian circles than in logical empiricist's circles. In this respect the present book is strongly influenced by the logical empiricists, witnessed among others by the topic of the present chapter, the distinction between descriptive and explanatory research programs in the first chapter,[4] and finally the fact that the next chapter along with Chapters 7 and 8 (on HD-evaluation) will reflect that explanation and prediction in explanatory programs primarily concerns (potential) observational laws, with the explanation and prediction of singular facts as derivatives.

Of course, Popper is well aware of the last point and when he is talking about explanation and prediction of facts (e.g. so-called 'novel facts'), he is mostly referring to general facts, hence, observational laws. Many natural scientists also make this elision, but they do not claim, as Popper does, that evaluation of theories is straightforwardly conducted in terms of singular facts. Although such facts may play a crucial role, it is an indirect role, viz. in testing the observational laws predicted by the theory. And even Popper does not always stick to his doctrine. When the evaluation of theories is the main subject, basic statements by definition deal with singular facts, but when talking about basic statements in other contexts, it is often impossible to conclude otherwise than that such statements also include observational laws. This ambiguity, however, has a fortunate aspect, for what is more plausible than to call all statements that can be formulated in basic terms, basic statements?

From the foregoing we conclude that Popper, starting from his own premises, would have improved his analysis considerably by introducing and exploiting the law-distinction.

PART II

PATTERNS OF EXPLANATION AND DESCRIPTION

INTRODUCTION

Guided by descriptive and explanatory programs, scientists generate observational laws, proper theories, and explanations of laws by theories as the products of their efforts. In Chapter 3 we present an analysis of the explanation of laws by theories, starting from three examples, the explanation of Galilei's law of free fall, Mendel's law of interbreeding, and Olson's law about collective goods. The result is a decomposition model according to which there occur in the explanation of laws by theories one or more applications of five well distinguished steps. Moreover, on the basis of the literature, starting with Nagel's seminal Chapter 11, it is argued that three of these rather different steps are apparently the reason for speaking of the reduction of the law. This diagnosis suggests a systematic explication of several distinctions made in the literature. The model of reductive and non-reductive explanations of laws and these distinctions are illustrated by briefly indicating a large number of examples.

Although Chapter 3 primarily deals with the explanation of laws by subsumption under a theory, it can easily be extended to a similar subsumption explanation of individual events. However, there seem to be many explanations that do not fit into the subsumption scheme. In particular, it will be argued in some detail in Chapter 4 that contrary to the claims of Nagel and Hempel, a subsumption reconstruction of intentional explanation of actions and functional explanations of biological traits, if possible at all, does not do justice to scientific practice. It will be argued that such explanations satisfy a general pattern, called explanation by specification, in terms of which it is easy to indicate the fundamental differences between intentional and functional explanation and to nevertheless reconstruct the relevant thought processes of scientists in a similar way. The general scheme is also shown to fit certain types of causal explanation, viz. explanations of individual events that select 'the cause' out of the set of causal factors. In the three subtypes, scientists are guided by a corresponding searchlight principle, viz. intentionality, functionality, and (specific) causality, respectively. Although, the results obtained in this way are typically called explanations (by specification), the programs in the context of which they are generated are usually called descriptive programs. In other words, the various patterns of explanation by specification may also be called patterns of description.

CHAPTER 3
EXPLANATION AND REDUCTION OF LAWS

INTRODUCTION

The standard view on the explanation of individual facts, such as events and states of affairs, is that of (deductive-)nomological explanation or explanation by *subsumption* under a law or theory in combination with one or more so-called initial conditions. This standard of explanation is also called the covering-law model of explanation. Formally, the underlying argument uses the rule of inference called universal instantiation (UI), followed by the rule called Modus Ponens (MP). That is, the most elementary form of explanation by subsumption is given by the valid argument:

(1)	*For all x if Ax then Bx*	law premise
(2)	*Aa*	initial condition premise
(3)	*if Aa then Ba*	from (1) by UI
(4)	*Ba*	from (2) and (3) by MP

According to the subsumption view, the search for and the establishment of the law and the initial conditions form the synthetic part of the explanation. The resulting valid argument forms the analytic part: when the real work has been done, the purported train of thoughts is placed in the opposite direction, leading to the alleged underlying argument.

The explanation of an observational law, or even a theory, can be reconstructed along the same subsumption lines, of course with some plausible modifications. Moreover, besides the deductive versions there are probabilistic and corrective versions of the subsumption model. In our opinion, together these various kinds of explanation by subsumption account for a large part of the explanation in the empirical sciences.

As far as explanation by subsumption is concerned, the emphasis in most elementary textbooks is on deductive and probabilistic explanation of individual events. Such explanations are not only relatively simple, there is also the question whether they form the core of scientific explanation. It is tempting to say that after an explanation of an individual event by subsumption under a law the really important issue then is to explain this law. In our opinion the core of explanation lies in the explanation of observational laws by subsumption under a theory, in short, theoretical explanation of (observational) laws. After a successful theoretical explanation of a law, we get as an extra bonus a theoretical explanation of the individual events fitting into that law.[1]

In this chapter explanation by subsumption is dealt with, and strong emphasis is placed on deductive and corrective theoretical explanation of laws.

Although the paradigmatic examples all concern the explanation of an observational law by a proper theory, the model to be presented can also be applied to other laws and theories.

The formal part of the model can also be used as a model for the prediction of laws by a theory. In this respect the model to be presented provides a detailed analysis of the first step in the refined hypothetico-deductive argument, the macro HD-argument, to be presented in Section 7.2.

Despite the cogency of many explanations by subsumption, the indignant criticism of interpreting certain types of explanation along the subsumption lines has never ceased. In particular, intentional explanation of human actions, as in history for example, and functional explanation of biological traits have frequently been mentioned as examples of explanation that do not conform to the subsumption pattern. In Chapter 4 we will deal with these objections and present an alternative main form of explanation, to be called explanation (or description) by specification, with explanation by intentional, functional and causal specification as special cases.

We begin this chapter with a number of famous theoretical explanations of observational laws in Section 3.1. They will be presented in a stepwise way, such that it becomes clear what typical steps occur and what specific kinds of auxiliary hypotheses are required. Then we will present in Section 3.2 a general decomposition model for explanation of a law by a theory. We will distinguish five standard steps that may occur in such an explanation. Finally, we will determine in Section 3.3 in what cases a theoretical explanation is a reductive explanation, i.e. in what cases one speaks in the literature about the reduction of a law. It turns out that there are at least three different reasons.

One of these types of law reduction essentially presupposes the reduction of concepts, using type–type identities as distinguished from causal correlation of concepts. This will be the subject of Chapter 5. Of course, reduction of laws and concepts is the crucial target of radical and moderate reductionistic strategies of interaction between two research programs, an issue dealt with in Chapter 1. The concepts and laws to be reduced belong to one research program, the reducing theory and the identities belong to another. It will be indicated at the relevant places when reduction essentially leads to the correction of the reduced laws and concepts, and when it leads to their further consolidation and even reinforcement, with or without some correction. It should be stressed in advance that the reduction of concepts does not presuppose the reduction of laws. Hence, when laws are not available in the primary field of study, one may concentrate on the weaker aim of concept reduction.

3.1 EXAMPLES OF EXPLANATIONS OF OBSERVATIONAL LAWS

Introduction

Guided by a special interest in generalizable steps, the following examples of explanation will be presented: the law of free fall by Newton's theory of gravitation, the ideal gas law by the kinetic theory of gases, a qualitative and a quantitative interbreeding law by Mendel's theory of heredity, and, finally, Olson's quasi-law about collective goods in utility theory. We will conclude this section with a general model of utilistic explanation, but postpone the presentation of the encompassing model of explanation of a law by a theory till Section 3.2.

3.1.1 *The Law of Free Fall*

A typical example of a theoretical explanation is the explanation of Galilei's law of free fall LFF by Newton's theory of gravitation TG (Examples (a*) and (a) in Chapter 2). The law states that a freely falling object p gets a constant acceleration a_p:

$$\text{LFF:} \quad a_p \text{ is constant}$$

It is clear that LFF is an observational law. Measurement of the acceleration of an object presupposes measurement of time and position, but these measurements do not (have to) presuppose the validity of the law.

The theory of gravitation TG consists of the three laws of motion and the law of gravitation, and it presupposes that the concepts of mass and force are TG-laden. Only the second law of motion is needed in the explanation. It states that the total force F_p operating on an object p equals the product of the (inertial) mass m_p of that object and its acceleration a_p:

$$\text{N2:} \quad F_p = m_p \cdot a_p$$

The law of gravitation states that an object q exerts on object p a (gravitational) force F_{pq} proportional to both gravitational masses, already known by Galilei and Newton to be empirically proportional to their inertial masses, and inversely proportional to the square of their distance d_{pq}:

$$\text{LG:} \quad F_{pq} = \tau \cdot m_p \cdot m_q / (d_{pq})^2$$

where τ indicates the so-called universal constant of gravitation. (Note that it follows immediately from N2 and LG that $a_p = \tau m_p / (d_{pq})^2$.)

The explanation can be decomposed into two steps. In the first step, the application step, these two laws of TG are applied to object p at the height $h(t)$ at time t freely falling onto the earth E with mass m_E and radius R, using the auxiliary hypothesis that the earth's gravitational force is the only force operating on p. This leads to the following TG-laden expression for the acceleration:

$$(*) \quad a_p = \tau m_E/(R + h(t))^2$$

From (*) we immediately get a prediction of the true observational law of free fall, a *corrected version* of LFF:

$$\text{CLFF:} \quad a_p \text{ is inversely proportional to } (R + h(t))^2$$

According to CLFF the acceleration is not constant. To obtain LFF we need a second step, an *approximation step*, on the basis of the auxiliary approximation hypothesis that $h(t)$ will normally be much smaller than $R(h(t) \ll R)$, and can hence be neglected. Hence, a_p is, by approximation, a constant, so we have LFF.

Scheme 3.1 summarizes the argument:

	TG: notably N2 + LG
application	AH: no other forces
	CLFF
approximation	AH: $h(t) \ll R$
	- - - - - - - - - -
	LFF

Scheme 3.1.

AH indicates one or more auxiliary hypotheses. A continuous line indicates a deductive step, a broken line a non-deductive step, in particular an approximation step.

Unfortunately, although we may be inclined to also infer from (*) another law known to Galilei, viz. that a_p is independent of p, this would be a circular explanation.[2] The reason is that LG is already laden with the empirical (general) fact that inertial and gravitational mass are proportional. If the conceptual distinction would be made explicit, the independence of a_p from p would not be derivable. One of the impressive explanatory successes of Einstein's general theory of relativity is the proportionality of inertial and gravitational mass, and hence indirectly of the p-independence of a_p.

The surplus value of the explanation LFF is twofold. First, using the (inertial/gravitational) mass proportionality, we have a theoretical interpretation of the 'constant' g (9.8 m/sec^2), that is, the object-independent acceleration of free fall near the surface of the earth. In particular, we may obtain a specification in terms of a universal constant and two measures of the earth (it is easy to check that this is $\tau m_E/R^2$). Second, the explanation predicts precisely how much the acceleration will decrease at high altitude, i.e. it predicts the corrected version CLFF of LFF. As is well-known, TG explains (and partially corrects) several other observational laws, such as Kepler's three laws of planetary motion and the law of the pendulum.

As we already noted, LFF is an observational law, the terms it presupposes (time, position, velocity and acceleration) are not LFF-laden. On the other hand TG is a proper theory, in particular the concepts of (inertial and theoretical) mass and force are TG-laden. Hence, TG is epistemologically stratified, and the application step drops theoretical terms.

3.1.2 *The Ideal Gas Law*

Another typical example of a theoretical explanation of an observational law is the explanation of the ideal gas law IGL by the kinetic theory of gases KGT. This explanation can be decomposed into three steps, of which we have already met one in the foregoing example, viz. the application step.

IGL states that for a mole (a standard amount) of gas in a container the product of volume V and (macroscopic) pressure P is proportional to the empirical absolute temperature T, such that the proportionality constant is the same for all gases; viz. the ideal gas constant R. Hence,

$$\text{IGL:} \quad PV = RT$$

According to KTG an isolated amount of gas consists of molecules which move and collide with each other and with the wall in accordance with Newton's laws of motion. In the application step these laws are applied to one molecule colliding with the wall. In this step we use the auxiliary hypothesis that the collision is elastic. The result is the 'individual law' which states that the momentum exchange q equals $2mv_w$ (m: mass; v_w: velocity in the wall direction):

$$q = 2mv_w$$

The second step, the aggregation step, is an ingenious aggregation, using some statistical auxiliary hypotheses, of the momentum exchange of Avogadro's standard number of molecules N in a mole of gas, leading to the 'aggregated law' which states that the product of the resulting kinetic pressure p on the wall and the volume V is equal to $(2/3)N\bar{u}$:

$$pV = (2/3)N\bar{u}$$

where $\bar{u}$ indicates the mean kinetic energy, i.e. the mean value of $(1/2)mv^2$ (v: the total velocity of a molecule).

In the third and last step, the *transformation step* or, more specifically, the *identification step*, two identity hypotheses are introduced:

$$p = P \qquad \bar{u} = (3/2)(R/N)T$$

It is easy to see that they enable us to derive IGL immediately.

Again we summarize the explanation in a scheme:

$$\begin{array}{ll} \text{application} & \underline{\text{KTG AH: elastic collisions}} \\ & \quad q = 2mv_w \\ \text{aggregation} & \underline{\text{AH: statistical hypotheses}} \\ & \quad pV = (2/3)N\bar{u} \\ \text{identification} & \underline{\text{AH: } p = P\,\bar{u} = (3/2)(R/N)T} \\ & \quad \text{IGL}: PV = RT \end{array}$$

Scheme 3.2.

In Kuipers (1982) it is shown that the second, quantitative identification can essentially be reduced to the following qualitative identification: being in the same thermal state (the basic qualitative concept for the notion of empirical absolute temperature) is identical to having the same mean kinetic energy (for some more information, see Appendix 1 of the previous chapter). The kinetic theory explains several other laws, notably Avogadro's hypothesis, see Section 1.7, as is not difficult to verify. Note that Avogadro's hypothesis is an observational law with respect to the kinetic theory, but of course not with respect to the atomic theory.

Returning to the explanation of the ideal gas law, it is clear that KTG is epistemologically as well as ontologically stratified. The aggregation step amounts to a jump to a higher ontological level, and the identification step amounts to the replacement of theoretical terms by observation terms.

3.1.3.1 *A Qualitative Interbreeding Law*

The explanation of the interbreeding laws by classical Mendelian genetics CMG is also well-known and very instructive. We will start with the explanation of a qualitative interbreeding law, thereby illustrating the fact that the explanation of laws is not at all restricted to quantitative laws. By systematic experiments Mendel found out first of all that certain pairs of mutually exclusive features F and G of a certain (sexually reproducing) species S, such as the color of pea plant flowers (e.g. green and white) and the shape of peas (round and wrinkled), are inherited in the following systematic way:

$$\text{L}: \qquad F \times F \to F \qquad G \times G \to G \text{ or } F \qquad G \times F \to G \text{ or } F$$

in which case G is called the dominant and F the recessive feature. If L holds for features F and G of species S it is evidently a qualitative observational law for that triple.

For the explanation we introduce what might be called Mendel's zero-th law and the qualitative version of Mendel's first law:

M_0: there are sets of competing factors such that each individual has one combination of two factors
M_1: each descendant gets one factor from each parent

Mendel called these factors also genes. In modern biology they are called alleles, reserving the term gene for a pair. In the *application step* M_0 and M_1, together with the auxiliary specification hypothesis stating that there are only two relevant factors *f* and *g* for the features *F* and *G*, lead to the following (qualitative, factor) genetic law, presented in a self-explanatory matrix form:

$$\mathrm{GL}: \quad \begin{array}{cccc} & ff & fg & gg \\ ff & ff & ff \text{ or } fg & fg \\ fg & ff \text{ or } fg & ff \text{ or } fg \text{ or } gg & ff \text{ or } fg \\ gg & fg & fg \text{ or } gg & gg \end{array}$$

The combinations *ff* and *gg* are called homogeneous and the combination *fg* is called heterogeneous.

In the *transformation step* or more specifically the (causal) *correlation step*, we combine GL with the appropriate auxiliary correlation hypothesis relating factor combinations to features, according to the following causal correlation matrix:

$$\mathrm{CH}: \quad \begin{array}{ccc} & f & g \\ f & F & G \\ g & G & G \end{array}$$

It is easy to check that L is an immediate consequence of GL and CH.

The resulting scheme reads:

CMG : $M_0 + M_1$

application AH: only two competing factors *f* and *g*

GL

correlation AH: CH

L

Scheme 3.3.

According to CH the asymmetry between the recessive feature *F* and the dominant feature *G* is due to the asymmetric causal role of the corresponding factors; of course, like *F*, the factor *f* is called recessive and *g* dominant, like *G*.

It is also clear that the notion of a factor is CMG-laden and hence that M_0 and M_1 contain only such CMG-laden terms and not the CMG-independent feature terms. On the other hand CH relates CMG-laden terms to CMG-independent terms. Although this transformation role of CH is similar to that of the identity hypotheses in the ideal gas law case, identities are nowhere stated: whole chains of causal processes are involved. In sum, Mendel's theory is epistemologically stratified, and the correlation step replaces theoretical terms with observation ones. No ontological stratification seems to be involved.

3.1.3.2 *A Quantitative Interbreeding Law*

For the promised quantitative example we introduce first the notion of a population being pure with respect to feature F, viz. if '$F \times F \rightarrow F$' holds for that population. If the qualitative law applies to features F and G of species S then each F-population (population of only F-individuals) is by definition pure, but not each G-population. Experimentally, it turns out that a (practically) pure population can usually be obtained by selecting across several generations only individuals with the relevant feature for the production of the next generation. The fact that in this way we get in the long run a pure population, can be explained by the aforementioned laws, in particular GL and CH.

The quantitative interbreeding law states that the first generation resulting from random crossing of a pure F- and a pure G-population is a (100%) G-population and that the second generation resulting from random crossing within the first generation contains 75% G-individuals and 25% F-individuals. In short we get the quantitative two-generation observational law:

$$\mathrm{L}^{2*}: \quad \text{pure } F \times \text{pure } G \quad \rightarrow \quad 100\%\ G \quad \rightarrow \quad 75\%\ G/25\%\ F$$

In order to explain this law we have to introduce Mendel's first law, the law of segregation, in its full quantitative strength:

M_1^*: each descendant gets one factor of each parent,
with equal probability for both factors of each parent

Applying M_1^*, M_0 and the same auxiliary (correlational) hypothesis as before, i.e. CH, we get now a quantitative version GL^* of GL. If two combinations are possible they have equal probability, i.e. 1/2 and in the single case of three possibilities the probabilities are 1/4 for *ff*, 1/2 for *fg* and 1/4 for *gg*. Applying GL^* for two generations in the described way starting with the auxiliary hypothesis that the two populations at the beginning have homogeneous factor combinations, i.e. *ff* and *gg* respectively, we get the quantitative genetic two-generation-law:

$$\mathrm{GL}^{2*}: \quad 100\%\ ff \times 100\%\ gg \rightarrow 100\%\ fg \rightarrow 25\%\ ff/50\%\ fg/25\%\ gg$$

Finally, it is easy to verify that GL^{2*} can be transformed into L^{2*} by applying the same correlation hypothesis CH as before. Hence, the explanation essentially follows the same kinds of steps as represented in Scheme 3.3 for the qualitative case.

The instructive feature of the quantitative example is that it clearly illustrates that the explanation of a statistical law is purely deductive. It is not at all something like a probabilistic transition from the premises to the conclusion. Of course, that one of the premises, i.e. M_1^*, is probabilistic is an essential but completely different matter. A point of historical interest is that Mendel was aware of L before he formulated his theory, but that he predicted L^{2*} with that theory.

For completeness we add Mendel's second law, the law of independent assortment, which did not play a role in the foregoing exposition. It states that the probability of inheriting a certain factor from one factor combination is independent of the possibility of inheriting a factor from another factor combination (related to a different type of feature) of the same parent.

Mendel's theory is in many respects a paradigm theory. First, in contrast to Mendel's cautious instrumentalistic attitude towards the factors, a realistic interpretation turned out to be defensible. Second, and this should also be stressed, the presented version of Mendelian genetics is nevertheless classic, for the theoretical laws, in particular the second one, have proved to have only restricted validity and had to be concretized in several respects.

3.1.4 *Olson's Quasi-law About Collective Goods*

The last example is of a economic-sociological nature. It is a paradigm case of a so-called methodological individualistic explanation. Moreover, it has in many respects the same structure as the explanation of the ideal gas law. It concerns the explanation by utility theory (also called the theory of rational choice) of the social law or, more cautiously, Olons's (1965) social quasi-law, stating that the chance (or level) of realization of a collective good decreases when the size of the relevant group increases:

LO: the larger the group the smaller the chance of realization

An extensive treatment of the example can be found in Kuipers (1984).

A collective good is by definition a good which can be consumed by all members of the relevant group, whether or not one has contributed to its realization. People who belong to the second category are called free-riders. A typical example would be the potential results, such as wage-increases, for all workers in a field obtained by the relevant trade union(s), assuming that membership in the union is free.

The utilistic explanation of the quasi-law takes the decision to participate or not to participate in the attempt to realize the collective good as a point of departure. Utility theory UT claims that people choose that action, out of a set of alternative actions, of which they expect the highest utility. In the present context the relevant form of this principle of utility maximization concerns the choice that the members of a group have to make between participation and non-participation in the realization of the collective good at variable group size. In short:

PUM: participation or non-participation depends on what alternative has the highest expected utility

In the *application step* PUM is combined with some auxiliary specification hypotheses. One of them concerns the individual utility assignments, in particular:

AHU: the subjective utility of the good and the subjective costs (negative utility) of participation are constant

The other specification hypothesis concerns the individual probability assignments, in particular:

AHP: the subjective estimation of the increased chance of realization by an extra participant is independent of the number of already participating individuals, and decreases with increasing group size

A direct result of applying PUM in the light of AHU and AHP is the following 'individual law':

Lin: individuals all have their own 'switch group size' (sgs), i.e. the size where they switch from participation to non-participation

In the *aggregation step* we assume the auxiliary statistical hypothesis that the relevant groups are statistically comparable as far as both mean and variance of this individual switch group size is concerned:

ASH: the relevant groups have equal mean and variance with respect to sgs

In this way we get the 'aggregated law':

Lag: the larger the group the smaller the degree of participation, i.e. the proportion of participants

Finally, in the *transformation step* we use the transformation, or more specifically, the correlation hypothesis:

CH: the smaller the degree of participation the smaller the chance of realization of the collective good

In this way we can easily derive Olson's (quasi-)law LO.

The scheme summarizes again the explanation:

$$\text{application}\quad \frac{\text{PUM}\quad \text{AHU/AHP}}{\text{Lin}}$$

$$\text{aggregation}\quad \frac{\text{ASH}}{\text{Lag}}$$

$$\text{correlation}\quad \frac{\text{CH}}{\text{LO}}$$

Scheme 3.4.

Of course, the present example of a social law as well as its explanation are highly idealized. Hence, both should be considered as 'ideal-typical' points of departure for concretization. The second part of AHU, i.e. the assumption that the subjective costs of participation do not depend on the size of the group is perhaps the most vulnerable auxiliary hypothesis, for at least the objective costs of participation will in many cases depend on the size of the group. This qualification leads to the recognition of a clear-cut disturbing factor for the manifestation of the mechanism. Other factors may compensate the free-rider inclination, e.g. positive incentives for participation or negative incentives for non-participation. Finally, and more fundamentally, it is well known that the core principle of utility maximization (PUM) has been concretized by Herbert Simon to a principle of bounded rationality in terms of satisfying choice behavior.

The example has also an interesting rationality problem, which is not directly related to the lastmentioned concretization of PUM. The probability assignments of the individuals may well be adequate. Hence, all individuals may be said to act rationally in the light of their utility assignments. However, it is perfectly possible that participation by all would lead to a greater realized utility for all. Hence, something is irrational. The only source of this irrationality seems to be the unrestricted application of the principle of utility maximization. In economics and sociology this phenomena is well known. It is a good reason to question the name 'theory of rational choice' as an alternative to the more neutral term 'utility theory'.

The similarities with the explanation of the ideal gas law are as striking as the differences. The main differences are the following. First, in the present case neither the law to be explained, nor the intermediate laws in the explanation, i.e. the individual and the aggregated laws are laden with the explanatory theory. They do not contain utilities and probabilities, they are observational or, more specifically, behavioral laws: Lin is an individual behavioral law and Lag a collective behavioral law. Second, the auxiliary transformation hypothesis is by no means an identity, but a causal correlation. As in Mendel's case, complex causal processes are involved between participation and realization. However, in other cases of similar explanations of a social law the transformation step might be an identity, perhaps in the sense of an institutional identity. For example, the democracy rule that a law proposal gets the status of law as soon as a majority in the parliament has voted for the proposal is a typical example of an institutional identity. Although it is not something like an ontological identity, there is also no causal process involved between the majority vote result and the status of law.[3]

	hypothesis of utility maximization
application	specification hypotheses
	individual behavioral law
aggregation	statistical hypotheses
	collective behavioral law
transformation	transformation rules
	social law

Scheme 3.5.: Model for utilistic reduction

This brings us to a general model, see Scheme 3.5., for what we like to call utilistic explanation (or even reduction, see below) of a social law.

In the scheme the transformation step may be an identification or a correlation.

Utility theory evidently is already epistemologically stratified when applied to one individual. Especially in the application step, theoretical terms, subjective utilities and probabilities are dropped. Although utility theory as such is not ontologically stratified, the aggregation step amounts to an interlevel step of an ontological nature, viz. from individuals to collectives. Finally, the transformation step may at first sight seem to be epistemologically interesting, i.e. from collectives to collective goods. However, since the concepts for describing collective behavior and collective goods as such are not laden with utility theory, the transformation step is not a replacement of theoretical terms by observational ones, relative to utility theory.

Finally, with respect to the debate about the status of utility theory, we would like to refer to Franssen (1997, Chapter 3) for a lucid survey of this debate, and we confine ourselves to the remark that the status of classical particle mechanics as an empirical theory is, for similar reasons, not unproblematic (see Chapter 12).

3.2 A DECOMPOSITION MODEL FOR THE EXPLANATION OF LAWS

The foregoing examples are all in agreement with the general claim that any explanation of a law by a theory can be decomposed into five standard steps specified in the following *five-step-model*:

'Theory *X* explains law L' if

formal condition: there are auxiliary mutually consistent hypotheses *A*1, ..., *A*5 such that *L* can be derived strictly or approximately from *X* using one or more times one or more of the five steps depicted in Scheme 3.6 for a paradigmatic case in which all five steps occur precisely one time in the indicated order:

(1) application *X* ———————— *A*1

(2) aggregation *L*1 ———————— *A*2

(3) identification *L*2 ———————— *A*3

(4) correlation *L*3 ———————— *A*4

(5) approximation *L*4 - - - - - - - - - - - *A*5

*L*5=*L*

Scheme 3.6.

empirical condition: there are good reasons for accepting theory *X* and the required auxiliary hypotheses *A*1, ..., *A*5 as approximately true

The empirical condition is in a sense necessary for a 'genuine' explanation, but it will only be assumed when we say so explicitly. We do this for practical reasons, which are not related to the fact that the condition might suggest that explanation by this condition becomes a kind of 'reducing to the familiar, to what is already known' as the classical dictum has it. Of course, in most cases this is certainly not the case at the birth of the explanation. To be sure, the empirical condition hides a world of problems, in particular problems of idealization and circularity, confirmation and (approximate) truth. The first two problems have already been addressed in Chapters 1 and 2, the third and fourth problems are given full attention in ICR.

In the following stepwise elucidation of the formal condition, the unessential but useful assumption is made of a paradigmatic case, that is, a case in which all five steps occur precisely one time in the indicated order.

Ad (1) application: *X* is tailored to the kind of object, system or situation that *L* is, implicitly or explicitly, about. This type of modeling usually requires a lot of auxiliary hypotheses, in particular specification hypotheses concerning variables occurring in *X*. Together with *X*, this leads to the formulation of some law, which may, in the face of an aggregation step, be called an 'individual law'.

Ad (2) aggregation: the total effect of the individual law for many objects is calculated by a suitable addition, or composition (or synthesis) if more than one type of individual law is involved. The required auxiliary hypotheses are primarily of a statistical nature, but 'macroscopic constraints' may also be involved. The resulting law is called an 'aggregated law'.

Ad (3) identification: the aggregated law is transformed with the aid of one or more identity hypotheses, ontologically identifying (object and predicate) terms of *L* with terms (definable in terms) of *X*. The resulting law need not yet be completely stated in terms of *L*.

Ad (4) correlation: the resulting law is transformed with the aid of some causal hypotheses, correlating terms of *L* with (defined) terms of *X*. In the next chapter the distinction between (ontological) identities and (causal) correlations[4] will be discussed; here we presuppose it as an intuitively clear distinction.

Ad (5) approximation: the deductively derived law is now simplified by some mathematical or logical approximation device, in particular counterfactual idealization, and justified by an auxiliary hypothesis stating implicitly or explicitly, for instance, that some relevant term is relatively small. In this way, the last step is a final opportunity to get rid of terms not occurring in *L*, viz. terms disappearing when they assume some extreme value.

This then, was the stepwise elucidation. The indicated order of steps seems to be the rule, but deviations are not excluded. Moreover, not all steps need to occur while others may occur more than once. The only claim with respect to the number and order of steps is that reconstruction of any particular case of a generally accepted explanation leads to a definite number and order of steps.

Up to now no examples of explanations of laws that did not fit well in the model have been found. However, if the model turns out to be too narrow to cope with new examples, it is unproblematic to extend the model as long as the number of kinds of steps remains small.

A few general remarks will be useful.

In order to get a uniform presentation we have called all hypotheses that were added in the course of the explanatory argument 'auxiliary hypotheses'. It should be stressed that these hypotheses frequently have an idealizational character, i.e. not only in case of the approximation step.[5] Moreover, in many cases some of these hypotheses would certainly be counted as belonging to the core (bridge) principles of the theory in the sense discussed in Chapters 1 and 2. A good example concerns the identity hypothesis for temperature in kinetic theory.

Steps (3) and (4) are more generally called *transformation steps*, and the auxiliary hypotheses (both identities and correlations) *transformation rules*. Following Nagel (1961) and Hempel (1966), and unlike the *internal principles* of the theory, these transformation rules might also be called *rules of correspondence* or *bridge principles*. However, both authors associate these terms with relations between different epistemological and/or ontological levels, whereas the name 'transformation rule' is intended to be neutral in this respect. Recall that we have reserved the term 'bridge principle' in the preceding chapter for linking ontological levels. In a reduction the decomposition of such a principle may well amount to the combination of an aggregation hypothesis and an identity hypothesis.

In the present context, the term 'law' need not be restricted to its strict sense of a (approximately) true (universal or statistical) lawlike statement. The five-step-model can also be used for so-called quasi-laws, i.e. laws which are only roughly or globally true (e.g. trends and the like) and that require further concretization.

The model can of course be used for the prediction of potential laws. More precisely, without the empirical condition it is a model for making formally correct explanations and predictions, irrespective of the truth-value of the theory and its auxiliary hypotheses. As suggested before, the model opens the black box covering the first step in the HD-evaluation of theories, as will be described in Chapter 7, viz. the derivation of general test implications from a theory. In that chapter the 'explanatory success' of a theory will be interpreted as its ability, irrespective of its truth-value, to formally and correctly explain and predict observational laws, which were known beforehand or became known afterwards.

In an explanation of a law by a theory, two 'vocabulary jumps' may occur. Application and approximation often imply a jump to a sub-vocabulary: the vocabulary of the resulting law concerns a part of the vocabulary of the preceding stage. On the other hand, transformation concerns a heterogeneous vocabulary jump: identification and correlation introduce new terms, replacing terms occurring in the preceding stage.

The law explained by a theory may, but need not be, an observational law with respect to the theory in the sense of Chapter 2, and the theory need not be a proper theory.[6] All examples exposed in Section 3.1. are however obviously of this epistemologically stratified kind. In such cases, application and approximation may amount to the elimination of theoretical terms, and both types of transformation to their replacement by observation terms, relative to the explanatory theory.

If the theory is also, or only, ontologically stratified, the explanation will include an ontological step to a higher 'aggregation level', notably, the aggregation step. Such a step does not introduce genuine new terms but 'aggregate' terms, such as 'total mass', 'kinetic pressure', 'mean kinetic energy', 'degree of participation' and the like.

We have so far restricted our attention to the explanation of a law by a theory. The following generalization is plausible. 'Theory X explains theory X^*' if and only if all laws constituting X^* can be explained by X and mutually compatible auxiliary hypotheses. In the literature (cf. in particular Kemeny and Oppenheim, 1956) there are also discussions about partial or instrumentalistic or observational explanation (and reduction) of one theory by another. It is obvious to explicate 'Theory X explains X^* observationally' if and only if X, using a consistent set of auxiliary hypotheses, can explain all observational laws independent from but explained by X^*.

3.3 REDUCTION OF LAWS BY THEORIES

Introduction

Scientific practice is almost by definition a matter of reduction, either in the transformational sense of inferring something (deductively or otherwise) from something else, or in the idealizational sense of concentrating on some aspects and provisionally neglecting others, to be accounted for later. The second sense of reduction is the 'reculer pour mieux sauter'-sense, that is, idealization and (subsequent) concretization dealt with Chapter 1. Obviously there may be connections. For example, in the following exposition the regaining (in a non-deductive way) of an idealized law, originally obtained by idealizational reduction, from a concretization of that law can and will be presented as a form of reduction in the transformational sense.

In the case of transformational reduction one may think primarily of the reduction of phenomena or the reduction of relations between them. Or, reformulated in terms of the conceptualization of phenomena and their relations, the reduction of concepts or the reduction of laws may be the primary subject of interest. In Chapter 5 the reduction of concepts, which is presupposed in one type of law reduction, will be dealt with.

3.3.1 *Three Kinds of Reduction*

In this section the reduction of laws is the subject. For many years there has been a discussion in the literature about the nature of reduction in science and of the (transformational) reduction of laws and theories in particular. However, on the basis of our model of explanation we can discern three different ways of explaining laws, each of which has been labeled a reduction. Moreover, this diagnosis enables precise definitions of distinctions occurring in the literature: homogeneous versus heterogeneous reduction, deductive versus corrective reduction, and micro-reduction versus other types of reduction.

Our pluralistic terminological diagnosis states that the concept of reducing a law has three essentially different meanings in the literature:

> 'Theory *X* reduces law L'
> or '*L* can be reduced to/by *X*'
> if and only if
> *systematic condition*: *X* explains *L* in accordance with the five-step-model, with at least one of the steps being aggregation (2) (provided it is non-trivial), identification (3) or approximation (5),
> *temporal condition*: *L* has been established before *X*.

We will not pay further attention to the temporal condition since its structural importance is of course limited.

The authors to whom we are alluding include among others: Adams (1959), Bunge (1977a), Causey (1977), Girill (1967), Hempel (1967), Kemeny and Oppenheim (1956), Mayr (1976/1981), Moulines (1984), Nagel (1961, 1979), Nickles (1973), Oppenheim and Putnam (1958), Pearce and Rantala (1985), Pluhar (1978), Popper (1982), Schaffner (1967), Simon (1969), Sklar (1967) and Sneed (1971).

It is evident that of the four examples of explanation only the explanation of the interbreeding laws is not an example of law reduction.

On the basis of our diagnosis it is possible to make the following explication of the terminology used in the literature. 'Micro-reduction' is reduction with an aggregation step or, simply, reduction by aggregation. Reduction without aggregation, hence with identification or approximation, might be called 'iso-reduction'. Reduction with an identification step or, simply, reduction by identification, is so-called 'heterogeneous reduction'. Reduction without such a step, hence with aggregation or approximation, and without a correlation step, is so-called 'homogeneous reduction'. Finally, reduction based on an approximation step or, simply, reduction by approximation, might be called 'approximative reduction', without such a step, hence with aggregation or correlation, 'deductive reduction'. However, we prefer to speak of 'corrective reduction' as opposed to 'deductive reduction', for it is clear that only reduction by approximation essentially leads to the correction of the reduced law, and its replacement by a corrected version. Hence, if no approximative step is involved, but only aggregation and identification, reduction is always deductive. The deductive reduction of a law amounts to its consolidation and reinforcement as soon as the reducing theory is accepted as (approximately) true. Of course, strictly speaking, corrective reduction of a law means only that the corrected law is consolidated when the reducing theory is accepted, but in these cases the reduced law is roughly speaking also reinforced, albeit in a revised form.

The explanation of the law of free fall is a clear case of corrective homogeneous reduction. The explanation of the ideal gas law represents a case of deductive heterogeneous reduction by aggregation and identification.

Note that reduction by aggregation or approximation containing also a correlation step but no identification step is a complicated case. The explanation of Olson's law is such a case. Calling it homogeneous reduction is certainly inadequate. However, calling it heterogeneous reduction is misleading, for the correlative transformation step may be the reason to speak of 'heterogeneous' explanation, it is not the reason to speak of reduction or, more briefly, there is no reduction by correlation. It is important to note, as the Olson example illustrates, that the presence of a correlation step in an explanation is no reason to stop seeing it as a reduction, provided at least one of the three 'reductive steps' is present. More generally, a reductive explanation may well need one or more auxiliary hypotheses, e.g. macroscopic constraints, that use concepts belonging to the domain of the reduced law. In the reduction of the ideal gas law, for instance, a hidden assumption in the aggregation step concerns the volume of the gas container, viz. that the volume is fixed. Moreover, the notion of volume explicitly occurs in the resulting aggregated law before the identification step.

It is clear that the explanation of interbreeding laws by Mendelian genetics is not a case of reduction according to the above criterion: only application and correlation occur. Of course, it may turn out to be possible to reduce the relevant correlation hypotheses by some theory about gene-expression in observable features. As a matter of fact, biologists are already rapidly moving in that direction. The possibility raises the interesting question whether a law may not be reducible but nevertheless explainable by some micro-theory due to the fact that the required transformation rules are irreducible correlations, constituting a complex causal bridge theory, as suggested by Maull (1977). Addressing this issue leads us to consider non-reductive interlevel research in general, a subject which will be dealt with in Chapter 6.

But let us return to cases satisfying our diagnosis. Again we have to note that the reduction may, but need not, concern an observational law with respect to a proper theory in the sense of Chapter 2. However, all given examples of reduction were of this kind.

There is also again a plausible generalization: 'Theory X reduces theory X^*' if and only if X explains X^* (see the preceding subsection) such that all laws constituting X^* can be reduced by X in at least one of the three distinguished senses. Finally we can define, of course: 'Theory X reduces X^* observationally' if and only if X explains X^* observationally such that all observational laws following from X^* can be reduced by X in at least one of the three senses. Observational reduction of a theory is frequently called eliminative reduction. It should be stressed that all kinds of reduction of a law discussed above are non-eliminative in this sense, the law undergoes at most some correction, namely in the case of an approximative step.

Some historical remarks also need to be made. The two famous formal conditions formulated by Ernest Nagel (1961) for 'heterogeneous reduction' of a law by one theory by another can be precisely stated: there must be transformation rules (A3 and A4, but also see below) (*condition of connectability*) and other auxiliary hypotheses (A1 and/or A2) such that the laws of the first theory can be derived from the second theory and the auxiliary hypotheses (*condition of derivability*). These two conditions became the semi-hard core of an explicative research program around the notion of reduction. Nagel was mainly thinking of examples in which later authors (Sklar (1967), Schaffner (1967), Causey (1977), among others) agreed that the transformation rules were identities. With this further restriction of the condition of connectability the two conditions specify precisely what is needed, according to the above exposition, for deductive reduction by identification. Note that the condition of derivability is not specifically related to reduction, for it is the condition required for deductive explanation in general. Unfortunately, Nagel was not very clear about formal conditions for homogeneous reduction, except for some kind of derivability without connectability, but his examples, the reduction of the law of free fall to the theory of gravitation (see the first section) and the reduction of rigid body mechanics to classical particle mechanics fit well in our model (see below).

Although Schaffner (1967) already pointed out that in many cases of reduction only a 'corrected version' can be derived, it was Nickles (1973) who very convincingly pointed out the distinction, in our terms, between reduction by aggregation and/or identification on the one hand and reduction by approximation on the other. The latter he called 'domain-preserving reduction', apparently only when occurring without reduction by identification and aggregation; hence it corresponds to our 'pure type' of 'corrective reduction'. What he called 'domain-combining reduction' is in our terms reduction by identification or reduction by aggregation followed by a correlation step, hence, in sum, deductive reduction with a transformation step.

In the scientific literature the terms 'epistemological and ontological reduction' are also used. Several explications are possible, one may equate 'epistemological reduction' with heterogeneous reduction, but one might also use it to indicate any kind of reduction of knowledge claims. It is further plausible to say that micro-reduction is a kind of ontological reduction, but the question is whether there are other kinds of ontological reduction. However, it would be strange to call the pure type of reduction by identification ontological reduction, since the identities involved are supposed to be of an ontological nature. Hence, a pure reduction by identification is an epistemological reduction by ontological identification of concepts.[7] However, as we will see in Chapter 5, it is not altogether clear whether there are pure cases of reduction of laws by identification (which will not prevent us from suggesting an explication of the idea of the ontological reduction of concepts).

We like to conclude this subsection with a brief discussion of Bickle's (1998) approach. Following Hooker (1981) and Churchland (1990), Bickle has elaborated, in structuralist terms, the idea that one first derives an analog or a relevantly isomorphic version of the law or theory to be reduced in the vocabulary of the reducing theory. In order to exclude artificial links, Bickle requires that the remaining reductive link between the two vocabularies should be an 'ontological reductive link' in the sense of Moulines (1984), that is, a link according to which the entities of the theory to be reduced are in some way or other composed of entities of the reducing theory. This approach enables Bickle to start with theory reduction and then to discuss the ontological consequences. This 'new-wave reductionism', as he calls it, leads him (p. 30) to an illuminating double ordering of the main historical cases of reduction: an epistemological ordering along a line from 'perfect smoothness' to 'limit bumpiness', ontologically corresponding to the line of 'perfect retention' (identity) to 'total replacement' (elimination). It is easy to verify that this approach fits very well in our five-step-model. The analog of the law to be reduced corresponds with the 'law' that has been derived after the application step and the aggregation step. The ontological reductive links correspond with the identities used in the identification step, in particular, for the reduction of 'thing-predicates'; hence, the reduction of 'attribute-predicates' seems to be neglected (see Chapter 5). However this may be, Bickle's and our approach are variations on Nagel's starting point, focussing on the varieties of reduction that actually occur.

3.3.2 *Some Further Examples*

We will now give a brief sketch of some other examples of reduction, sometimes including remarks of a more general interest.

The restricted validity of Newtonian mechanics can be explained by approximation, assuming low velocities and starting from an application of Einstein's special theory of relativity to the relevant system of objects. As already noted by Nickles (1973), physicists frequently say that the special theory reduces to classical mechanics, whereas philosophers tend to say just the opposite. The latter is reflected in our terminological diagnosis. It is perhaps generally the case that scientists say that the concretized law or theory can be reduced, by (simplifying) approximation, to its idealized predecessor. Be this as it may, the present example is based on application (step (1)) and approximation (5) and hence illustrates a case of *corrective, homogeneous iso-reduction*.

The reduction of rigid body mechanics to classical particle mechanics is a paradigm example of *deductive, homogeneous micro-reduction*: the laws of motion of a rigid body are obtained by aggregation (2) of the result of application (1) of the laws of motion to the constituting particles, of which the mutual distances are assumed, by definition of a rigid body, to be constant (Adams, 1959; Nagel, 1961; Balzer et al., 1987).

Looking back to the example of the ideal gas law in the first section, it is clear that two of the three steps occur that are each sufficient in order to talk about reduction. With a detour via the Van der Waals's law we get an interesting example in which all three steps occur, leading to the qualification corrective, identificatory micro-reduction. With the aid of integrated application (1) and aggregation (2) of the kinetic theory one first derives the kinetic version of the Van der Waals's law, next one obtains, using the same identifications (3) as in the ideal gas case, the standard macro-version of this law. Hence, at this point we have obtained a deductive, identificatory micro-reduction of the Van der Waals's law. Finally, by approximation (5) (neglecting the so-called attraction and volume constant) one obtains the ideal gas law. In sum, a corrective reduction of the ideal gas law occurs, along with a deductive reduction of the Van der Waals's law.

The present example is particularly interesting from the 'idealization and concretization' perspective, since (the macro-version of) the Van der Waals's law is the paradigm for the concretization of a law, i.e. the ideal gas law. If concretization is really an improvement, as in the present case, an adequate explication of the notion of truthlikeness should count it as a step in the direction of the truth. In ICR (Section 10.4) we show in detail in what sense this is the case. Moreover, and this is the main point in Kuipers (1985), the indicated deductive reduction of the Van der Waals's law by the kinetic theory is a concretization of the deductive reduction of the ideal gas law in the following precise general sense. The explanation of a concretization of an idealized law is said to be a concretization of the explanation of the idealized law if the former can be obtained from the latter by concretization of one or more of the principles of the explanatory theory or of the auxiliary hypotheses. The paper mentioned shows that the naive kinetic explanation of the Van der Waals's law as well as the standard textbook explanations are presented as a concretization of the kinetic explanation of the ideal gas law in a fundamentally wrong way. Moreover, it is shown that a conceptually decent concretization, satisfying the mentioned condition, is possible if one follows the I&C-heuristic systematically.

The suggested explanations by Dalton's theory of the atom (as revised by Avogadro and Cannizaro) of the five observational laws of reaction in Chapter 1 are easily seen to use application (1); aggregation (2); and identification (3), and no approximation. Hence, they provide cases of deductive, heterogeneous micro-reduction. Note also that the explanation of the fourth law discussed, the Gay–Lussac law of combining volumes, uses in the transformation step, besides the core identities of the atomic theory, also Avogadro's hypothesis which correlates the (aggregate) micro-concept of the number of molecules with the macro-terms volume, pressure and temperature. As we have already remarked, Avogadro's hypothesis can be explained by the kinetic theory of gases.

In Hettema and Kuipers (1988) it is shown, among other things, that the explanation of Mendeleev's periodic law (table) by quantum atomic theory uses precisely all five steps of our model. Hence this is an example of corrective, heterogeneous micro-reduction. It is particularly interesting because it shows that transformation by identification (here, the atomic number is equal to the number of electrons or protons) and by correlation (here, chemically similar behavior is caused by analogical atomic structure) can occur concurrently. An example in which they occur sequentially, i.e. with an intermediate, disappearing term, would be hardly surprising.

Looijen (1987, 1998/2000) shows that an explanation, which has previously been called into question, of the physiological so-called Bohr-effects concerning the in- and output of oxygen of the blood is a case of deductive, heterogeneous micro-reduction of these effects to the theory of chemical bonding, via the theory of allostery. The following steps occur in making this reduction: application, aggregation, correlation, application, simple aggregation and identification, which make it a nice illustration of an explanation where certain types of steps occur more than once. The specific goal of Looijen is to show that a sophisticated explication of the idea of emergent properties does not exclude micro-reduction of the lawlike occurrence of such properties.

The explanation of the two constitutive laws of Mendelian genetics with the aid of the principles of molecular genetics is in any case an example of *corrective, identificatory reduction*, because, after application (1), there is certainly identification (3) of something (roughly speaking, a gene is identified with a piece of a DNA-molecule) and approximation (5). In particular the fact that there results after identification a *corrected*, i.e. concretized, version, of Mendelian genetics, which has still to be adapted (approximated) to obtain the classical laws, has drawn much attention in the philosophy of biology. Whether something like aggregation occurs, and hence whether it is a case of *micro*-reduction, needs further research.

Janssen and Kuipers (1989) have shown that the claim of (special) collective demand theory, viz. that goods will be exchanged at equilibrium prices in equilibrium amounts, can be derived from the claim of (special) individual demand theory. The steps application (1) and aggregation (2) are needed for this purpose. Hence, it is an example of *deductive homogeneous micro-reduction* in economics.

Our last example illustrates three of the five use values of cognitive structures mentioned in the introduction of this book: explaining a controversy as a deviation from the ideal course, suggesting concrete new research and suggesting, more generally, a new research policy for the relevant research field. In economics there is a debate about the proper micro-foundation of macro-economics. Janssen (1987, 1993) distinguishes two approaches of this goal: the aggregation and the representativety approach, and explains on the basis of the five-step-model presented here why these approaches are unsatisfactory. For the representativety approach this is easy to summarize: one tries to put the cart before the horse, namely to aggregate first and then to apply utility theory to the so-called representative subject. Janssen (1987) also shows that and how deductive homogeneous micro-reduction of macro-economics is possible in a straightforward way by presenting the utilistic reduction of a (probabilistically) concretized version of the standard macro-consumption function, i.e. a linear relation between national income and national consumption demand. This utilistic reduction uses application (1) and aggregation (2) and resembles in many respects the example of Olson's hypothesis. Of course, the standard version of the consumption function can be obtained by approximation (5) after the aggregation step.

CONCLUDING REMARKS

The models of reductive and non-reductive explanation of laws by a theory presented in this chapter may not only contribute to (dis)solving disputes about explanation and reduction among philosophers of science, and be used for didactic purposes, they may, among other things, also play a heuristic role in both ongoing and new research. In particular in the context of 'interlevel research', where at least two levels of an epistemological and/or ontological nature are involved, they may be useful. However, before we can concentrate on different types of interlevel research, we first have to deal with other types of (mono- and multi-level) explanation, viz. explanation by specification, as opposed to explanation by subsumption, which will be done in the next chapter. Moreover, in Chapter 5 we will focus specifically on the reduction and correlation of concepts, and close with a global survey of reductionistic and holistic strategies. In the final chapter of Part III (Chapter 6) we will distinguish several types of interlevel research, in particular mind-body research, in which the previously distinguished types of reductive and non-reductive explanation will find a place.

CHAPTER 4
EXPLANATION AND DESCRIPTION BY SPECIFICATION

INTRODUCTION

There is no doubt that explanation by subsumption in general and theoretical explanation of observational laws in particular are very important in the empirical sciences. However, it is not at all evident that all respectable kinds of explanation occurring in the scientific literature obey the basic pattern of explanation by subsumption, as dealt with in Chapter 3.

Among historians and philosophers of history for example there has been much dispute about whether or not the intentional explanations that are frequently used by historians are of a different nature than the nomological explanations occurring in the non-life sciences. But also in other areas the discussion about different types of explanation has been very time-consuming. In particular one may think of functional and teleological explanations in biology and psychological explanations in psychology and psychiatry. Among the scientists who use such explanations, and most of them do, many have a bad conscience about it, because such explanations are not easily seen to meet the standards of explanation by subsumption. The same holds for many causal explanations in everyday life, technology, medicine, etc.

In this chapter we will show that such explanations are indeed of a different nature, but nevertheless have a sound logical structure and a substantial information content. In contrast to most defenses of this claim we do not, however, argue that these explanations compete with eventual nomological explanation for the same facts. On the contrary, although they may be highly informative, they are not so pretentious.

More specifically, we will present the main objections to nomological or subsumption reconstructions of intentional explanation of actions, of functional explanation of biological traits, and of a particular kind of causal explanation. For each of these three types of explanation we will show that a totally different reconstruction is possible, which is much closer to scientific practice and which demonstrates that these types of explanation, though different from nomological ones, are sound and informative. It will be shown that these three types have a lot of structure in common, but that it is also possible to indicate clearly where the differences are. The common structure is however sufficiently substantial to distinguish (besides explanation by subsumption) another main type of explanation: explanation by specification. The common structure turns out to be quite different from that of explanation by subsumption.

Within the general pattern of explanation by specification there is room for new special cases, besides the special cases that get the main emphasis in this chapter, viz. explanation by intentional, functional and causal specification. Section 4.1. deals with intentional explanations, Section 4.2 with functional explanations, and Section 4.3 with causal explanations of a special kind. Finally, Section 4.4 will indicate some extrapolations and present some further speculations.

It should be stressed in advance that our explications can be conceived as minimal explications. Although the presentation will frequently suggest that these explications present the whole story, it is perfectly possible to interpret and evaluate them as minimal expositions that can be strengthened.

4.1 INTENTIONAL EXPLANATION OF ACTIONS, GOALS AND CHOICES

Introduction

Explanation of actions in terms of conscious or at least semi-conscious beliefs and desires plays an important role in everyday life, and also, for example, in historical research. Such explanations are frequently called 'intentional explanations'.

In psychology the idea of intentional explanation functions on the one hand as a kind of extreme model: explanations in terms of *un*conscious motives for example are grafted upon the idea of intentional explanation. On the other hand, actions which can be explained intentionally without problems are not interesting for psychologists: they have only to start working as soon as, and as long as, someone's actions cannot simply be explained intentionally. Combining the two considerations leads to the conclusion that explanations in psychology frequently are reasoned deviations from intentional explanations.

From the foregoing it follows that an adequate explication of intentional explanations is not only important for understanding everyday life and historical practices but also for psychology and psychiatry. Moreover, similar things, with the corresponding implication, could be said about sociology, anthropology and economics, in relation to the explanation of the behavior of institutional subjects as well as of the collective effects of individual behavior.

We will start with a sketch of the two well known competing explications of Hempel kand Von Wright and an indication of the main objections to both. Then we will present an entirely different explication. First we will give a meaning analysis of specific and unspecific intentional statements. In terms of such statements we will reconstruct the thought process of producing an intentional explanation, of which a specific intentional statement turns out to be the main product.

4.1.1 *The Standard Explications and their Objections*

The two standard explications of an action share a desire and a belief premise. Moreover, both require a formally similar general premise. Hence, both explications can be represented with one scheme:

$D(x, z)$	x desired goal z
$BN(x, y, z)$	x believed action y to be necessary to approach z
$G(NE/SE)$	for all x, y and z and all occasions if $D(x,z)$ and $BN(x, y, z)$ then $P(x,y)$
P(x, y)	x performed y

Scheme 4.1.

However, they deviate about the interpretation of the general premise. Note first that it is a valid argument scheme. It leads to (a simplified version of) the *nomological explication* (*NE*) of Hempel (1965, Chapter 12) if the general statement G is conceived as an empirical, hence falsifiable generalization. The scheme leads to the *semantic explication* (*SE*) of Von Wright (1971) if G is conceived as a meaning postulate, i.e. as a non-falsifiable meaning relation between the expressions 'to desire', 'to believe something to be necessary' and 'to perform an action'.

Hempel and Von Wright are mentioned as paradigmatic proponents of the two views. Cognate versions of the nomological view have been defended by for instance Gardiner (1952) and Nagel (1961). Versions of the semantic view have been defended by Dray (1957) and Anscombe (1957) and Von Wright's version was exposed in a benevolent way by Stegmüller (1975), along with that of Hempel.

The main objections to the two explications are partly overlapping and partly specific. Common objections to both are that there is no clear correspondence with research practice (a), that there is no explicit role for specific intentional statements (b), and that the belief premise is extremely strong (c). Specific objections to the nomological view are that the empirical generalization is intuitively speaking irrelevant (d) and that it wrongly implies symmetry between intentional explanation and prediction (e). The semantic view has been developed as an alternative to the nomological view, among others because of objections like (c) and (d). However, the main specific objection to the semantic view is that it is based on a meaning postulate with a magical character (f).

The following elucidation of these objections has to suffice.

Ad (a) A minimum condition of adequacy for an adequate explication of any type of explanation is that all ingredients that are present in informal versions should have a place in it and that it does not introduce elements that are alien to informal versions. Moreover, it is desirable that something of the order of the thought process is reflected in the explication. The latter is certainly not the case in both explications, even if we place them upside down. Objections (b) and (d) are instantiations of the first mentioned condition. Moreover, something like a principle of intentionality is not very visible.

Ad (b) In some way or other intentional explanations concern statements of the form 'x performed the action y intentionally' and 'x performed action y with the intention to approach goal z'. However, in both versions such statements play no role, let alone that their meaning becomes precise.

Ad (c) In most cases it is absurd to maintain that the actor believed the action to be *necessary* for his goal, for usually there are alternative known and possible actions that promote the same goal. Due to this extreme strength of the belief premise both explications apply only in rare cases and have to be, and have been, modified for the normal case. But such modifications can only explicate the explanation of a 'disjunctive action', i.e. that the actor performed one of the alternative actions suitable for his goal. This is highly unsatisfactory.

We come now to the two specific objections to the nomological explication.

Ad (d) According to *NE* there is in an intentional explanation an (at least implicit) appeal to the indicated empirical generalization *G*. However, the critic may ask, first, why *G* is seldom made explicit, and second and more important, what does it matter what one or another might do in similar situations, the question is why somebody did what he did on this particular occasion. Hence, *G* is felt to be irrelevant or at least superfluous.

Ad (e) It is easy to see that the premises according to *NE* are sufficient to predict (or retrodict) that the actor will perform the action. However, historians, for example, usually do not claim that they could have predicted an action of an historical subject on the basis of the explanation they offer for that action in terms of the actor's beliefs and desires. Only if they know a lot about his character, including the fact that his behavior is rather uniform, may they claim that they could virtually have predicted the action. Hence, in general, an intentional explanation does not imply the prediction or retrodiction of the action, or to put it in the standard phrase: the symmetry-thesis is false for intentional explanations.

It is easy to check that the objections (d) and (e) do not apply to the semantic explication. No empirical generalization is involved, and a (hidden) meaning postulate is as such no problem. Moreover, prediction of the action on the basis of the premises is impossible, for the meaning postulate interpretation of *G* excludes the possibility of testing the two other premises without testing the conclusion. However, we have already mentioned one specific objection to *SE*, viz. (f).

Ad (f) We have presented the semantic view as a valid argument. To reach validity, *G(SE)*; i.e. *G* interpreted as meaning postulate, is indispensable. It is our best proposal to make something of the so-called logical connection argument. Without *G(SE)* the argument is openly invalid, and the main objection is then that the absence of a meaning postulate is in conflict with the intuition that there is some valid argument involved. With *G(SE)* the problem is not that *SE* appeals to some meaning postulate but that the particular kind is highly problematic. We like to call the postulate magical, for it connects a specific action, as a consequence of meaning relations, to specific mental states. To be sure, the problem is not that some primarily mental concepts have behavioral connotations, e.g. 'hot-tempered' certainly has such connotations, but that statements using only such concepts, even if they are quite specific, might imply a specific action.

A number of objections are not new. In particular (c), (d) and (e) have been raised in the literature in many different forms. Both standard explications have been modified in order to cope with objections like those above. However, these attempts have not been convincing, for none of the two has received general acceptance. Hence, there is room for a third alternative.

In the criticism of the nomological explication by non-analytic philosophers the reference to Dilthey's concept of 'Verstehen' has been dominant. There is good reason to assume (Van Nierop, 1988) that Dilthey only wanted to formulate so-called transcendental conditions for the possibility of knowledge about human affairs. However, he has frequently been interpreted as providing a methodological recipe for the acquisition of such knowledge, i.e. 'Verstehen' or 'Einfühlen' as a necessary condition for a proper understanding (as opposed to mere explanation). This modification of Dilthey's position not only raised the issue of who is able to explain the behavior of someone like Hitler, but it even brought people to believe that such 'Verstehen' could be a sufficient condition for proper understanding, without further testing. The explication of intentional explanations to be presented can also be interpreted as a proposal for a more adequate methodological counterpart of Dilthey's primarily transcendental objectives.

4.1.2 *Explanation of Actions by Intentional Specification*

Let us start with a clear-cut example of an intentional explanation. 'Why did Jane open the window?' One possible adequate explanation is that she opened the window in order to let the room cool down. Readers are advised to keep such a simple example in mind during the following exposition, for they should realize that an adequate explication of intentional explanations should be able to deal with such simple everyday life examples. For in everyday live intentional explanation evidently plays a central role and it is very effective. Hence, if such an example is not in agreement with a certain explication, the disagreement poses a problem for the explication, not for the example.

The two leading intuitions of our explication are, first, that the search for an intentional explanation essentially is of some hypothetico-deductive nature and, second, that, though there may be a valid argument involved, the standard conclusion that the actor performed the action in question (i.e. $P(x,y)$) does not need to be the conclusion of that argument.

The claim of the explication is, of course, that it captures the kernel of simple, normal examples, leaving room for modifications, complicated cases and new, related 'why- and how-questions'. One particular modification should already be mentioned. We will start to concentrate on the explanation of actions, but later it will be shown that everything can be modified in terms of the explanation of goals, that is, in terms of an answer to the question of why somebody pursued a certain goal.

The explication consists of two steps: meaning analysis, followed by a reconstruction of the thought process.

4.1.2.1 *Meaning Analysis*

Consider a *specific intentional statement* of the form:

$IP(x, y, z)$ x performed action y with the intention of approaching goal z

In order to exclude trivial explanations we assume that goal z in $IP(x, y, z)$ is always different from the *internal goal* of action y, that is, the goal of action y according to the description used for it. For example, the internal goal of 'opening the door' is 'having the door open'. The reason for excluding the internal goal from being z is that we consider it as a trivial meaning component of $P(x,y)$ that x performed y with the intention of approaching or even realizing the internal goal of y. Hence, we concentrate further by definition on explanatory goals that are not as a matter of meaning related to the relevant action. On the other hand, it will be argued later, in order to explain an action intentionally, that it is usually sufficient to explain the internal goal of that action intentionally.

The following meaning postulate is a first approximation of the meaning of $IP(x, y, z)$ by way of a meaning decomposition:

$$\text{MP-1}_i \; IP(x, y, z) \approx P(x, y) \text{ and } D(x, z) \text{ and } BU(x, y, z)$$

where the following abbreviations have been used:

$P(x, y)$	x performed action y
$D(x, z)$	x desired goal z
$BU(x, y, z)$	x believed y to be *useful* to approach z

In this so-called (meaning-)key to specific intentional statements we distinguish, beside the action component $P(x, y)$, two mental components: a desire component $D(x, z)$ and a belief component $BU(x, y, z)$. Notice that the desire component corresponds to the first premise in the scheme of the standard explications, but that the belief component is much weaker than the second premise in that scheme. Instead of 'necessary' it only requires that the action is believed to be 'useful' for approaching the goal.[1] By the way, 'x desired goal z' should not be interpreted as if x by definition considers z as a pleasure. It is assumed to be a neutral expression in this respect, as is evidently the case in the standard expression 'beliefs and desires', which is the reason to use 'desired' instead of 'wanted'.

MP-1_i is a first approximation in the sense that the three components do not exhaust the meaning of $IP(x, y, z)$. For instance, some time conditions are plausible, in particular to exclude hindsight 'rationalizations': $D(x, z)$ and $BU(x, y, z)$ may not 'start' later than $P(x, y)$. Such conditions will be implicitly assumed. More importantly, it is highly defensible to add a *causal* component stating that the belief and desire component were causally effective in the sense that the combination of the two mental states played a significant causal role in the formation of the plan to perform the action. However, in order to simplify the presentation let us first stick to the weak, in fact minimal, interpretation as expressed by MP-1_i. Later we will consider the suggested stronger version.

Consider now the *unspecific intentional statement* of the form:[2]

$IP(x, y)$ $\quad$ x performed action y intentionally

There are two plausible interpretations of such a statement. The first is that x performed the action y with the intention of realizing the internal goal of the action. We have already said that we consider this to be a trivial meaning component of $P(x, y)$. The second, non-trivial, interpretation is that the action was directed at some goal different from the internal goal of the action, without specifying this goal. Hence, we introduce the following non-trivial meaning postulate:

MP-2_i $\quad$ $IP(x, y) \approx$ there is a goal τ such that $IP(x, y, \tau)$

Note that, MP-2_i entails that $IP(x, y, z)$ validly implies $IP(x, y)$, by so-called existential generalization (EG).

Let us return for a moment to the example of Jane's opening of the window. The relevant specific and unspecific intentional statements read: 'Jane opened the window in order to let the room cool down' and 'Jane opened the window intentionally'. The relevant belief and desire statements read: 'Jane believed that opening the window would probably cool down the room' and 'Jane wished that the room would cool down', respectively. Notice at least the following: she does not have to believe that the opening of the window is necessary or sufficient for cooling down the room. Moreover, a totally different explanation could have been adequate, e.g. that she wanted to let out her tame canary. Finally, additional why- and how-questions are easy to imagine. But these questions need not be raised, let alone be answered, before we can accept the cooling down explanation as a satisfactory and adequate explanation of her particular action.

4.1.2.2 *Thought Process*

Following the foregoing meaning analysis the question is how the crucial intentional statements function in the process of searching for an intentional explanation and in the result of that process. First of all, the heuristic-methodological principle dominating the research in an 'intentional context' is of course:

PoI $\quad$ *principle of intentionality*: if $P(x, y)$ then $IP(x, y)$

that is, if someone performs (or has performed) an action, he will do (have done) that intentionally. Here the notion of 'intentionality' is not used in the philosophical sense of 'aboutness', but in the straightforward sense of consciously aiming at a certain goal. (See Dennett 1996, Chapter 2, for this distinction.)

It should be stressed that PoI does not state the analytic fact that anyone who performs an action, according to a certain description, must have the intention of realizing the internal goal of that action according to that description. The point is that a goal other than the internal goal is aimed at. For example, when one opens the door this is supposed to have still another goal than just having an open door.

As in many other methodological principles, PoI should be conceived as a principle that fundamentally leaves room for exceptions. Such exceptions can only improperly be called falsifications. That is, the research is guided by the idea that the qualification 'intentional' is adequate, but the research may fail to substantiate this claim. Although it is impossible to have conclusive reasons for concluding that an action was not intentionally performed, the researcher may provisionally come to this conclusion, that is, for the time being.

Such methodological principles essentially are heuristic variants of default rules of inference in the sense of non-monotonic logic (see Tan, 1992, and Marek, W. and Truszczynski, 1993); as long as one has no good reasons to the contrary a particular instance of a certain type is supposed to have a feature that most of the instances of that type have.

Thought-Process

(1)	Verified action statement	$P(x, y)!$
(2)	*question*	why $P(x, y)?$
(3)	unspecific intentional statement	'by PoI'
	as hypothesis	$IP(x, y)?$
(4)	specific intentional statement	'by idea'
	as hypothesis	$IP(x, y, z)?$
(5)	non-trivial implications to be tested	hence, by MP-1_i
	desire hypothesis	$D(x, z)?$
	belief hypothesis	$BU(x, y, z)?$
(6a)	falsification of one or both,	
	hence, go back to step (3)	not-$IP(x, y, z)!$
(6b)	or, no clear results,	$IP(x, y, z)?$
	go to step (3) or (8)	
(6c)	or, verification of both,	
	hence *answer*: verified specific	
	intentional statement	$IP(x, y, z)!$
(7)	now, conclude first as a side-step:	by MP-2_i and EG
	verified unspecific	
	intentional statement	$IP(x, y)!$
(8)	then go to new related	
	why- and how-questions	why?/how?

Scheme 4.2.

Now we present a stepwise reconstruction of the thought process governing an intentional explanation. Questions and hypotheses are indicated by question marks, verified statements by exclamation marks. It should be stressed that 'verification' is used here in the pragmatic sense in which it is not irrevocable.

Scheme 4.2. summarizes the following train of thought. Starting from the established fact (1) that somebody x has performed a certain action, which includes that x wanted to achieve the internal goal of that action, the question (2) is raised why x did so. The search for an answer is governed by the unspecific intentional hypothesis (3) that x will have done it intentionally, that is, directed by some goal which has still to be specified. As soon as some specific goal, and hence a specific intentional hypothesis (4), has been thought of as a serious candidate, its non-trivial meaning components have to be tested (5). Such testing will be elucidated separately. Here we restrict our attention to the two extreme cases of the test results. If at least one of the two implications is convincingly falsified the total specific hypothesis is falsified (6a) and we have to go back to step (3) and invent a new specific hypothesis, i.e. a new serious goal-candidate. If both implications are convincingly verified and if we neglect hidden meaning components, the total specific hypothesis is also verified (6c). In this case we may first conclude, by way of a side-step, that the unspecific hypothesis is also verified (7). The main road, however, leads to new, related why- and how-questions (8), which are suggested by the verified intentional hypothesis (see below).

We leave it as an exercise to transform the search for an explanation of Jane's opening of the window into an example of this scheme.

The claim is of course that the process of producing an intentional explanation is characterized by the above scheme, that the main product is the corresponding specific intentional statement, and that an interesting by-product is given by the valid implication of the corresponding nonspecific intentional statement.

It is clear that the steps (3)–(6c), leading to the main product, represent the synthetic part of the explanation, which may be called the process of *intentional specification*. The step (6c)–(7), leading to the by-product, represents the analytic part, to be called *intentional generalization*. Moreover, it is plausible to decompose the process of intentional specification into an heuristic phase (3)–(4) and a test phase (5)–(6).[3]

Let us now concentrate on the test phase, which is directed at testing the two non-trivial meaning components, i.e. the hypotheses that x desired the goal and thought the action to be useful for that goal. Of course, whether and how such hypotheses can be tested depends heavily on the context and our standards for such a context.

It is instructive to compare the present test phase with hypothetico-deductive evaluation of theories, in terms of the general and individual test implications presented in Part IV. Of course, here there is also a clear hypothetico-deductive structure: a hypothesis is put forward, and implications to be tested are derived. One important difference between the present application of the hypothetico-deductive method and its application to theories is that the hypothesis is now not of a general, if not modal-nomological, nature but is some complex individual statement. A second difference is that the implications to be tested are of a fixed type and small number (even if we include a causal component, see below); they exhaust the hypothesis, which is almost inconceivable in the case of theories. Of course, each of the two (or three) implications amounts itself to a complex statement. For example, '*x* believed *y* to be useful in approaching *z*' is far from a simple ternary fact. In particular, the implications to be tested, being concerned with mental states, can certainly not be called observational relative to the main hypothesis. Hence, they are not 'test implications' in the specific observational sense discussed in Chapter 2. But the situation is not as dramatic as it might seem.

Let us speak of 'strong predictions' (including strong retrodictions) of a hypothesis when their eventual falsehood is observable (relative to the hypothesis) and falsifies the hypothesis, at least in principle; hence they are observable necessary consequences. The general and individual test implications in Chapter 7 are of this nature. Let us speak of 'weak predictions' when their eventual truth is observable (relative to the hypothesis) and verifies the hypothesis, at least in principle, hence they are observable sufficient conditions.

Now it is easy to ascertain that the two implications to be tested, i.e. the belief and the desire hypothesis, primarily yield such weak predictions. Neglecting the general distrust towards human utterances, think, for example, of the evidence for the desires of historical figures provided in their private notebooks. Under certain conditions, however, it is also possible to falsify them, on the basis of strong predictions. Think of the explicit denial of a desire in the notebook.

Consequently, the intuitive feeling of historians that their research is frequently of a hypothetico-deductive nature can be explained in terms to the implications to be tested on the basis of weak and strong predictions, without assuming interest in nomological hypotheses. Such a way of testing seems characteristic for testing (complex) individual hypotheses. Hence, we might speak of the hypothetico-deductive method for individual hypotheses.

With respect to the step (6c)–(7) leading to the by-product, i.e. the intentional generalization (as a case of existential generalization) of the unspecific intentional statement, it should be stressed that this step does not seem to be important in practice. However, it has been the conviction of many philosophers that there is some valid argument involved in intentional explanation. More specifically, it is usually assumed that the conclusion of this argument should be $P(x, y)$. In our explication the natural argument involved is the existential generalization step (7) leading to the conclusion that the action was intentional, that is $IP(x, y)$. Notice that this view confirms what we have already stated as a hypothesis in step (3). In the meantime new information has been obtained implying that this hypothesis may be accepted as true, In our opinion, this argument explains our intuitive feeling that everything fits, that the puzzle is solved.

To be sure, it follows directly from MP-1_i and MP-2_i that $P(x, y)$ is a conjunctive meaning component of $IP(x, y)$. Hence, $P(x, y)$, the standard conclusion, forms a part of our conclusion $IP(x, y)$. But it is easily seen to be a trivial part (in fact the only one), for $P(x, y)$ is already, according to MP-1_i, a conjunctive meaning component of the premise of the existential generalization, that is, of $IP(x, y, z)$.

4.1.2.3 *New Related Questions*

As already indicated in step (8) a successful intentional explanation suggests new related questions. The first one is certainly whether the belief and desire were causally effective in the sense that the combination of the two mental states played a significant causal role in the formation of the plan to perform the action. As already suggested we could have added this condition as a meaning component of $IP(x, y, z)$. Hence, the related question is equivalent to the question about the consequences of replacing the weak interpretation of $IP(x, y, z)$ by the strong one obtained by adding the *causal effectiveness component*:[4]

$CE(x, y, z)$ $D(x, z)$ and $BU(x, y, z)$ were causally effective for $P(x, y)$

As an implication to be tested, it is of a rather different nature than the other two. It was less difficult to more or less verify the belief and desire component than to falsify them. In contrast, although one may succeed in falsifying the causal effectiveness component more or less convincingly, it is as yet impossible to verify it in some straightforward way. However, here it is plausible to make use of a default rule of inference to the effect that, as long as there is no evidence to the contrary, the belief, desire and action component together are supposed to imply the causal effectiveness component:

CE-default if $BU(x, y, z)$, $D(x, z)$ and $P(x, y)$ then $CE(x, y, z)$

The resulting thought process for a strong intentional explanation is obtained by inserting the relevant causal effectiveness component at step (5), by adding the possibility of its falsification at step (6a), and by inserting the CE-default as an argument in step (6c).

It is important to note that the difference between the weak and the strong version of an intentional explanation is not that the former still can constitute a 'rationalization' of the action by the actor. This possibility was already excluded by the implicit time conditions. The difference it makes only concerns the relatively rare possibility (justifying the default rule) that one particular appropriate combination of belief and desire were in fact not causally effective, whereas some other pair was. This point was stressed by Davidson (1963). Since this possibility is relatively rare, or so it seems, we have chosen to start with the weak version for reasons of simplicity.

The other main questions raised by a successful intentional explanation (in the weak sense) seem to be the following. Why did x aim at z? Did x aim with y only at z? Why did x think that y was useful for z? Why did x choose y among the available alternatives? The first question brings us directly to the problem of intentional explanation of goals, the last one to the problem of intentional explanation of choices. Both will be dealt with in separate subsections. As to the second question, whether or not z was the only goal, it is evident that one may aim at different goals with the same action. This may be in the sense that the one goal is an intermediate goal between the action and the other goal, which brings us back to the first question, for it creates the possibility of an intentional explanation of that intermediate goal in terms of the other goal. But the possibility of having two (causally effective) 'parallel goals' with the same action is not excluded: Jane may open the window in order to cool down the room and in order to let out her canary.

Finally, with respect to the third question about the actor's grounds for his belief, it will be clear that this question may be interesting in particular cases, but it is of course a subject in itself. In the present context it may only serve to illustrate a terminological remark. Intentional explanations are also frequently called 'rational' explanations. This terminology is however highly misleading, for it is likely to confuse two things: reasons and good reasons. The term 'reason-giving-explanations' does not have this ambiguity. The crucial point is that it should be possible to call an action intentional but dispute its rationality. For example, atheists will not dispute that the praying of a believer may be intentional, but they would be unwilling to call it rational. In general, beliefs and even desires may or may not be rational, notwithstanding the fact that they may constitute the crucial components of an intentional explanation.

Let us now look back at the objections raised against the standard explications. These objections do not apply to our explication. The objections (a) and (b), concerning the correspondence with research practice and the role for specific intentional statements, are clearly met. Objection (c) does not apply because the belief component in our explication is considerably weaker than the one in the standard explications: 'useful' instead of 'necessary'. Objection (d) does not apply because there is no appeal to an empirical generalization, let alone to an irrelevant one. Objection (e) does not apply because there is no prediction claim with respect to $P(x, y)$, hence there is a strong asymmetry between prediction and (intentional) explanation of $P(x, y)$. Finally, objection (f) does not apply because the two meaning postulates are not magical in the sense that they do not connect a specific action, as a meaning consequence, to (specific) mental states.

4.1.2.4 *Intentional Explanation of Goals*

Let us start by mentioning Nagel's favorite example of what we call an intentional explanation:

> 'Why did Henry VIII of England seek to annul his marriage to Catherine of Aragon? A familiar explanation for this historical occurrence consists of imputing to Henry a consciously entertained objective ... Thus historians often account for Henry's efforts to annul his marriage to Catherine by citing the fact

that, since she bore him no son, he wished to remarry in order to obtain a male heir. ... his efforts at obtaining an annulment are explained as deliberate means instituted for realizing a conscious goal (or end-in-view).' (Nagel, 1961, p.19)

It is clear that several related actions and goals are involved in the example. What is exactly explained? The goal to divorce, the efforts to divorce, the actual divorce? A first important point is to realize that in everyday life an intentional explanation of an action and an intentional explanation of the internal goal of that action are more or less equivalent. For example, in order to explain Henry's actual divorce it is sufficient to explain his goal or desire to divorce. Importantly it is not being implied that people are assumed always to perform an action when they desire the internal goal of that action. The only assumption is that if someone has performed a certain action, the interesting question usually is why he wanted to achieve the internal goal of that action and not why and how he managed to perform the corresponding action.

It is plausible to set up the general explication of intentional explanations of goals in line of this relation between explanation of an action and explanation of the internal goal. Such an alignment automatically occurs as far as possible if we make our inquiry analogous to our explication of the intentional explanation of actions. It is easy to see that this is possible with the following abbreviations (replacing $P(x, y)$, $IP(x, y)$, $IP(x, y, z)$):

$A(x, z)$	x aimed at goal z
$IA(x, z)$	x aimed intentionally at goal z
$IA(x, z, z*)$	x aimed at goal z with the intention to approach goal z^*

Moreover, $D(x, z*)$ and $BU(x, z, z*)$ and the relevant meaning postulates are defined completely analogously to those in the case of actions. Then it is easy to confirm that an intentional explanation of an action can be transformed into an intentional explanation of the corresponding internal goal. Such ease of confirmation is also true in the other direction, provided the relevant goal can be conceived as the internal goal of an action, which is not of course possible for all conceivable goals.

A second point that is clear from the example of Henry VIII is that different but related goals and actions may be involved. To say that Henry wanted to divorce in order to remarry is correct, but so is the fact that he wanted to divorce in order to obtain a male heir, simply because he wanted to remarry in order to obtain a male heir. Hence, it is a typical example of a chain of actions and goals with a corresponding chain of intentional explanations. Of course, if we have established one link between actions and/or goals we may investigate new related why- and how-questions, as indicated in the thought process.

For a similar, but still more refined analysis of statements like $IP(x, y, z)$ and $IA(x, z, z*)$, the reader is referred to (Holmström-Hintikka, 1991). She distinguishes (p. 13) and extensively analyzes 'instrumental actions', 'purposive actions' and 'goal-directed wills'. The second type of action corresponds with our IP-type of action and 'goal-directed wills' with IA-type of goals. An 'instrumental action' is defined more broadly than an purposive action, viz. as 'seeing to it' that a certain goal becomes realized or remains realized, depending on what is the case.

4.1.2.5 *Intentional Explanation of Choices*

The last question suggested by a successful intentional explanation is why the actor did choose one particular action from many alternatives. In discussing the two standard explications we saw that both had trouble with alternative actions due to their strong belief premise that the action is thought to be necessary for the goal. As a result of our much weaker premise, the action is only thought to be useful for the goal, alternative actions do not play a role in our explication. However, it remains to be shown how alternatives are to be treated when they are brought into question.

Let us start by assuming the (weak) intentional explanation $IP(x, y_1, z)$ of $P(x, y_1)$, implying $D(x, z)$ and $BU(x, y_1, z)$. Suppose now that for the possible actions $y_2, \ldots, y_n$, holds also that $BU(x, yi, z)$. Let y_0 denote 'doing nothing' and let $BU(x, y_0, z)$ be false; note that this is a non-trivial assumption.

From these assumptions it follows that we would be able to explain all 'positive' actions in terms of the same goal z. But x did choose y_1, why?

In order to answer this question we will postulate quantitative utilities (u) and probabilities (p) for convenience, but only use comparative relations between them.

We begin with some obvious interpretations, leaving out the reference to x, in quantitative terms:

$D(x, z)$:	$u(z) > u(not\text{-}z)$	
$BU(x, y_i, z)$:	$p(z) < p(z/y_i)$	$i = 1, 2, \ldots, n$
$not\text{-}BU(x, y_0, z)$:	$p(z) \geq p(z/y_0)$	

In addition to these assumptions it is reasonable to assume that each of the positive actions costs more or less effort; i.e. they have, compared with y_0, lower utility:

$$u(y_i) < u(y_0) \qquad i = 1, 2, \ldots, n$$

A standard question in utility theory, more precisely utility oriented decision theory, now is whether or not the *expected utility* of y_i, for $i = 0, 1, 2, \ldots, n$, i.e. in the present context,

$$EU(y_i) = u(z) \cdot p(z/y_i) + u(not\text{-}z) \cdot p(not\text{-}z/y_i) + u(y_i)$$

was maximal for $i = 1$.

Suppose now that x made up his mind before performing y_1, considered the set of alternative actions $Y = \{y_0, y_1, \ldots, y_n\}$, attached utilities and probabilities as suggested above. Next he calculated the expected utility of each action and reached the conclusion that $EU(y_1)$ was maximal. Finally, he decided to use the maximization of expected utility as the criterion for choice-making, chose a_1 accordingly, and performed it. Now it is natural to say that the intentional explanation of $P(x, y_1)$ is still in terms of z, i.e. $IP(x, y_1, z)$, and that x's choice of y_1 out of Y was also intentional, but now not directed at z but at maximizing expected utility.

In sum, we have arrived at a natural distinction between an intentional explanation of an action and an intentional explanation of 'choosing one out of the alternative actions', i.e. of choice. Intentional explanation of choice will be called utilistic as soon as the actor uses some utilistic decision rule.

In our opinion the success of utility theory, which is of course mainly interested in utilistic explanations of (and devices for) choice, has unfortunately led to the theoretical neglect of the intentional explanation of actions. Standard presentations of utility theory usually do not start from a particular goal, but from alternative actions leading to incompatible possible outcomes, which obscures the fact that in many cases there is really one goal at stake, which will or will not be achieved. For instance, it may well be that scientific interest in particular actions of human beings, in historiography, is more concerned with intentional explanation of such actions than with utilistic explanation of the choice of them. On the other hand, in contexts where scientific interest is mainly occupied by the collective effects of individual actions, e.g. in economy, sociology, and political science, it is reasonable to expect that the emphasis will be on the utilistic explanation of choices.

These differences in emphasis will partly be due to the fact that there may be an intentional explanation of an action, but not an intentional, let alone utilistic, explanation of the choice of that action. For example, somebody may jump into the water to rescue someone else without any deliberation of alternative actions serving the same purpose. Conversely, in a complex decision situation, with many possible actions and incompatible outcomes, it is hardly possible to talk about the intention with which a certain action is performed, although the choice may be perfectly intentional in a utilistic sense.

We conclude this subsection with a link to Chapter 3. There we have seen that utility theory naturally leads to the possibility of utilistic explanation, even reduction, of social laws. The basic principle is that of utility maximization stating that people tend to choose the action with the highest expected utility. Such explanations usually do not strictly assume that the actors consciously choose actions according to that principle. Hence, in general it would be better to speak of 'as-if-utilistic' explanations.[5] However, it is of course possible in certain cases that the assumption of conscious choice is realistic for most people involved. It is plausible to explicate such an 'intentional interpretation' of the principle of utility maximization in terms of the present analysis of utilistic intentional choice (see Kuipers (1984) for a detailed analysis).

Concluding Remarks

We claim that our explications of the intentional explanation of actions, goals and choices are close to everyday and historical practice. Moreover, they show that these practices can be sound and informative, without appeal to laws and theories.

The reconstruction of the thought process may well be used as a heuristic checklist in order to evaluate and streamline these practices.

We would like to conclude this section with a few general points. So far we have not explicitly spoken of true and false intentional explanations of an action, goal or choice, but it is plausible to call an intentional explanation true when the two (or three) additional meaning components are true, and false when at least one of them is false. We need these terms in order to formulate a general question concerning intentional explanations as precisely as possible. If, as we have suggested, there are very many true intentional explanations in everyday life, why are they not considered as belonging to our stock of scientific knowledge? As with a lot of other (true) information – and true intentional explanations just provide a special type of information – its neglect must be due to the fact that this information is, in general, not considered to be scientifically interesting. An important exception is the science of history, where many intentional explanations are considered to be interesting, e.g. the divorce of Henry VIII. As is clear from our exposition, historians should not, in our opinion, worry about the well known nomological criticism. If they support their hypothetical intentional explanations with evidence they explain actions in a significant and sound way. Attempts to strengthen such explanations so that they constitute nomological explanations do not seem to result in the formulation of any interesting laws. This inability does not of course prevent the historian from being interested in statistical information, e.g. of the form: in that period and region most people doing such and such did this with intention so and so. Such information may suggest intentional explanations in particular cases. However, this use of statistical information should not be confused with the imperfect form of deductive-nomological explanation, viz. statistical explanation of an individual event (action) on the basis of a statistical law. Such explanations do not explain very much, as long as the laws underlying them are not explained by some theory.

It is also worthwhile to consider historiography in the light of Chapter 1 and ICR (Chapter 7). To begin with, historiography obviously is primarily a descriptive science, with research programs in which the Principle of Intentionality, is one of the main heuristic principles. In ICR (Chapter 7) we explicate the distinction between approaching the true description and the true theory. The nomological criticism can now be seen as based on the false claim that the science of history is, or should be, a theory-directed science, where nomological explanations play an important role. In contrast to this nomological point of view, one may argue that intentional explanations can be of crucial importance in historiography for it is easy to show that a true intentional explanation, conceived as the transition from $P(x, y)$ to $IP(x, y, z)$, is a (synthetic) step which brings us closer to the true description of an historical episode in a given vocabulary. As a consequence, the pattern of explanation by intentional specification essentially is a pattern of intentional description.

Let us conclude this section with a remark of a different nature. The rejection of the nomological explication of intentional explanation does not imply the rejection of a causal-nomological explanation of actions. In principle it is possible with the aid of causal laws and theories to enumerate sufficiently many causal factors, including mental components, on the basis of which it can be derived that somebody performed a certain action. These causal laws and theories will however be much more complex than *G*(*NE*). Moreover, such a causal-nomological explanation of a particular action will seldom be interesting. What is interesting in this connection is of course the general question: how do human beings succeed in transforming mental factors into actions? What is the mechanism? What we need for this is a general theory about the total system of mind, brain, body, and environment, with the central focus on the nature of our consciousness. Several such theories have been proposed, e.g. P.M. Churchland (1990), Dennett (1991), and Minsky (1985) to mention just a few. However, such theories are certainly not involved in our practice of explaining particular actions in terms of beliefs and desires. In Chapter 6 we will study various kinds of interlevel research in psychology, in which, among other things, intentional as well as functional explanations have a role to play. So let us turn to functional explanations.

4.2 FUNCTIONAL EXPLANATION OF BIOLOGICAL TRAITS

Introduction

In this section we will argue that functional explanations in biology have the same general structure as intentional explanations and that the important difference lies in the meaning decomposition of specific functional and intentional statements. That is, not surprisingly, the key meaning postulate for specific functional statements will turn out to be of a quite different nature than those of specific intentional statements.

To get the appropriate feeling for our present subject we mention four examples of generally accepted functional explanations in terms of the related specific functional statements. The function of lungs in animals is to facilitate oxygen supply by breathing. The function of chlorophyll in plants is to enable them to perform photosynthesis. The heartbeat in vertebrates has the function of circulating blood through the organism. The function of the fanning movement of sticklebacks is to supply the eggs with oxygen.

4.2.1 *Standard Explications and the Objections to Them*

First we will sketch the two standard explications of functional explanations (Nagel, 1961, Chapter 12.I; Hempel, 1965, Chapter 11) and survey the main objections to them. Both Nagel and Hempel reconstruct attempts by biologists to provide an explanation for the presence of a relevant trait. According to Nagel the trait, say chlorophyll, is present because it is causally necessary for another trait or process, e.g. photosynthesis. According to Hempel one has to show that the trait in question is causally sufficient for another trait that is, in its turn, necessary for the adequate functioning of the organism. These explications lead to two different underlying arguments, both with a nomological premise. In the case of Nagel it is valid, in the case of Hempel it is not.

The following schemes give schematic versions of these, inessentially simplified, explications, where MP indicates a meaning postulate and 'x-organisms' are organisms of a certain kind x.

Nagel's explication can be reconstructed as having the following underlying argument, which is easily seen to be valid:

$H(x,z)$	x-organisms have (perform/produce) trait (process/phenomenon) z
$N(x,y,z)$	trait y of x-organisms is *causally necessary* for trait z
MP	if $N(x,y,z)$ then: if $H(x,z)$ then $H(x,y)$
———————————	
$H(x,y)$	x-organisms (must) have trait y

Scheme 4.3.

Note that due to MP the second premise is a nomological premise.

According to Hempel we are basically concerned with the following *in*valid argument (indicated by '_ _ _'), where the second and the third premise are of a nomological nature:

$A(x)$	x-organisms function adequately
$A(x) \rightarrow H(x,z)$	trait z is a necessary condition for $A(x)$
$S(x,y,z)$	trait y of x-organisms is *causally sufficient* for trait z
MP	if $S(x,y,z)$ then: if $H(x,y)$ then $H(x,z)$
_ _ _ _ _ _ _ _ _ _ _ _ _	
$H(x,y)$	x-organisms (must) have trait y

Scheme 4.4.

The main objections to the two explications are again, as in the case of the two standard explications of intentional explanation, partly overlapping and partly specific. Some of the objections are similar to the ones in the intentional case, others are different.

Common to both explications is the fact that there is no clear correspondence with research practice (a), that there is no explicit role for specific and unspecific functional statements (b), and that there is no clear 'reproduction and survival' component (c). Nagel's explication has the additional objections that there is no adequate room for functionally equivalent traits (d) and that it makes no difference between explanation and prediction (e). The main specific objection to Hempel's explication is of course that his argument is invalid (f).

We will very briefly elucidate these points.

Ad (a) The lack of correspondence with research practice will become striking in comparison with our explication. Here it suffices to note that there is no clear role for something like a principle of functionality of traits and the objections (b) and (c).

Ad (b) Analogous to the case of intentional explanations, one wonders if there is no clear role for, nor an explication of functional statements, whereas such statements are crucial in the type of biological research we are considering.

Ad (c) Biologists are eager to show by their functional explanations that the trait in question has a direct or indirect survival value in terms of the theory of evolution, i.e. the theory of variation and natural selection.

Ad (d) According to Nagel's reconstruction, the possibility of functionally equivalent traits is excluded, except when one is willing to weaken the explanatory claim to a disjunction of traits. Note that Hempel's explication leaves room for more than one sufficient condition, so there is room for functionally equivalent traits. However, the requirement of causal sufficiency may also be deemed as being too strong for other reasons.

Ad (e) According to Nagel an adequate explanation would enable the biologist to claim that he could have predicted the presence of the trait in question. But biologists seeking functional explanations do not seem to be interested by such retrodiction claims. Due to the invalidity of Hempel's argument scheme there is no symmetry. Due to the possibility of functionally equivalent traits there is even no reason to assume that the premises make the presence of the trait (very) likely.

Ad (f) However, the invalidity of Hempel's scheme is an objection because biologists certainly do claim, or at least have the intuitive feeling, that their argumentation is basically sound, notwithstanding the fact that they feel uneasy about the nomological criticism.

4.2.2 *Explanation of Biological Traits by Functional Specification*

As announced above there are many similarities between our explication of intentional and functional explanations. Our leading intuition is again that there is a valid argument involved, but that the 'standard conclusion', in the present case that x-organisms (must) have trait y, is not the proper conclusion. Our claim is again that the explication to be presented captures the kernel of unproblematic cases, leaving room for modifications, complicated cases and new related 'why- and how-questions'.

The reader is advised to keep the example of the fan-movement of the sticklebacks in mind. Systematic observation has taught biologists that male sticklebacks, having fertilized the eggs in the nest of their female partner, start to make periodical intensive movements with their pectoral fins in the direct neighborhood of the nest. The example is very instructive, as it is not immediately clear why they do this. However, the biologist will be strongly inclined to believe that this behavior is functional. The question is, why precisely? The idea that turned out to be the correct answer, at least for the time being, is the already mentioned fact that the movement improves the oxygen supply. Having this idea, however, is one thing, to prove its correctness another. In our opinion the proof centers around the following two points. First, it should be shown that the fanning movement leads to oxygen rich water around the eggs. Second, it should be shown that the oxygen supply is favorable for the reproduction and survival of sticklebacks. It is easy to imagine relevant experiments that can be designed to test these two hypotheses.

If both hypotheses are confirmed, it is likely that the biologists will accept the functional explanation as having been established, which does not of course exclude the fact that there may be new questions of interest. For instance, is the supply of oxygen the only function of the fanning movement? Why is this trait realized rather than one of the functional alternatives? And did the behavior (phylogenetically) originate along the suggested lines? How to deal with the first question is likely clear. The second question is very difficult to answer. However, biologists tend to resort to a kind of utility maximization known as optimality analysis in explaining (the presence or selection of) a feature rather than its functional alternatives; hence, they make a clear distinction between the 'absolute' functional explanation of a trait and the 'comparative' explanation of it among alternatives. We will concentrate on the former and briefly come back to the latter in the concluding remarks.

As for the third question (viz. 'Did the behavior (phylogenetically) originate along the suggested lines?'), the answer may be straightforward (viz. that the behavior was selected along the suggested evolutionary lines), but need not be. There are several reasons for the latter possibility. One is that the present environment may have changed fundamentally, compared with the 'selection environment'; i.e. the environment in which the relevant trait originated in the organisms, was at that time in a certain overall state of development. To be sure, the trait may well be functional in the present environment, to be called its present function(s), which may explain why it remains in existence and did not become rudimentary. Wouters (1999) even argues convincingly, by discussing many examples, that biologists normally functionally explain the persistence of a trait in their present environment. If they want to functionally explain their selection, they explicitly refer to the selection environment. According to Wouters it is a misapprehension among most (general) philosophers of science and many philosophers of biology that biologists are primarily interested in selection explanations. Fortunately, in our general analysis we can and will stay neutral regarding this matter. In particular, in our meaning analysis we will explicitly refer to the present as well as to the selection environment.

To get a feeling for the fact that the present function of a trait frequently is not the original function, we would like to mention the example of the animal tail. According to present insights, the tail originated in fish (and whales) as a propulsive organ. For mammals the tail lost that function, but it remained as an organ with several other possible functions, e.g. as a grasping organ for many sorts of monkeys, as a support organ for kangaroos (third leg!), and as a fly-swatting device for cows and horses. As is well-known, the tail finally disappeared altogether in animals like frogs and apes.

There are reasons other than changes of environment that determine why a possible functional aspect may not have played a substantial role in the selection of the trait. In particular, the trait may have primarily come into being as a by-product of some other functional trait, which came into being at the same time. Or the trait may have another functional aspect, surpassing the identified functional aspect in survival value. The existence of such possibilities leads to a distinction between a weak and a strong version of functional explanations similar to what was the case for intentional explanations. First, we will concentrate on the weak version.

Whatever the origin of a trait may be, it is important to note that, when the evidence strongly suggests that a trait originated due to its fulfilling such and such function(s), it will usually not be very interesting to lay down the specific historical-nomological, i.e. evolutionary, explanation in great detail.

Let us now turn to the explication proper. Again the explication consists of two steps: meaning analysis, followed by a reconstruction of the biologist's train of thought.

4.2.2.1 *Meaning Analysis*

Consider a specific functional statement of the form:

$FH(x, y, z)$ the (or one) function of y of x-organisms is (was) its contribution to trait (process/phenomenon) z in the present (or selection) environment

We assume throughout that in such a statement y is different from z. As a first approximation we introduce the following meaning postulate:

MP-1_f $FH(x, y, z) \approx H(x, y)$ and $P(x, y, z)$ and $P(x, z, r\&s)$

where the following abbreviations are used:

$H(x, y)$ x-organisms have trait y

$P(x, \alpha, \beta)$ trait (process/phenomenon) α of (performed/produced by) x-organisms is a positive causal factor for trait (process or phenomenon) β in the present (or selection) environment

$r\&s$ reproduction and survival

The following remarks will clarify this meaning postulate. First, besides the descriptive component there are two causal-nomological components. Of the latter, assuming z unequal to $r\&s$, the first one, i.e. $P(x, y, z)$, may be called the *proximate* component and the second one, i.e., $P(x, z, r\&s)$, the *ultimate* component. Second, the ultimate component, stating that z is a positive causal factor for reproduction and survival, corresponds of course to what biologists call, in terms of the theory of evolution, (positive) *adaptive* or *survival value*. More specifically, in the 'selection-reading', the claim of the theory of evolution is that, as a rule, y became a trait of x-organisms due to selection along the lines suggested by the combination of the ultimate and proximate component. Third, there arise two types of specific functional statements: if z is unequal to $r\&s$, $FH(x, y, z)$ is called a proximate functional statement and $FH(x, y, r\&s)$ is called an ultimate functional statement. Fourth, and finally, assuming that $H(x, r\&s)$ and $P(x, r\&s, r\&s)$ are both tautological and that $P(x, \alpha, \beta)$ is transitive (regarding α and β, and fixed x), it is easy to derive that a proximate functional statement $FH(x, y, z)$ implies both corresponding ultimate ones, i.e. $FH(x, y, r\&s)$ and $FH(x, z, r\&s)$.

We would like to mention that our MP-1_f can be seen as a modified version of the mutually distinct proposals made by Ruse (1973) and by Van der Steen (1982). See Millikan (1993), Mackor (1997) and Section 6.2. for more refined distinctions in terms of proximal, distal and ultimate functions.

Consider now the unspecific functional statement

$FH(x, y)$ trait y of x-organisms is functional (in the present (or selection) environment)

The following meaning postulate seems inevitable:

MP-2_f $FH(x, y) \approx$ there is a trait (process, phenomenon) τ such that $FH(x, y, \tau)$

It is again immediately clear that $FH(x, y, z)$ implies by existential generalization (EG) $FH(x, y)$. Moreover, $FH(x, y)$ implies $FH(x, y, r\&s)$, i.e. y is ultimately functional. Finally, it is clear that it is possible that y is ultimately functional, but not proximately. The fact that plovers have a stable clutch-size of 4 (in their present type of environment) provides an example; i.e. there does not seem to be a demonstrable trait or process or phenomenon 'between' the particular clutch-size and reproduction and survival.

4.2.2.2 *Train of Thought in a Functional Explanation*

With the use of the presented meaning postulates it is possible to reconstruct the main lines of the researcher's thought process.

Again we can formulate a heuristic rule that plays a crucial role.

PoF *principle of functionality*: if $H(x, y)$ then $FH(x, y)$

that is, all common traits of the organisms of a certain type are functional (at least in the ultimate sense). It is clear that PoF, taken as descriptive generalization, is false. Prima facie counter-examples, that is, counter-examples relative to the present environment, are well-known: vestigial or rudimentary traits, such as tonsils, appendix, etc. However, they are not genuine counter-examples for they probably were functional in their selection environment. But there are also traits that are no more than causal by-products of functional traits, e.g. the heart sounds caused by the heartbeat.[6] As already explained, this accidental nature of such traits may even apply, despite a seemingly functional feature. Hence, PoF is to be considered as a heuristic default rule; though leaving room for exceptions, it governs the following thought process.

Thought process

(1)	Verified attributive statement	$H(x, y)!$
(2)	question	why $H(x, y)$?
(3)	unspecific functional statement	'by PoF'
	as hypothesis	$FH(x, y)$?
(4)	specific functional statement as hypothesis	$FH(x, y, z)$?
(5)	non-trivial test implications:	hence, by MP-1_f
	attributive hypothesis	$H(x, z)$?
	proximate causal-nomological hypothesis	$P(x, y, z)$?
	ultimate causal-nomological hypothesis	$P(x, z, rs)$?
(6a)	falsification of one of them, hence, go back to step (3)	not-$FH(x, y, z)$
(6b)	or, no clear results, go to step (3) or (8)	$FH(x, y, z)$?
(6c)	or, verification of all three, hence answer:	
	verified specific functional statement	$FH(x, y, z)!$
(7)	now, conclude first as a side-step:	by MP-2_f and EG
	verified unspecific functional statement	$FH(x, y)!$
(8)	then go to new, related why- and how-questions	why?/how?

Scheme 4.5.

This time we leave the stepwise elucidation to the reader, but a number of remarks should be made. Scheme 4.5 is designed for proximate functional explanations (i.e. where z is different from $r\&s$); for ultimate functional explanations (i.e. when $z = r\&s$), it can be simplified somewhat. The claim is of course again that the process of producing a functional explanation is characterized by the above scheme, that the main product is the corresponding specific (proximate or ultimate) functional statement and, finally, that the valid implication of the corresponding unspecific functional statement is an interesting by-product: this view explains the philosopher's conviction that there is, in some way or other, a valid argument involved.

The train of thought is, due to step (4), essentially of a hypothetico-deductive nature, much closer to the standard account of Chapter 7 than in the case of intentional explanations. In the present case, the main hypothesis is a concealed, complex general hypothesis, the implications to be tested are proper test implications in the sense of testable (e.g. by experimental imitation of the selection environment) general hypotheses.[7] The only special aspect is that the main hypothesis is equivalent to the conjunction of the descriptive component with precisely two well-defined test implications.

Again it is plausible to call the steps (3)–(6c), leading to the main product and representing the synthetic part of the explanation, the process of *functional specification*, and the step (6c)–(7), the analytic part that leads to the by-product, *functional generalization*. Moreover, it is tempting to decompose the process of functional specification into an heuristic phase (3)–(4) and a test phase (5)–(6).

Of course, a functional explanation is called true when all meaning components are true, and false when at least one component is false. After a successful proximate functional explanation (i.e. a proximate functional explanation considered to be true), there arise, as indicated in step (8), new related questions. We repeat the already mentioned ones and we add some others. Is z the only function of y? Why is the trait realized rather than one of the functional alternatives? Did y evolve along the suggested evolutionary lines? Other questions are: How do the proximate and ultimate causal-nomological components come true, i.e. how to explain them by subsumption under a theory and what are the mechanisms involved? One may even argue, as Wouters (1999) in fact does, that a proximate functional explanation presupposes that both explanatory questions have been answered in a satisfactory way. In order to allow equating a functional explanation in the present environment with the selection explanation, another question is whether the present environment is sufficiently similar to the selection environment? Moreover, one may ask, how does trait y ontogenetically evolve? Finally, is it also possible to give a proximate functional explanation for trait z, if that is itself a trait or process of x-organisms?

In the case of an ultimate functional explanation the last four questions apply also to some extent, but the primary question is now of course: Is it also possible to give a proximate functional explanation? As already suggested, this possibility need not exist, as exemplified by the example of the plover's clutch-size.

We will restrict our attention to the third question, assuming a selection reading of the relevant functional explanation, to be indicated by the subscript 's': did y evolve along the suggested evolutionary lines, i.e. did the identified functional aspect play a substantial role in the selection of the trait? The situation is rather similar as in the intentional case. Here we could have added this condition as a fourth meaning component of $FH(x, y, z)$. In this case the related question becomes equivalent to the questions about the consequences of replacing the weak interpretation of $FH_s(x, y, z)$ by the strong one obtained by adding a *historical evolutionary component*, that is, a selection version of a causal effectiveness component:

$CE_s(x, y, z)$ $P_s(x, y, z)$ and $P_s(x, y, r\&s)$ were evolutionary effective for $H(x, y)$

Again the implication to be tested is of a rather different nature than the other two test implications. It cannot be called a test implication in a straightforward sense. However, the possibility of a convincing falsification is not excluded, for example, when the trait evidently is a causal by-product of some other functional trait originating at the same time and surpassing the by-product in survival value. Or the trait may have another functional aspect, surpassing the identified functional aspect in survival value. Verification of the extra component certainly is still more difficult. However, again it is plausible to introduce a default rule of inference to the effect that, as long as there is no evidence to the contrary, the descriptive component and the two causal-nomological components together imply the historical evolutionary component:

CE_s-default if $H(x, y)$, $P_s(x, y, z)$ and $P_s(x, y, r\&s)$ then $CE_s(x, y, z)$

The resulting thought process for a strong functional (selection) explanation is obtained by inserting the causal effectiveness component at step (5), by adding the possibility of its falsification at step (6a), and by inserting CE_s-default as an argument in step (6c).

Of course, a persistence explanation also raises the related 'third' question about whether the trait persists due to the identified function in the present environment. This issue leads to 'presentist' versions of the causal effectiveness component and default, to be respectively indicated by $CE_p(x, y, z)$ and CE_p-default, and a corresponding stronger version of the explanation.

It is important to note that the difference between the weak and the strong versions of a functional explanation concerns the relatively rare possibility (justifying the default rule) that one particular appropriate combination of two causal-nomological components were, or are, in fact not causally effective for the selection, or persistence, of a trait, but some other pair were. For this reason and for the sake of simplicity we have chosen to start with the weak version.

It is also interesting to check that the objections to the two standard explications do not apply to our explication. The first three objections common to both explications are again clearly met: it is close to research practice, specific functional statements play an important role and there is a straightforward (causal-nomological) reproduction and survival component (viz. $P(x, z, r\&s)$), without troublesome 'effects as causes' problems. In the strong selection version, there is even an historical evolutionary component, viz. $CE_s(x, y, z)$. There is also room for functionally equivalent traits, which is of course only relevant for the case of proximate functional explanation. There is no prediction claim with respect to $H(x, y)$, hence there is an essential asymmetry between the prediction and the functional explanation of a trait. This is of course not in conflict with the fact that there are two fundamental prediction claims involved, viz. contained in the two causal-nomological components. Finally, there is a valid argument involved;

however, $H(x, y)$ is not the conclusion, but $FH(x, y)$, using existential generalization; $H(x, y)$ is an explicit premise, in view of MP-1_f, hence a trivial conclusion, even implied by $FH(x, y)$. Note that our analysis of functional explanations leaves room for another localization of the supposed underlying argument than the side-step application of existential generalization. As already noted, $FH(x, y)$, suggested by PoI, implies, according to our meaning postulates, that y is ultimately functional. The resulting functional explanation $FH(x, y, z)$ enables us to derive this ultimate functionality, due to the transitivity of the ternary causal relation, from its proximate and ultimate component. Of course, the presence of two sound arguments will strengthen the feeling that an underlying argument is involved.

Concluding Remarks

As far as it is defensible to talk about biological natural kinds, it is possible to interpret the above explication as dealing with the 'absolute' functional explanation of natural kind laws. Moreover, modified explications for related functional explanations can easily be made. For example, the *comparative* functional explanation of the persistence or selection of traits, rather than functionally equivalent alternatives, can be reconstrued along lines similar to the formal structure of intentional explanation of choices between alternative actions at the end of Section 4.1. Instead of utility maximization, some other kind of optimization has to be taken into account, e.g. optimal efficiency.

As has been stressed in general, our analysis of the functional explanation is, like the other analyses, a minimal analysis in the sense that it can be combined with stronger claims. For instance, it is possible to argue that, given a certain kind of organism, comparative functional explanations can be so convincing that theoretically conceivable functional equivalents of an actual trait cannot be taken seriously, which restores the possibility of deriving the conclusion that a certain kind of organism (must) have that trait. See Looijen (1998/2000, Chapter 6) for this strengthening of the foregoing analysis in so far as selection explanations are concerned. See Wouters (1999) for an essentially similar strengthening of persistence explanations: a functional explanation is only complete when it shows that an item with a certain characteristic performs a certain function in a better way than any conceivable alternative characteristic would do.

Our claims have been restricted to biology, but it may be worthwhile to modify our explication such that it becomes adequate for a functional explanation in sociology, anthropology, psychology and economics (see Section 4.4.). On the other side of the scientific spectrum similar adaptations will not be easily possible, e.g. minimality explanations in physics in terms of the shortest distance seem to lack something like an ultimate component.

4.3 SPECIFIC CAUSAL EXPLANATIONS

Introduction

Causal explanations of events are seldom mentioned as examples of a type of explanation that cannot be seen as a subsumption explanation. On the contrary, causal explanations are usually seen as explanation by subsumption par excellence, viz. subsumption under a causal law. However, in our opinion there is a certain type of causal explanation for which the subsumption interpretation is, though formally possible, not the most adequate explication. We are especially thinking of causal explanation in the context of e.g. everyday life, medicine, jurisdiction, insurance, technology, etc. ... In many cases one does not hesitate to call one of the causal factors *the cause*, whereas the subsumption view is essentially neutral with respect to the causal factors.

The following examples are all of the suggested kind, provisionally called 'specific causal explanations':

- Her child-bed fever was caused by 'cadaveric matter' (a well-known example from Hempel's report (1966) of Semmelweis's discovery).
- The catastrophe with the Challenger was caused by a defective packing.
- The fire was caused by a short-circuit.
- The cause of his death was a heart attack.
- The cause of the accident was that the driver drove through red traffic lights.

The main objections to the subsumption account are the following. It has no clear correspondence with research practice (a). There is no explicit role for specific and unspecific causal statements (b). It is neutral with respect to the causal factors (c). It makes no difference between explanation and prediction (d).

The following will elucidate these objections.

Ad (a) Causal research practice in many areas is not only everything but making an inventory of all causal factors, it is not at all reflected in the subsumption account.

Ad (b) None of the two statements mentioned occurs explicitly as (part of) a premise or the conclusion of standard nomological explications of causal explanation.

Ad (c) All causal factors are treated equally: they are reported as an initial condition and generally referred to in the antecedence of a law premise. Note that this neutrality-objection is new in this chapter. It only applies to the subsumption explication of specific causal explanation.

Ad (d) In many cases a specific causal explanation could not have been used for prediction due to incomplete knowledge of all relevant causal factors, and in other cases when complete knowledge of all relevant causal factors and laws is available, prediction is possible, but specific explanation only becomes possible with a selection device for selecting the cause among the causal factors.

In the light of these objections it is plausible to try to explicate specific causal explanations as a subtype of explanation by specification.

4.3.1 *Explanation by Causal Specification*

The following abbreviations will be used:

$O(x, y)$	event (factor) y occurred in system x
$CO(x, y, z)$	y occurred in x due to cause z
$CO(x, y)$	y occurred in x due to some specific cause

The five examples of specific causal explanation mentioned above are all formulated in terms of the second type of statement, i.e. $CO(x, y, z)$. For such specific causal statements we propose the following meaning postulate or key

MP-1$_c$ $CO(x, y, z) \approx O(x, y)$ and $AO(x, z)$ and there are factors $f_1, \ldots f_n$ such that $NO(x,f_1), \ldots, NO(x,f_n)$ and $CL(x, y, z, f_1, \ldots f_n))$

where the following new abbreviations are used:

$AO(x, z)$	$O(x, z)$ and z is an *abnormal* factor (event, intervention, condition) for x
$NO(x, f_i)$	$O(x, f_i)$ and f_i is a *normal* factor/condition for x
$CL(x, y, z, f_1, \ldots f_n)$	'if $O(x, z)$ and $O(x,f_1), \ldots, O(x,f_n)$ then $O(x, y)$' is a causal law in the strict sense.

A causal law in the strict sense is minimally taken in Nagel's sense (1961): an experimental law (Nagel's term for what we are calling an 'observational law') such that the factors z and f_i are space-and-time-contiguous and such that there is time-asymmetry between y and z: the latter precedes the former. One may or may not prefer to add a nomic or counterfactual connotation. That is, one may or may not assume in addition that the law excludes certain conceptual possibilities as physically or nomically possible (see Chapter 9, see also ICR). If one adds this condition it allows the determination of the truth-value of the corresponding counterfactual conditionals.

The meaning postulate for the corresponding unspecific causal statement reads of course:

MP-2$_c$ $CO(x, y) \approx$ there is a factor τ such that $CO(x, y, \tau)$

The presented key takes specific causal statements, and hence the corresponding explanations by causal specification, as primarily reporting the abnormal causal factor among all causal factors. Hence, they apply only if there is at least one abnormal factor. In this line, it is plausible to assume that the relevant principle of causality only applies to abnormal events:

PoC *principle of (specific) causality*:
if $AO(x,y)$ then $CO(x,y)$

Again it is a heuristic default rule; although restricted to abnormal events, it may have exceptions. For example, all causal factors as such may be normal, and only their joint occurrence may be highly exceptional. In this case, causal specification is impossible.

It is also clear that an explanation by causal specification can now be formulated in a way completely analogous to explanation by intentional and functional specification, with a similar train of thought, main product, by-product, synthetic specification phase (with heuristic and test subphases), analytic generalization phase, and (suggestions for) new questions after a successful explanation. It is even possible to introduce a stronger version, analogous to the case of intentional and functional explanation, with similar test problems and leading to similar adaptations of the train of thought. To be precise, strictly speaking our meaning decomposition only requires that $AO(x,z)$ was a possible causal factor, not that it was an effective one. The component to be added for this purpose is essentially a generalization of the 'causal effectiveness' components introduced for intentional and functional explanations:

$GCE(x,y,z)$ z was causally effective for $O(x,y)$

and the corresponding default rule states:

GCE-default if $AO(x,y)$ and $AO(x,z)$ and there are factors $f_1, \ldots f_n$ such that $NO(x,f_1), \ldots, NO(x,f_n)$ and $CL(x,y,z,f_1, \ldots f_n))$ then $GCE(x,y,z)$

Here it is of course again crucial that the first condition reads $AO(x,y)$, and not merely $O(x,y)$; only abnormal events may be assumed, by default, to be specifically caused by abnormal conditions.

A number of remarks about this analysis are in order. We do not claim that the abnormality view on specific causal explanation is new. What is new is the specification perspective. Moreover, our claim is only that the specification explication grasps the core of relatively simple examples of what we have called specific causal explanations, leaving room for complex cases.

It is easy to verify that all four mentioned objections to the subsumption interpretation are invalid in this case. In particular, the specification is non-neutral with respect to the causal factors; it is biased in the direction of the abnormal factor (if there is any). Moreover, due to the existential clause it will not be possible to use the explanation for prediction as long as the postulated law is not explicitly known. In this respect, and others, a diagnostic non-monotonic logical turn of the present approach to specific causal explanation (Tan, 1992) is very illuminating.

From MP-1_c it immediately follows that an explanation by causal specification implies the possibility of providing an explanation by causal subsumption, if the particular causal law is explicitly known, but not the converse. More precisely, the converse holds if and only if (the particular causal law is known and) just one of the causal factors is abnormal. In this respect an explanation by causal specification is quite different from explanation by intentional and functional specification. In the latter cases there is no reference to an underlying law, let alone to a known one, which could provide the basis for the corresponding subsumption explanations, notwithstanding the fact that in the case of functional explanations there certainly are laws involved.

In most cases, 'the cause' z will be, relative to $f_1, \ldots, f_n$, a so-called INUS-condition in the sense described by Mackie (1965): an Insufficient but Necessary part of a condition which is itself Unnecessary but Sufficient (for the occurrence of y). More precisely, 'U' and 'S' are already guaranteed, but 'I' and 'N' need not always be satisfied. Note that an INUS-condition typically is a possible causal factor, not necessarily an effective one.

Finally, an explanation by causal specification of an individual event can be generalized to an explanation of an empirical regularity (e.g. the Semmelweis example).

Concluding Remarks

It is important to note that not only is the distinction between normal-abnormal factors of course vague, but also that in many contexts and cases it is easy to reach agreement on what the abnormal factor was; or what the abnormal factors were, for there may of course be more than one. However, the question 'Why $O(x, y)$?' is usually raised because the event to be explained, i.e. y, is abnormal ($AO(x, y)$), but this is no meaning component of $CO(x, y, z)$.

As already mentioned, the relevant principle of causality should leave room for exceptions. Compare, e.g. dying from a heart attack or 'due to old age'. In the second case there is usually not supposed to be an abnormal factor. Hence, the methodological causality principle should not be confused with some kind of universal determinism, i.e. something like 'for every event there are a number of causal factors that are jointly sufficient'. The principle of specific causality is compatible with determinism as well as with (occasional) indeterminism.

Finally, as announced in a note, we will see in Section 6.2 that an explanation by intentional specification can also be reconstrued as an explanation by causal specification, where the mental states of desiring and believing function as the 'abnormal' or at least special causal factors, viz. mental factors.

4.4 EXTRAPOLATIONS AND SPECULATIONS

Explanation by a certain type of specification automatically leads to a corresponding type of description, in particular classification. Organs performing the same function, though in quite different ways, are primarily classified according to that function. Actions are classified according to an internal goal that may be ascribed to the relevant behavior. Hence, in general we may distinguish a *functional style* and an *intentional style* of description and explanation. Besides these we distinguish the *causal style*, that is, causal description and explanation in the general sense of identifying causal factors as well as specific causal description and explanation. Hence, in the causal style we combine explanation by subsumption under causal laws and explanation by causal specification, along with corresponding types of description. Moreover, all styles are also supposed to include relevant conceptual and ontological claims, such as identity and part-whole relations.[8] Roughly speaking, the styles correspond to Dennett's intentional design and physical stance (Dennett, 1987).

In this final section we will make some general remarks about the scope of the three styles of description and explanation, with emphasis on the functional and intentional style. In Chapter 6 we will elaborate some of these points and some other related points.

From the account in Section 4.2 it is clear that the functional style is at least applicable to all kinds of organs and processes in plants and animals. In the case of processes, e.g. breathing, one is perhaps more inclined to speak about the teleological style, but the structure of the underlying argument is the same: this process is directed to that particular goal; hence, this process is goal-directed. Occasionally some intermediate causal correlation may be distinguished giving rise to an intermediate so-called *distal* function. A case in point is the color adaptation mechanism of the chameleon (see Section 6.2). As exemplified with the stickleback example the analysis also applies to instinctive behavior of animals. If instinctive behavior happens to seem dysfunctional, it is likely that the current environment differs in some relevant aspect from the environment in which the natural selection operated. The same holds true for rudimentary organs, etc.

There also seems to be no fundamental difference between functional description and explanation of instinctive behavior and learned (or 'operant') behavior, apart from the ontogenetic origin of the latter, as opposed to the phylogenetic origin of the former. More specifically, learned behavior presupposes a mechanism of learning: the law of effect, the show-piece of behaviorism. This law of effect is itself a functional product of evolution, and it produces behavior which is, as a rule, functional. However, the decomposition analysis becomes somewhat more complicated, the same holds for forms of representational behavior, whether innate or learned. In Chapter 6 we will show in detail how our analysis can be applied to such types of behavior.

In human beings (and higher animals?) all this has culminated in the possibility of conscious functional behavior, i.e. intentional (or 'intelligent') behavior, again a functional product of evolution. Among other things, it includes the possibility of intentional learning and intentional problem solving. The capacity of intentional behavior obviously constitutes the kernel of human freedom. As we have seen before, intentional description and explanation of actions does not compete with nomological explanation at all. Hence, human freedom does not compete with nomological explanation. Instead of resulting from indeterminism of one kind or another, the core of human freedom is that beliefs and desires operate as causal factors, together with other ones. In Chapter 6 we will see in detail how this *prima facie* dualistic interactionism can be interpreted in a monistic approach of a biophysicalistic nature.

We do not, of course, want to suggest that all non-instinctive human behavior is intentional. On the contrary, the law of effect remains to produce unconscious functional behavior constituting the so-called second nature, to be distinguished from the first nature of physiological processes and instinctive behavior. But, unfortunately, not all second nature behavior is merely functional, especially in cases where 'unconscious goals and motives' are in conflict with conscious goals. This is the domain of psychological explanation and therapy. Whereas behavioral therapy relies on the law of effect to stop or to change problematic second nature behavior, psychotherapy attempts to make beliefs and desires conscious in order to increase the amount of undisturbed intentional behavior.

It is tempting to extrapolate the scope of explanation by specification to the social level. To be sure, in the context of so-called institutional subjects, like firms and states, it generally makes sense to explain their behavior intentionally. But apart from that, there is also the phenomenon of collective effects of individual intentional behavior, for example utilistic choice behavior. Such effects may or may not be functional on the level of groups or whole societies, in the sense that the persistence of the individual behavior producing such effects may be caused in one way or another by the beneficial character on the social level of the effect. However this may be, as far as the effect itself is concerned, we arrive again at a type of phenomena also occurring in non-life sciences and population biology, where straightforward explanation by subsumption is feasible and interesting, namely, the collective results of similar behavior involving many individual entities of one kind or another. The explanation of Olson's hypothesis presented in Chapter 3 is a typical example of this kind. As we have noted before, it combines the causal style on the social level with the functional, if not intentional, style on the individual level. Such explanations normally do not postulate that goals or functions operate on the social level. Hence, these explanations are not explanations by specification, at least not by intentional or functional specification, but they may be explanations by causal specification.

The scope of explanation by specification may range from biology to sociology and economics, but it should be stressed once more that there is no substantial but only a formal analogy in such explanations: the meaning components of specific intentional and functional statements are rather different. However, the work of Ruth Millikan (1984, 1993) seems to provide a substantial integration of the intentional and (the selection version of) the functional style which is compatible with the analysis presented here. In Chapter 6 we will elaborate this point. Here we will just give a few clues. Millikan starts with the notion of the 'proper function' or 'teleofunction' of a biological trait in terms of its 'normal performance' under 'normal (internal and external) conditions'. This can be seen as a refinement of our meaning postulate for specific functional statements. Next Millikan argues that beliefs and desires have also proper functions. Roughly, the proper function of desires is to help cause their own fulfillment, and the proper function of beliefs is to help the fulfillment of the proper functions of other (momentary or more permanent) features of organisms, such as desires. This distinction amounts to a kind of second-order functional specification of the belief and desire components in our analysis of specific intentional statements. The result is a conceptual unification of biology and psychology, which may be very useful for some forms of research in these fields, for instance stress research (Guichard, 1997). Moreover, the unification can be extended to sociology by analyzing (meaningful and) rule-guided behavior in terms of proper functions, leading to a new naturalistic perspective on the spectrum of disciplines and their possibilities of co-operation (Mackor, 1997). As said, this substantial integration of the styles, which originates with Millikan, will be elaborated in Chapter 6.

Returning to explanation by subsumption and by specification, it is important to note that, although there are many facts which can be explained only by subsumption and not by specification, this is not due to some incompatibility between these two types of explanation. In our view the situation is as follows. In certain contexts of scientific interest we happen to focus on certain types of explanation. Besides interest in explanation by subsumption in many cases, one is frequently also or only interested in explanation and description by specification. For example, put in our terms, biologists are often primarily interested in explanation and classification by functional specification (frequently called 'functional biology'), historians in explanation and classification by intentional specification and insurance companies in explanation and classification by causal specification.

By way of a slogan one might say: although explanation by subsumption may in principle always be possible, it is certainly not always interesting; on the other hand, explanation by (some relevant) specification is certainly not always possible, but, as long as the answer is not well known, it is always interesting when it is possible. What holds for explanation, holds *eo ipso* for the corresponding kind of description, i.e. classification, viz. classification according to structure, function or intention. Hence it holds for the styles of description and explanation in general. In particular, explanatory-cum-descriptive research programs may partly or even completely be governed by a style, more precisely, by a core heuristic-methodological principle.

Subsequent to this line of thought we would once more like to draw attention to the scope of explanation by specification. In particular, we want to address the question of whether there are other subtypes of explanation by specification. Specifically, our explication raises the question of whether there are 'clear and distinct' subtypes of explanation by specification between intentional explanation on the one hand and functional explanation on the other. To verify this conjecture, further research may lead to proposals for the corresponding *keys* of these subtypes. In psychology one might think of a key for explanation by (specification of) unconscious motives.

In sociology and anthropology there may also be specific keys for functional explanation, which are not strictly biological. The fact that the latter disciplines use such explanations is illustrated by Pettit (1996), who gives a couple of well-known examples. Religious rituals have the function of promoting social solidarity (Durkheim); common ideas of time and space, cause and number serve to make mental contact and social life possible (Durkheim); peacemaking ceremonies serve to change the feelings of the hostile parties to one another (Radcliffe-Brown); social stratification in modern societies makes it possible to fill socially indispensable but individually unattractive positions (Davis and Moore). Pettit further discusses the crucial problem for a social analogue of functional explanation in biology, viz. a mechanism for selection and persistence of a feature comparable to (promoting) reproduction and survival seems to be missing. He proposes to extrapolate his earlier suggestion (Pettit (1993)) of a (realistic) 'resilience' interpretation of rational choice (utilistic) behavior to functional social phenomena. The former amounts to the claim that as soon as there are threats of straightforward 'as-if-utilistic' behavior, people tend to make utilistic calculations. By analogy, one may hypothesize that people tend to take measures as soon as straightforward socially functional behavior is seriously threatened. In this way, resilience becomes the mechanism for explaining the persistence rather than the origin of phenomena, comparable to equilibrium explanations in population genetics. Pettit even argues that this resilience criterion should be conceived as the core of a 'significant' research program based on the question: 'Which are the more or less passing ephemera and which the phenomena that are deeply embedded in the society?' Unfortunately, he uses the example of the golf club as a (relatively) deeply embedded phenomenon, since it serves networking purposes for wealthy people; apparently he is not aware of the increasing 'democratization' of the golfing public. However this may be, the resilience criterion raises at least two fundamental questions. Resilience may be a practical criterion for separating socially functional features from non-functional ones, but this does not answer the question, in a non-circular way, why the resilience response to threat occurs. Moreover, it does not answer the question what the specific key, in our sense, would be for such a socially functional explanation. As long as no convincing proposals have been formulated for such a key, we believe the alternative to be the case: there is merely a diffuse continuous path from biological functional explanation to intentional explanation.

CONCLUDING REMARKS

Interest in a certain style is not only a matter for individuals, concrete cases and research programs. Even a scientific community may favor a certain style, in which case it is likely to form a substantial part of the style of thought in the research tradition to which the group adheres. More awareness of such, in our opinion, defensible differences in thought style between communities working in roughly the same area may not only elucidate many communication problems between these communities, but it may also be a useful viewpoint in the sociology of science.

PART III

STRUCTURES IN INTERLEVEL AND INTERFIELD RESEARCH

INTRODUCTION

In this part we will explore some new patterns, notably concerning reduction and correlation of concepts, and then we will apply all patterns distinguished so far to mind-body research.

A necessary step for one type of reduction, though by no means sufficient, amounts to the (type-type) reduction of concepts by using type-type identities. Chapter 5 deals with concept reduction. The very idea requires a distinction between ontological identities and causal correlations, which is intuitively clear but difficult to explicate. Inspired by Causey's extreme claims, a detailed analysis will be given, mainly guided by the reduction of substance terms to molecular terms, showing that concept reduction is technically not as simple as is frequently suggested. Moreover, it suggests distinguishing three kinds of type-type concept reduction: singular, multiple and quasi-reduction, which turn out to be in line with some of Kim's ideas about type-type and token(-token) reduction. Besides dealing, in addition, with the intermediate notion of multiple concept reduction and the relation between concept and law reduction, special attention is paid to the radical and moderate reductionistic and holistic research strategies that were already indicated in Chapter 1. Moderate strategies, in particular by means of co-operating research programs, recognize, first, that for reduction of concepts and laws there should be something to reduce, and, second, that reduction of concepts and laws has by no means to imply the elimination of them.

The second chapter of this part, Chapter 6, extends the treatment in Chapter 1 of symmetric and asymmetric co-operation between research programs, with emphasis on mind-body research. Programs may or may not take different levels of aggregation into account and may even belong to different disciplines. Moreover, they may or may not use the same or different styles of description and explanation, i.e. causal, functional, or intentional. The result is a matrix of four kinds of so-called interfield research, each of which gets separate attention. First we will deal with interstyle (and primarily monolevel) research, more specifically the relations between the styles will be explicated. Then we will focus in two sections on interlevel, biophysical, mind-body research, first restricted to two levels and one main style, then the scope will be broadened to examples and models of interlevel and interstyle interfield mind-body research. We will conclude with some examples of monolevel and monostyle research of an interdisciplinary nature. This chapter will use several means developed in the previous chapters, either in a straightforward or in a new way.

In Chapter 3 we dealt already with the reduction of laws by theories. In the following two chapters reduction will even be the dominant focus of attention. This is not a popular point of view in the philosophy of the cognitive sciences in general and the philosophy of mind in particular. We like to conclude this introduction with a dramatic expression by Kim (2000, p. 89) of this even much more general but regretful situation:

> 'Expressions like ''reduction'', ''reductionism'', ''reductionist theory'', and ''reductionist explanation'' have become pejoratives not only in philosophy, on both sides of the Atlantic, but also in the general intellectual culture of today. They have become common epithets thrown at one's critical targets to tarnish them with intellectual naivete and backwardness. To call someone ''a reductionist'', in high-culture press if not in serious philosophy, goes beyond mere criticism or expression of doctrinal disagreement; it is to put a person down, to heap scorn on him and his work. We used to read about ''bourgeois reductionism'' in left-wing press; we now regularly encounter charges of ''biological reductionism'', ''sociological reductionism'', ''economic reductionism'', and the like, in the writings about culture, race, gender, and social class. If you want to be politically correct in philosophical matters, you would not dare come anywhere near reductionism, nor a reductionist. It is interesting to note that philosophers who are engaged in what clearly seem like reductionist projects would not call themselves reductionists or advertise their work as reductionist programs.'

In a note, Kim mentions work of Jerry Fodor and Fred Dretske, in which they speak of 'naturalization', for that term 'doesn't offend, it seems, in the way 'reduction' does'. Be this as it may, with Kim and Bickle (1998), we like to continue to call the subject by its plausible name.

CHAPTER 5
REDUCTION AND CORRELATION OF CONCEPTS

INTRODUCTION

In the philosophy of science literature produced since 1960, emphasis has fallen on the reduction of laws, notwithstanding the fact that something like concept reduction frequently appears as a necessary intermediate link.

As in the case of the reduction of laws, the notion of general or type-type identities, as opposed to merely causal correlations, plays an important role in the reduction of concepts. The distinction between identities and correlations will be explained, and defended as a working hypothesis in Section 5.1. In Section 5.2 we will analyze in detail some crucial variant forms of concept reduction, using elementary physical examples. The correlation of concepts will be studied at the end of this section simply by substituting correlations for identities. Section 5.3 is devoted to the relation between reduction of laws and concepts, the importance of 'multiple reduction', and to research strategies, from radical reductionistic to radical holistic ones, along with the mixed strategies in between.

5.1 TYPE-TYPE IDENTITIES AND CORRELATIONS

In his prominent exposition of heterogeneous reduction, Nagel (1961) assumed that the required rules of transformation are empirical laws or rules of correspondence. Soon after its publication it was argued (Schaffner 1967; Sklar 1967; Girill 1976; Pluhar 1978) that these rules of transformation should not merely be correlations, but should be identities, and hence that there is something like concept reduction involved in heterogeneous reduction. This discussion forms the background for the introduction of two transformation steps in the general decomposition model for the explanation of laws in Chapter 3: a correlation step, based on some (causal) correlation hypothesis and an identification step, based on an identity hypothesis. In line with this analysis, the occurrence of an identification step in the explanation of a law was claimed to be a sufficient reason to speak about reduction of that law in the literature. The important question is, however, what precisely is the difference between (causal) correlations and (ontological) identities, and are there usable criteria to distinguish them? The paradigm example of a synthetic or informative identity is the discovery: 'The morningstar is the eveningstar'. What were first considered to be two names for two different objects, turned out to be two different names for the same object, namely, the planet Venus. This example has already occasioned considerable brain-racking by philosophers, but it concerns only an individual identity, whereas in the case of reduction general or type-type identities are involved.

Turning to type-type relations, it is important to begin by noting that, whereas a 'causal correlation' between two concepts is, to say the least, not necessarily of a deterministic nature, it may well be of a statistical nature, it is difficult to imagine how an 'ontological identity' could be of a statistical nature. Hence, we may focus on deterministic type-type relations.

The most far-reaching claims seem to have been formulated by Robert Causey (1977). According to him general identities can be represented as biconditionals satisfying the surface feature that, just like correlations, they need empirical support, but, in contrast to correlations, they are not subject to causal explanation: the quest for a causal (-nomological) explanation of, e.g. 'morningstar = eveningstar' or 'water = aggregate of H_2O-molecules' is absurd, because there is no mechanism in between. Although this surface feature is manageable in practice, it is essentially circular, because not being subject to causal explanation seems in the end only justifiable by claiming that the relevant proposition states an identity.

Among other reasons, Causey proposes in the light of this circularity a more profound criterion: the substitution criterion. In fact this criterion is just an extrapolation to general identities of the more or less generally accepted qualified version of Leibniz's definition of individual identities: *a* is identical to *b* if and only if *b* can be substituted for *a*, without changing the truth-value (*salva veritate*), in all statements in which no personal attitude or other intensional expression occurs. This depth criterion is not circular, but, in contrast to the surface criterion, it is not manageable in practice, simply because it is impossible to check for every statement whether substitution is allowed or not. To be sure, it can nevertheless be used to conceive an identity claim as a general hypothesis to be tested. If it gets falsified, the relation between the concepts is at most merely a causal correlation. But as long as it has not been falsified, the judgement 'identity' has to be provisional.

The main difference between Causey and the more cautious advocates of identities as rules of transformation consists of the fact that Causey does not hesitate to leave not only room for so-called *thing-identities*, but also for *attribute-identities*. Thing-identities are general identities in which types of objects, belonging at first sight to different domains, are identified: 'light consists of electromagnetic waves'; the bridge principles in the atomic theory, with specific instances such as 'water is an aggregate of H_2O-molecules'; 'genes are aggregates of DNA-molecules', etc. Attribute-identities are general identities in which properties, relations or functions are identified. In the reduction of the ideal gas law, presented in Chapter 3, there occur two nice examples. In the first place the usually hidden identification 'the macroscopic pressure is the same as the kinetic pressure'. In the second place, the identification of temperature and mean kinetic energy, about which we have already noted that the latter identification of two functions can be based on the identification of two relations: 'equal thermal state is the same as equal mean kinetic energy'. The identification, by subatomic physics, of the atomic number of chemical elements with the number of protons of an atom, is another example of an attribute-identity.

It is easy to imagine different sorts of criticism directed at Causey's views. One point of criticism, however, can easily be put aside in the light of our diagnosis in Section 3.3, where we note that there are three different reasons of speaking about the reduction of laws. As soon as a critic provides examples of reduction in which the transformation rule is undeniably a correlation (in which case it often is not even of biconditional form), then it is only necessary to show that at least one of the other two crucial steps, aggregation or approximation, occur in the decomposition of the explanation. Put in a different way, in the light of our diagnosis, the Causey's general claim should be sharpened as follows: if a certain explanation of a certain law is usually called a case of reduction and if there occurs a transformation step in that explanation, but no aggregation or approximation step, then the transformation rules will be identities.

More important than the above criticism of Causey's unconditional claim are the uttered and conceivable objections against the distinction between identities and correlations itself, against the proposed criteria for it, against the distinction between thing- and attribute-predicates, and finally, if one accepts this distinction and is prepared to admit thing-identities, against the idea of attribute-identities. Any discussion of all these objections would lead us too far afield. We confine ourselves to the remark that there seems no reason to be very impressed by objections against the distinctions between identities and correlations and between thing- and attribute-predicates based on the observation that there are examples which are difficult to classify. Moreover, it often occurs that explication of intuitively appealing distinctions involves a lot of work in formal logic, of which the successful completion is anticipated by others, taking the risk for granted. A nice historical example is the distinction between ordinal and cardinal numbers.

Note that besides ontological identities and causal correlations we may distinguish 'ontological correlations' in the sense of 'part-whole' relations. For example 'all humans have hearts' is a law of an ontological nature, but does not express an 'ontological identity'. Although ontological identities and correlations resemble causal correlations insofar as they also require 'empirical correlation' for empirical support, the former cannot be explained, whereas the latter ask for explanation. However, although this diverging criterion between causal and ontological correlations is again circular, it can not now be easily replaced by a more sophisticated substitution criterion.

In Chapter 3 the hypothesis that the distinction between correlations and identities can be successfully explicated, was used to disentangle the explanation of laws in general and their reduction in particular. In the next section the same working hypothesis is presupposed in the analysis of concept reduction: identities will play a crucial role. Moreover, by way of precaution we will start with thing-predicates. In the second instance, however, it will turn out that the reduction of attribute-predicates occurs in a completely analogous way. By consequence, the present analysis does not provide an argument against attribute-identities and in favor of thing-identities.

5.2 ANALYSIS OF REDUCTION AND CORRELATION OF CONCEPTS

Introduction

The analysis of reduction (and correlation) of concepts that follows has a more technical character than most of the expositions so far. However, the general patterns have been uncovered by analyzing two, relatively simple, examples in detail. Hence, the technically somewhat complicated character of the patterns shows first of all that the standard presentations of (kinds of) type-type reduction are rather naive, not to speak of weaker kinds of concept reduction.

All concepts to be considered will be predicative. Hence, we will casually talk about the reduction of predicates, in particular of thing- and attribute-predicates. In the first subsection we will deal with the reduction of the thing-predicate 'water', being a 'macro-type', to the corresponding 'micro-type', which is defined in micro- or base-terms. In the second subsection we will argue, using the temperature function as the leading example, that the reduction of attribute-predicates (representing properties, relations and functions) can take place essentially along the same lines, except that a time parameter now has to be introduced.

Next, a formal survey of concept reduction will be given, followed by a survey of the ontological and epistemological levels of reduction that are involved. Finally, in the fifth subsection it will be argued that all this can be adapted for the correlation of concepts.

Although the analysis is directly suggested by some examples from elementary physics, it seems plausible that it gives a generally adequate analysis of different forms of reduction and correlation of concepts. We hope at least that they provide instruments with which, together with such other instruments as the model for reduction of laws, it is possible to deal with more complex, but at the same time at least as interesting, reduction questions. The main challenge in this respect seems to be the mind-body problem, and a primary use-value may be to suggest new research for a materialistic or biophysicalist account of mental concepts and laws, and the mind-body relation. In this connection it is relevant to stress again that, whereas reduction of laws presupposes reduction of concepts, reduction of concepts does not presuppose reduction of laws; hence it does not presuppose the existence of laws at the side to be reduced. Concept reduction only requires concepts at the side to be reduced which is, of course, supposed to imply that these concepts are relatively stable and intersubjectively applicable.

We will need the notion of a set-theoretic or relational *structure*. It is an ordered sequence of sets, starting with one or more 'domain- or base-sets', followed by one or more relations (including properties and functions), defined on the basis of the sets. For instance, a city can be represented by a structure specifying, for example, the base-sets of inhabitants, houses, and streets, and some of their properties (e.g. male/female), relations (e.g. houses in the same street) and functions (e.g. the house in which an inhabitant lives).[1] Moreover we will need the notion of a *family* of predicates for a certain domain. A family is a set of (unary) predicates which are pairwise exclusive and together exhaustive for the members of the domain. Such a family is said to lead to a *partition* of the domain in mutually non-overlapping and together exhaustive subsets.

5.2.1 *Reduction of Thing-Predicates*

We will take substance terms, such as 'water', as our leading example of thing-predicates. Consider the statement '*a*o is water', that is, the claim that a certain isolated amount of substance *a*o is (a token of) water. It is plausible to call the ontological level of the 'object' *a*o the *macro- or aggregate-level* and the level of molecules the *micro- or base-level*.[2] We sharply distinguish micro- or base-objects from aggregate-objects, that is, aggregates of a (large) number of base-objects of one or more kinds, which will just be called aggregates, for simplicity. Hence, aggregates essentially belong to the ontological macro-level, and aggregates dealing with just one base-object form improper extreme cases. Moreover, all macro-objects can be conceived as aggregates of one kind or another.

In the following we will introduce three kinds of representation of the aggregate-level: macro-representation, and proper and quasi micro-representation. Micro-representation will be completely (proper) or partially (quasi) based on the representation of the base-level in terms of the base-vocabulary, that is, the base-representation. Figure 5.1 depicts this outline of what follows.

Our exemplary statement '*a*o is water' gives a partial representation of a macro-object in macro-terms. More specifically, *a*o is assigned one *macro-type* out of a

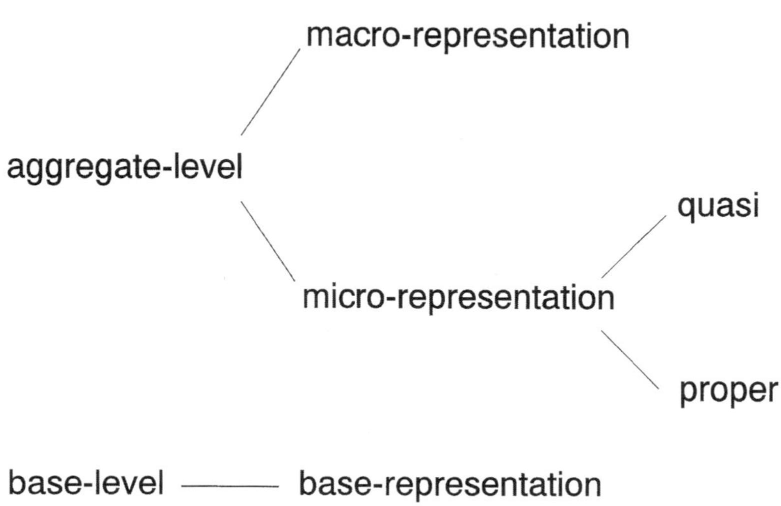

Figure 5.1 Outline of representations.

coherent set of macro-types, i.e. the family of substance terms. Hence, it will be called the (relevant) *macro-type representation* of *ao*.[3]

To generalize the example, let A indicate the relevant set of macro-objects of a certain kind, i.e. samples of substances, and let $F = \{P1, P2, \ldots, Pm\}$ indicate the relevant family of macro-types, i.e. substance terms (including terms for mixtures), with W for water. If a provisional F is non-exhaustive with respect to A, it is of course easy to add an extra type, by definition covering the remaining cases, such that a genuine family arises.[4] Now we assume that there is a function R, called the *macro-type representation function*, which specifies, for all a in A, the (unique) macro-type of a in F. Our elementary statement of interest becomes in this way '$R(ao) = W$'.

Macro-objects will be conceived as aggregates (of base-objects, see above). However, usually not all aggregates are easily conceived as macro-objects, let alone known as such. Hence let A^* indicate the set of all aggregates, having (the set of macro-objects) A as a subset. Let the variable 'a' range from now on over the entirety of A^*. However, regarding the macro-type representation, it remains that we need only to assume that it applies to members of A.

Reduction of substance terms means of course that we relate them to a suitable representation of macro-objects in terms of their base-objects and (some of) their properties, relations and functions. Let B indicate the set of all relevant base-objects. In our example B is the set of all molecules. Assuming a certain 'base-vocabulary' to represent base-objects and their properties and relations, let Mp indicate the set of all conceptually possible micro-structures or simply *structures* corresponding to that vocabulary.[5] In our example of molecules and kinds of molecules, Mp is of the following kind: one base-set, to be used for representing a set of molecules, and a set of subsets forming a partition of it, to be used for representing the partition of the molecules in chemical kinds, e.g. the subset of H_2O-molecules.[6]

Let d be the function from B to A^* indicating for all b in B from what $a = d(b)$ in A^* b is a part of. Hence, $d(b) = a$ states that b is a part of a. So we assume that the division of aggregates is such that they do not share parts. Let the subset Parts(a) of B indicate the set of base-objects being part of aggregate a, i.e. $\{b \text{ in } B/\ d(b) = a\}$. Now we will assume that all members a of A^* can be assigned a (up to isomorphism) unique structure x in Mp, with base-set Parts(a). That is, we assume that there is a function s, called the *structure representation function*, which specifies for all a in A^* its base-set and its further structure. Let $s(ao) = xo$ for our token of water.

So far one minimum condition for reduction has already been taken into account, viz. that every a in A has a macro-type as well as a structure representation, with the appropriate base-set. This may be called the *Token Identity Hypothesis* (Token-IH) with respect to $\langle A, F, Mp \rangle$ (cf. Kim 1996, p. 59). In our specification and notation it amounts to: for all a in A, there is a Pi in F such that $R(a) = Pi$, and there is a x in Mp such that $s(a) = x$ and the base-set of x is Parts (a). In other words, a is represented in two ways, by a macro-type and by a (related micro-)structure.

The idea of type-type (TT-)reduction is not that macro-types are to be identified with structures, but with 'base-induced' types of aggregates, called (proper) *micro-types*.[7] Micro-types will directly apply to structures and indirectly to aggregates. Hence, let $G = \{C1, C2, \ldots, Cn\}$ indicate a family of types for Mp, that is, a family of predicates which is explicitly defined in terms of the base-vocabulary, that is, in elementary base-terms or aggregate base-terms. In the present context it is plausible to define a micro-type as specifying that certain 'molecular kind-subsets' are non-empty and the others are empty. Let $g(x)$, for x in Mp, indicate the (unique) micro-type which applies to x; g is called the *micro-type classification function*. For instance, let $g(x\mathrm{o}) = Cw$ in G indicate the micro-type according to which the 'H2O-subset' is the only non-empty cell of xo. Let rp indicate the function assigning to each a in A^* the same micro-type as g assigns to the structure representation $s(a)$ of a. Hence, $rp(a) =_{\mathrm{def}} g(s(a))$. In this way, micro-types indirectly apply to aggregates and even form a family for A^*. We will call rp the (proper) *micro-type representation function*. Note that we are able to derive the micro-type representation '$rp(a\mathrm{o}) = Cw$' from the structure representation '$s(a\mathrm{o}) = x\mathrm{o}$' and the micro-type classification $g(x\mathrm{o}) = Cw$.

The core of TT-reduction of concepts now amounts to the *Type-Type Identity Hypothesis* (TT-IH) (cf. Kim 1996, p. 59) with respect to $\langle A, F, Mp, G\rangle$:

> for all Pi in F there is a (unique) Cj in G such that 'for all a in A $R(a) = Pi$ iff $rp(a) = Cj$' is an 'ontological equivalence'.

An 'ontological equivalence' is an equivalence on ontological grounds, here in the specific sense that the macro-types of macro-objects are ontologically identical, in the sense of the first section, to micro-types of aggregates. For our example we assume of course that 'for all a in A $R(a) = W$ iff $rp(a) = Cw$' is such an ontological equivalence. Note that it is not implied that there corresponds to every micro-type a macro-type. Think, e.g. of wave length intervals outside the visible area of the spectrum.

TT-IH enables the truth-preserving interchange of corresponding macro- and micro-type attributions to macro-objects (at least in transparent, non-intensional contexts). It is the possibility of deriving macro-type attributions from micro-type attributions to which the term 'type-type (TT-)reduction' specifically refers. Still more specifically, it refers to the derivation of the macro-type attribution from the structure representation and its micro-type classification. Both points will be illustrated by our example. We may derive the macro-type attribution '$R(a\mathrm{o}) = W$' from the micro-type attribution '$rp(a\mathrm{o}) = Cw$' and the supposed ontological equivalence 'for all a in A $R(a) = W$ iff $rp(a) = Cw$'. Earlier we saw already that the micro-type attribution '$rp(a\mathrm{o}) = Cw$', by its very definition, can be derived from the structure representation '$s(a\mathrm{o}) = x\mathrm{o}$' and its supposed micro-type classification '$g(x\mathrm{o}) = Cw$'. Hence, together with the ontological equivalence we can derive the macro-type attribution from the structure representation and its micro-type classification.

The ontological equivalence 'for all a in A $R(a) = W$ iff $rp(a) = Cw$' may be said to express that W is ontologically identical to Cw, an example of Dalton's core idea, viz. (being) water is (being) an aggregate of water-molecules. In general, TT-IH for the example of substance terms, restricted to pure substances, amounts to a specification of the first bridge principle in Dalton's theory, presented in Subsection 1.1.7 (B.1 for pure substances).

Figure 5.2 represents (perfect, singular, see below) TT-reduction in some detail. It illustrates that, assuming the appropriate identity, the macro-type $R(a)$ of a may be derived from its micro-type $rp(a)$, which in its turn may be derived from its structure representation $s(a)$, and the micro-type classification $g(s(a))$ of the latter.

As suggested, micro-types may not only be based on straightforward base-types, such as 'being a H2O-molecule', but also on so-called 'aggregate concepts', such as mean kinetic energy, or even 'aggregated concepts', if some kind of standard aggregation is involved, such as is the case with 'kinetic pressure'. Aggregate concepts are by definition essentially defined in terms of base-types of base-objects for aggregates of such base-objects.[8]

Instead of a perfect link between macro- and micro-types, there may be an approximative link. It may be that for most cases a macro-type attribution goes together with a certain micro-type attribution. Here we come upon *approximative* forms of TT-reduction. The approximative Type-Type Identity Hypothesis, $\text{TT-IH}^{\approx}$, states that there is a function relating macro-types to micro-types, such that their denotations in A approximately coincide. From now on we will indicate the perfect TT-IH by $\text{TT-IH}^{=}$, to distinguish the corresponding kind of *perfect* TT-reduction from its approximative version.

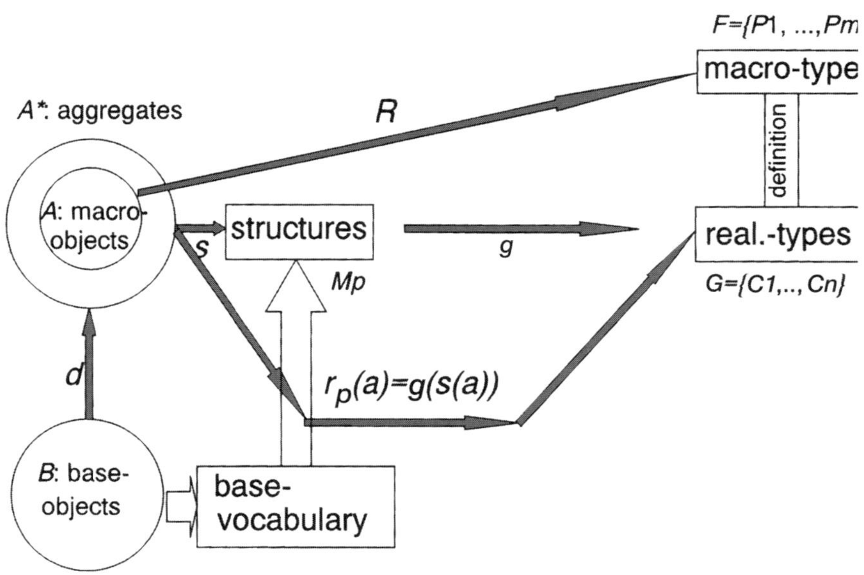

Figure 5.2 Type-Type reduction (perfect, singular).

In the case of the corrective reduction of laws the reduced law was essentially replaced by a corrected version. However, in the case of the approximative reduction of concepts one might, but need not, replace the macro-types by the approximately identical corresponding micro-types. Keeping both in honor only runs the risk of more or less incidental deviations.

It is easy to verify that TT-$IH^{=}$ implies the (Supervenience or) *Realization Hypothesis* (Kim, 1996, p. 224):

> for all a and a' in A if $s(a) \cong s(a')$ ($\cong$: the isomorphy relation on Mp) then $R(a) = R(a')$ on 'ontological grounds'

That is, as far as A is concerned, an aggregate 'ontologically realizes'[9] its macro-type by its having a certain structure or, simply, the structure realizes the macro-type. It is also said that, within A, Mp realizes F.

Note that we did not at all assume that the macro-type of a macro-object uniquely determines its structure. On the contrary, as a rule, many structures will realize the same macro-type. Hence a macro-type is paradigmatically assumed to be 'multiply realizable' in this sense, that is, by different structures. However, this should not be confused with the more problematic phenomenon of multiple realizability of a macro-type by a number of different micro-types. That is, it may well be that there is no micro-type that can play the role indicated above for (perfect or approximative) TT-reduction, but that formally the same role can be played by a disjunction of (mutually exclusive) micro-types, assuming that this disjunction is composed of such diverse micro-types, that it cannot (as yet) be conceived as another, encompassing kind of micro-type of aggregates.

In this case the corresponding macro-types are said to be *multiply realizable* by the original micro-types. The corresponding reduction, which requires a multiple version of the Type-Type Identity Hypothesis, may be called *multiple or one-many TT-reduction*, as opposed to the first kind of TT-type reduction, which may now more precisely be called *singular or one-one* TT-reduction.

Let us now look more closely at the Realization Hypothesis itself. As we will see, it allows the definition of ('subvenience-' or 'quasi micro-' or) 'realization-classes' of structures. These classes give rise to corresponding 'realization-types' of aggregates, leading to a realization-type classification function for structures and to a realization-type representation function for aggregates. The realization-class of a macro-type amounts to the set of all structures representing the aggregates that realize the macro-type, that is, aggregates to which the macro-type applies. Let Pi be a member of F. The Realization Hypothesis guarantees that the following definition:

$$K^*(Pi) = K^*i = \text{def}$$
$$\{x \text{ in } Mp / \text{ there is } a \text{ in } A \text{ such that } R(a) = Pi \text{ and } s(a) \cong x\}$$

leads to non-overlapping subsets of Mp, and even a partition of Mp, if we take K^*0 to refer to the set of remaining members of Mp. K^*i is called the *realization-class*[10] (of structures) of Pi.[11]

Let Ki indicate the corresponding *realization-type*, a classification primarily for structures, viz. 'belonging to K^*i'. Let $H = \{K0, K1, K2, \ldots, Km\}$ and let h indicate the *realization-type classification function* assigning to each x in Mp its realization-type. Finally, let rq indicate the *realization-type or quasi micro-type representation function* assigning to each aggregate a in A its realization-type according to the definition $rq(a) =_{\text{def}} h(s(a))$, that is, the realization-type classification of the structure-representation of a.

We should sharply distinguish between the formal definition of the realization-classes and realization-types, allowed by the Realization Hypothesis, and their empirical determination, that is, their determination in base-terms on the basis of empirical research.[12] In other words, realization-type attribution to a macro-object may be a matter of definition, viz. when we start from a macro-type representation, or a matter of empirical research, assuming that the realization-types Ki's have been empirically determined in base-terms. From now on we will mention it when we presuppose that the Ki's have been empirically determined in this way as well the actual attribution $rq(a)$ of some a in A^*.

Without a (one-one or one-many) TT-IH proper TT-reduction is not possible, but as soon as we may assume the Token-IH and the Realization Hypothesis, we may speak of *quasi type-type (quasi-TT-)reduction* in the sense that a macro-type attribution $R(a) = Pi$ may be derived from the corresponding realization-type attribution $rq(a) = Ki$ (and the latter from the structure representation and its realization-type classification). Note that there are no approximative or multiple versions of quasi-TT-reduction, due to its definitorical nature. Hence, quasi-TT-reduction is by definition perfect and singular. Figure 5.3 represents quasi-TT-reduction.

Despite the rather formally similar way in which they were introduced, there is a fundamental difference between realization-types and micro-types. The former are by definition, hence apriorily, ontologically identical to macro-types, whereas the latter may only turn out to be so aposteriorily. The latter discovery is the core of (singular) TT-reduction, which can now also be expressed in terms of realization-types: for all Pi in F (with corresponding Ki in H) there is a (unique) Cj in G such that 'for all a in A $rq(a) = Ki$ iff $rp(a) = Cj$' is an ontological equivalence, or, equivalently, such that '$Ki = Cj$' is an 'ontological equality', that is, an equality on ontological grounds, or, again equivalently, such that not only Pi and Ki are ontologically identical, by definition, but also Pi and Cj and hence Ki and Cj as a matter of ontological fact. Similar things can be said about approximative TT-reduction. Figure 5.4 combines quasi-TT-reduction and proper (perfect, singular) TT-reduction.

5.2.2 *Reduction of Attribute-Predicates*

'Water' is a typical thing-predicate, more in particular a substance term. Assuming that the foregoing subsection gives an adequate analysis of the reduction of thing-predicates, the question is whether the analysis is also applicable to the reduction of

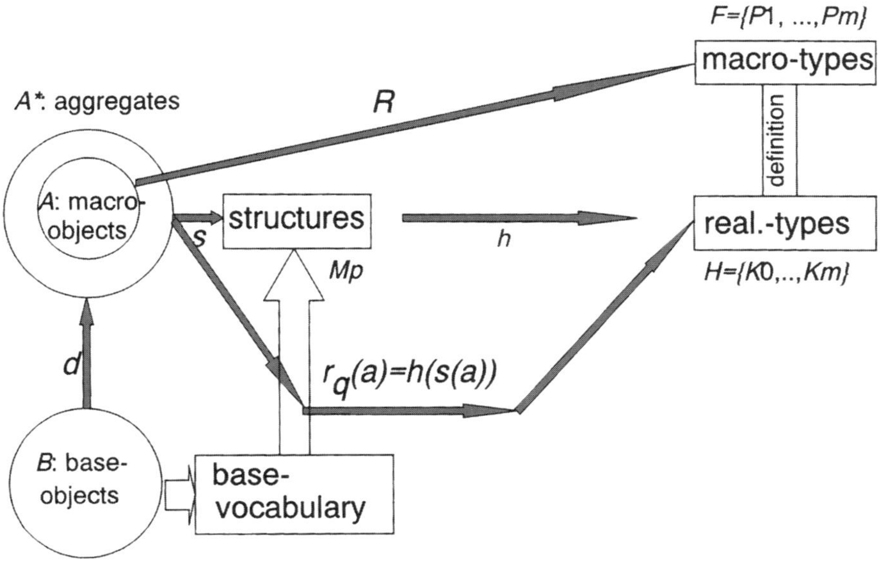

Figure 5.3 *Quasi-Type-Type-reduction (by definition perfect and singular).*

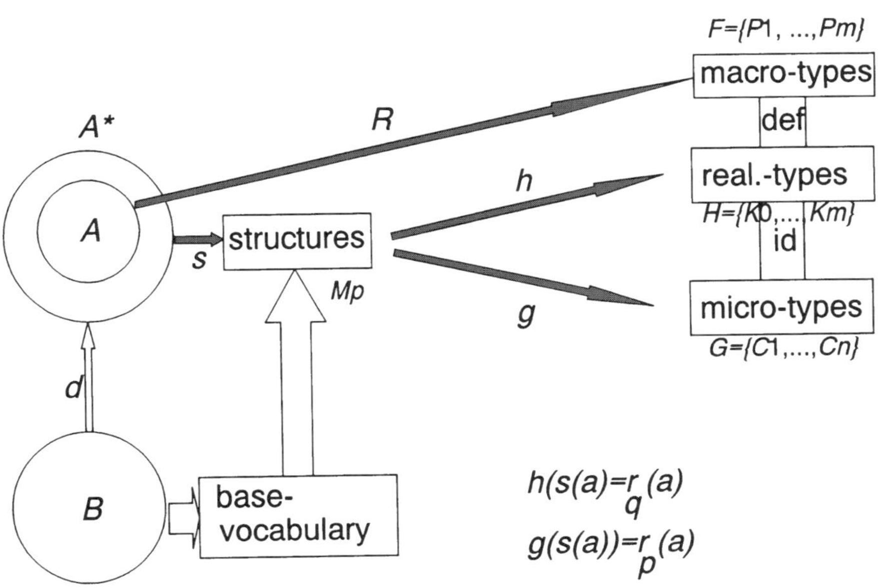

Figure 5.4 *Quasi-TT-reduction combined with proper TT-reduction.*

attribute-predicates. One usually distinguishes three types of attributes: properties, relations and functions. In view of the fact that properties are indicated by unary predicates, just as thing-predicates, the reduction of property-predicates has precisely the same form as the reduction of thing-predicates. The water example could have been analyzed as a combination of thing-with-property: water in liquid form. It is indeed not difficult to see that the predicate 'fluid' and the predicate 'water' in the sense of 'water in one of the three phases' can be reduced in a similar way. In the new example of 'liquid' the relevant structures should have to specify that their base-sets concern molecules, moreover, their kinds would not be required, but their mass and (relative) position and some other properties, relations and functions would be required. The relevant micro-types would be defined by specifying the relevant base- and micro-conditions for the solid, liquid and gaseous phase of a substance. The only thing which immediately becomes clear is that it is now advisable to introduce time as a variable.

Let us now direct attention to the reduction of a (quantitative) functional attribute-term: temperature. Consider the statement that the (absolute empirical) temperature Θ of an isolated amount of gas (e.g. water-vapor!) *ao* at time *t*o is equal to *T*o: '$\Theta(ao, to) = To$'. The macro-types now are of course the possible temperature values, i.e., the temperature scale, and Θ is the macro-type representation function.

The required structures now just have to specify a base-set of molecules, together with their mass and temporary position, such that we can calculate their velocity, momentum and kinetic energy. The relevant structure attribution statement is, apart from the time parameter, again of the same type: the structure representation '$s(ao,to) = xo$'.

The standard way of linking temperature to structures is by means of the aggregate concept of 'mean kinetic energy' $\bar{u}(x)$, which is explicitly defined for any structure x in (the adapted) *Mp*. The micro-types of structures are defined by having a certain mean kinetic energy, e.g. $Cu(x)$ iff $\bar{u}(x) = u$, with the corresponding micro-type classification $g(x) = Cu$. The corresponding micro-type representation $rp(a) = Cu$ is by definition determined by the condition that $s(a)$ belongs to Cu or, equivalently, $g(s(a)) = Cu$. The familiar functional relation $\bar{u} = (3/2)kT$ (where k is Boltzmann's constant) specifies the function postulated by the TT-IH=, by linking the macro-type T with the micro-type Cu as ontologically identical only in the case that $u = (3/2)kT$.

Turning now to relational attribute-terms, recall that it was reported in Appendix 1 of Chapter 2 that the present identity between the two quantitative concepts can be obtained from the identification 'equal thermal state is the same as equal mean kinetic energy', where the macro-attribute ('equal thermal state') is a qualitative (equivalence) *relation*. Now it is not difficult to check that the above sketched reduction can be transformed into the reduction of this relation, by which we have indicated an example of reduction of the third category of attributes, beside properties and functions, viz. relations.[13]

By now it has been sufficiently argued that the reduction of attribute-predicates can take place in a completely analogous way to the reduction of thing-predicates. Only the introduction of the time parameter has now become unavoidable.

5.2.3 *Formal Survey of Concept Reduction*

By way of summary this subsection first gives a survey of the formal aspects of the foregoing two subsections, leaving out all example related notation, and including the time parameter throughout. The survey will essentially be restricted to monadic, including functional, predicates. But it is not difficult to adapt the formal aspects to genuine relations.

General notations and assumptions

A	set of macro-objects (variable: a)
A^*	set of aggregate-objects, including A (variable a)
T	time axis (variable: t)
F	family of macro-types for A (variable: Pi)
Mp	set of conceptually possible structures (variable: x)
B	the set of base-objects
d	$B \times T \to A^*$, where $d(b,t) = a$ states that b is, at t, a part of a
Parts	$A^* \times T \to Po(B)$, where Parts(a,t) indicates the set of base-objects being part of a at time t, and $Po(B)$ the powerset (i.e. the set of subsets) of B

Representation assumptions: there are functions R and s such that

R	$A \times T \to F$, *macro-type representation function*
s	$A^* \times T \to Mp$, *structure representation function*, such that $s(a,t) = x$ with base-set Parts (a,t)

implying

Token-IH (*Token Identity Hypothesis*):
for all a in A and t in T there is a Pi in F such that $R(a,t) = Pi$, and there is a x in Mp such that $s(a,t) = x$ with base-set Parts(a,t)

Type-type (TT-)reduction (proper)

G	a family $\{C1, C2, \ldots, Cn\}$ of *micro-types* for Mp (and indirectly for A^*) defined in base-terms
g	$Mp \to G$, the *micro-type classification function*, such that $g(x) = Ci$ iff Ci applies to x
rp	$A^* \times T \to G$ such that $rp(a,t) = g(s(a))$, *(proper) micro-type representation function*

TT-IH$^{=}$ (perfect TT-IH, i.e. perfect *Type-Type Identity Hypothesis*):
for all Pi in F there is a (unique) Cj in G such that 'for all a in A and t in T $R(a,t) = Pi$ iff $rp(a,t) = Cj$' is an 'ontological equivalence' (Pi and Cj are ontologically identical as far as A is concerned)

or, equivalently, there is a 1–1-function $I^{=}$ from F to G such that Pi and $I^{=}(Pi)$ are ontologically identical[14]

TT-reduction (singular, perfect): deriving a macro-type attribution from the corresponding micro-type attribution, represented by Figure 5.2

$\approx$ $U \approx V$: set U approximately coincides with set V

TT-IH$^{\approx}$ (approximative TT-IH): there is a 1–1-function $I^{\approx}$ from F to G such that the approximate equalities $\{a \text{ in } A/R(a) = Pi\} \approx \{a \text{ in } A/rp(a) = I^{\approx}(Pi)\}$ are of an ontological nature

Approximative TT-reduction (singular): deriving a macro-type attribution from the approximately corresponding micro-type attribution

Multiple versions: one-many versions of (perfect and approximative) TT-reduction can be based on corresponding multiple versions of the Type-Type Identity Hypothesis

Quasi-type-type (quasi-TT-)reduction

Realization (or Supervenience) Hypothesis
for all a and a' in A and t and t' in T, if $s(a,t) \cong s(a',t')$ then $R(a,t') = R(a',t')$ (' $\cong$ ' is the isomorphy relation on Mp)

H the family of (macro-induced) *realization-types* $\{K0, K1, K2, \ldots, Km\}$ to which the *realization-classes* give rise, that is, the partition of Mp, such that, for $i = 1, \ldots, m$, $K^*(Pi) = K^*i =$ def $\{x$ in $Mp/$ there is a in A such that $s(a,t) \cong x$ and $R(a,t) = Pi\}$ is the realization-class of Pi and K^*0 is the rest-class of structures not realizing a macro-type; realization-types are defined as follows: x is of realization-type Ki iff it belongs to the realization-class K^*i

h $Mp \to H$, the *realization-type classification function*, such that $h(x) = Ki$ iff Ki applies to x

rq $A^* \times T \to H$ such that $rq(a,t) = h(s(a))$, *realization-type or quasi micro-type representation function*

quasi-TT-reduction: deriving a macro-type attribution from the corresponding realization-type attribution, represented by Figure 5.3 (there are no approximative or multiple versions of quasi-TT-reduction)

quasi-TT-reduction combined with proper TT-reduction

> According to TT-IH$^{=}$ (which implies the Realization Hypothesis) the macro-types, and hence the realization-types, are ontologically identical to the corresponding micro-types. This allows the derivation of the macro-type attributions from the corresponding realization-type attributions (quasi-TT-reduction), which may in their turn be derived from the corresponding micro-type attributions (TT-reduction); Figure 5.4 represents the combination of reductions.

5.2.4 *Levels of Reduction*

Both examples of concept reduction are clear cases in which two related ontological levels play a role, namely, the level of macro-objects and that of micro- or base-objects, i.e. molecules. This distinction of levels is based on the fact that the base-objects are in the most literal sense parts of the macro-objects, as soon as the macro-objects are conceived as aggregates of base-objects. Of course, the ontological level distinction is relative. Molecules are composed of atoms, and they in their turn are composed of elementary particles. A base-level may or may not be restricted to one level in some strict sense, or include one or more of its own base-levels, e.g. only molecules, or molecules, atoms and elementary particles.

Aggregates, although being an ontological category, do not constitute a separate ontological level, but merely belong to the macro-level. In fact, the macro-level may be said primarily to house aggregates, for which reason it is better to speak of the aggregate-level, despite the fact that many aggregates, but not all of them, are better or only known as macro-objects.

Unlike the foregoing two ontological levels, it is less easy to distinguish conceptual or epistemological levels. Recall for this purpose that macro-objects are primarily represented by macro-types, representing macro-objects, properties of macro-objects and relations between them. Similarly, base-objects are represented by base-types, representing base-objects, properties of them and relations between them. However, macro-objects can be represented in a second way, by conceiving them as aggregates of base-objects, micro-aggregates, to be represented by micro-structures. As long as we neglect macro-aggregates and -structures,[15] we just speak of aggregates and structures, meaning micro-aggregates and -structures.

Aggregates need not be conceived as macro-objects. Structures are intended to represent aggregates in terms of micro-types, drawn from a certain base-vocabulary. It is important to note that structures do not represent the structure of base-objects, but the structure of aggregates of base-objects, including macro-objects, conceived as such aggregates.

The set of structures may be partitioned by various types of structures. The crucial point of concept reduction is that there may be base-induced micro-types, which are ontologically related to macro-types. In the case of TT-reduction the macro-types are ontologically identical to certain micro-types.

As soon as the Realization Hypothesis is assumed we can also introduce macro-induced classes of structures, called realization-classes, and hence macro-induced types of structures and hence of aggregates, called realization-types.

Hence, at least the following five kinds of concepts are in any case involved in the question of conceptual levels: macro-types, base-terms, structures, micro- and realization-types of structures and aggregates. The five kinds of concepts are members of *F*, the base-vocabulary, *Mp*, *G* and *H*, respectively. They all occur in Figure 5.4. Whereas the micro-types are essentially 'base-induced', the realization-types are essentially 'macro-induced', viz. derived from the macro-induced partition of the set of structures in realization-classes.

One may speak of a conceptual base-level and an aggregate-level and situate on the base-level the concepts that only deal with base-objects (hence base-terms) and on the aggregate-level those dealing in one way or another with aggregates (hence macro-types, structures, micro-types and realization-types). However, now it is also plausible to introduce one or more sublevels of the aggregate-level. To begin with, the sublevel of macro-types on the one hand, called the macro-(sub)level, and the sublevel dealing with structures, the micro-(sub)level, on the other. The micro-(sub)level may be said to have two epistemological sides, the base-(induced) side of (proper) micro-types and the macro-(induced) side of realization-types, that is, quasi micro-types.

In sum, we have distinguished four conceptual (sub-)levels and two conceptual sides, as indicated in Figure 5.5.

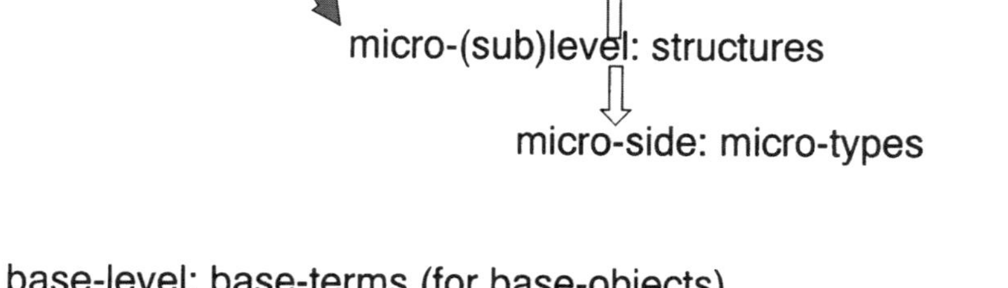

Figure 5.5 Conceptual levels, sublevels and sides.

It is also tempting to formulate a plausible explication of the idea of *ontological reduction* of concepts: the TT-reduction of *thing*-predicates, i.e. macro-types representing (types of) macro-objects, to micro-types of aggregates. Our first example, the case of the substance term water, is a case in point. Although this explication deviates formally quite a lot from the one provided by Moulines (1984), it is close to it in spirit. Generalizing our water example, the atomic theory is a paradigm example of ontological reduction, i.e. the reduction of substance predicates to that of the micro-types of aggregates of molecules. The reduction of molecules to aggregates of atoms and the reduction of genes to aggregates of DNA-molecules are others.[16]

5.2.5 *Correlation of Concepts*

In contrast to the reduction of laws, the reduction of concepts is usually not seen as a form of explanation, although this is of course just a matter of terminology. However this may be, TT-reduction can directly be transformed into *type-type (TT-)correlation*, that is, deriving macro-type attributions from causally correlating micro-type attributions, by substituting appropriate (causal) TT-correlations for TT-identities. In contrast to TT-identities, TT-correlations may not only be of a deterministic nature, but also of a statistical nature. These correlative arguments are generally useful. Hence, even for somebody who not only rejects attribute-identities but also thing-identities most of the foregoing analysis was at least of some use.

Consider for the suggested transformation instead of the perfect and approximative versions of the Type-Type Identity Hypothesis the corresponding versions of the *Type-Type Correlation Hypothesis*:

> *TT-CH*$^{=}$: for all Pi in F there is a (unique) Cj in G such that 'for all a in A and t in T $R(a,t) = Pi$ iff $rp(a,t) \in Cj$' is a 'causal equivalence'
> (F and G are causally correlated as far as A is concerned)
>
> or equivalently, there is a 1–1-function $C^{=}$*from* F *to* G *such that the correlation between* $C^{=}$ (Pi) *and* Pi *is of a causal nature*[17]
>
> *TT-CH*$^{\approx}$: there is a 1–1-function C. from F to G such that the approximate correlation between $C^{\approx}(Pi)$ and Pi is of a causal nature

It is plausible to assume that besides these singular versions, multiple versions would also make sense.

The two main questions of the first section concerning correlation and identities can now be summarized: is there only correlation of concepts or is reduction (by some kind of type-type identification) also possible, and if so, is there only reduction of thing-predicates or also of attribute-predicates? As we have already seen in Chapter 3, genetics provides a good example of causal correlations.

Biological types of organisms are causally correlated with the combination of genetic factors, the geno-types, of the fertilized egg, the zygote.

Although there are TT-correlation variants of (perfect-or-approximative) singular and multiple TT-reduction, there is no meaningful correlation variant of (perfect-or-approximative) quasi-TT-reduction, due to the latter's definitorical character. Recall that for the same reason there was also no approximative version of perfect quasi-TT-reduction.

We conclude this section with a table of the different kinds of reduction and correlation of concepts, ordered in degrees of increasing strength. '*X*' indicates that there are examples, and '–' that examples are conceptually impossible.

Table 5.1 Degrees of type-type reduction and correlation of concepts.

degree	1.1/1.2 perfect/approx	2.1/2.2 perfect/approx	3.1/3.2 perfect/approx
	quasi (one-one)	multiple (one-many)	singular (one-one)
identities	X/–	X	X
correlations	–/–	X	X

Examples of the three degrees of perfect reduction of concepts are the following:

- Perfect reductions of the third degree (singular/one-one) are that of, roughly speaking, water to aggregates of H_2O-molecules and temperature to mean kinetic energy.
- Perfect reductions of the second degree (multiple/one-many) are that of, for example, uranium to pure and mixed aggregates of (molecules of) its three isotopic atoms[18] and, on closer inspection, temperature to mean kinetic energy of translation of the molecules in the case of gases and to mean kinetic energy of vibration of the molecules in the case of solids.
- A perfect reduction of the first degree (quasi) is that of color (or at least color classification) to its corresponding wave length interval (for there is no natural partition of wave lengths on that level).

Examples of the two (!) degrees of perfect correlation are the following:

- A perfect correlation of the third degree (singular/one-one) is a recessive phenotype and its corresponding homozygote genotype (which can only arise from the parents in one way).
- A perfect correlation of the second degree (multiple/one-many) is a dominant phenotype and its two corresponding genotypes, a homozygote and a heterozygote (where the latter, but not the former, can even arise from the parents in two ways).

5.3 THE RELATION BETWEEN CONCEPT AND LAW REDUCTION, MULTIPLE CONCEPT REDUCTION, AND (NON-)REDUCTIONISTIC STRATEGIES

Introduction

First we will investigate the relation between the reduction of laws and concepts, to get an overview of the possibilities of reduction. Next we will pay some attention to the importance of multiple concept reduction. Finally, we will formulate and discuss reductionistic and holistic research strategies in relation to reductionistic and holistic philosophical positions.

5.3.1 *The Relation Between the Reduction of Concepts and Laws*

It is clear that none of the presented variants of concept reduction presupposes (some form of) the reduction of laws. Hence, the interesting question concerns the other direction: does the reduction of laws presuppose concept reduction? Let us answer this question by considering the five steps in the decomposition model of the explanation of a law as outlined in Chapter 3. This procedure will show that a number of mistakes are easily made when explanations are not decomposed.

To start with, it is again clear that the application step and the (causal) correlation step, which on their own do not justify our speaking of a reduction of the law, have nothing intrinsically to do with concept reduction.

Consider next the approximation step, the occurrence of which was one of the reasons to speak in the literature of the reduction of the law in question. It has already been mentioned that this step is related to a reduction in the idealizational sense: the idealized law was originally obtained by consciously or unconsciously neglecting certain factors. By the approximation step we regain that idealized law from a concretized version, by neglecting already accounted factors. Hence, although this step does not involve concept reduction in one of the transformational senses, it does involve concept reduction in some idealizational sense, i.e. a kind of re-idealization: some aspects, and hence the concepts representing them, which were first neglected, are later accounted for; going backwards, they may again be deleted in the approximation step.

It is fairly evident that the two remaining steps, aggregation and identification, have something to do with concept reduction in the transformational sense. Note first that the identification step presupposes by definition some kind of Type-Type Identity Hypothesis, where it is plausible to think of the kinds that we have introduced in the context of concept reduction. More specifically, the following claim seems rather plausible: if an aggregation step is followed by an identification step, the latter consists of carrying out all relevant type-type reductions of concepts, more precisely, all 'aggregate concepts' occurring in the aggregated law are replaced by their macro-counterparts, while the relevant Type-Type Identity Hypotheses guarantee intrinsic preservation of truth-value of the law during this transformation.

Now what if there is identification but no aggregation? As a matter of fact, it still is a challenge to elaborate proper examples of the corresponding pure type of reduction of a law. If they exist, then, again, it seems plausible that type-type concept reduction is involved. One particular possible form will be mentioned here. There may be a macro-law for which we can find and establish an aggregate-version, that is, a law formulated in terms of micro-types, which can be transformed into the original macro-law by the relevant type-type concept reductions. But this does not imply that the aggregate law is an *aggregated law*, in the sense that it is the result of an aggregation of the effect of an individual law applied to a collection of objects. The aggregate law is only also an aggregate*d* law if it turns out to be possible to micro-reduce it by such an aggregation step. Hence, we may look in particular for a macro-law which can be deductively explained by an aggregate law (solely) on the basis of identification, whereas it is not yet known whether the aggregate law itself can be micro-reduced. If the indicated explanation is called reduction, we have a pure example of reduction by identification, although we should not exclude the fact that calling it reduction may still be an implicit anticipation of a successful micro-reduction. Examples of this kind might exist around the ontological identity of light rays and electromagnetic rays or the ontological identity of inertial mass and gravitational mass. But these examples still have to be analyzed in detail.

In both cases discussed so far, that is, identification with or without aggregation, there is of course also room for multiple and approximative variants. That is, the role played by micro-types may also be played by disjunctions of micro-types, leading to a multiple type-type reduction of concepts, or the fit between the classes corresponding to the micro-types and the realization-types may not be exact but approximative. The case of multiple type-type concept reduction will be addressed further in the next subsection.

Let us finally consider the aggregation step. At first sight it may seem a plausible idea that the aggregation step in itself involves some kind of concept reduction due to the fact that aggregation presupposes the definability of 'aggregate(d) concepts', i.e. of micro-types. This implies in its turn the very possibility of reduction of macro-concepts. But this does not imply that this possibility is realized in the aggregation step. In sum, there is no good reason to relate homogeneous micro-reduction of a law to type-type reduction of concepts.

5.3.2 *Multiple Concept Reduction*

We have already remarked that it is possible that a macro-type cannot be identified with one micro-type but only to a disjunction of micro-types, assuming that these disjunctions are composed of so diverse (usually non-overlapping) types of structures and that they cannot (as yet) be conceived as micro-types themselves. In this case the corresponding macro-types were said to be *multiply realizable* with respect to the micro-types, for which reason the corresponding reduction was called a multiple or a *one-many TT-reduction*. Recall the examples of isotopes (certain substances are realizable by aggregates of (molecules of) atoms of different isotopes) and temperature (realized by different kinds of molecular kinetic energy in gases and solids, viz. of translation and vibration). We have also already noted that a macro-type apparently is already multiply realizable *with respect to structures* as soon as its realization-class contains more than one element.

In the philosophy of mind it is generally thought that multiple realizability is an obstacle to reduction. However, as our detailed analysis makes clear, such is at least not generally the case, for the (one–one) TT-reduction in the examples of physics typically concerned cases where the macro-types, as well as the micro-types, evidently are multiply realizable in terms of structures. Hence, the problem seems to be restricted to the meaningfulness of one-many TT-reductions. As our examples illustrate, in the physical sciences such reductions are generally considered as unproblematic. In particular, the case of temperature has been put forward earlier (Hooker, 1981; Sklar, 1993; Bickle, 1998) in arguing against 'multiple realizability' as a blockade of successful reduction. If one finds such reductions nevertheless problematic, the following three considerations should be taken into account. First, one frequently gets the impression that reservations originate from thinking that the relevant micro-type disjunct should be derivable from the macro-type and the relevant identity hypothesis. However, the direction of reductive derivation of macro-types, and macro-laws for that matter, is from micro to macro. Hence, it suffices that the relevant micro-type disjunct together with the relevant partial implication of the multiple version of the TT identity hypothesis enable the derivation. Second, it may well be that the relevant disjunction of micro-types can be seen, on closer inspection, as a genuine, more encompassing micro-type.[19] Examples are (aggregates of (molecules of) atoms having) the same number of protons and (aggregates of (molecules of) atoms having) the same mean kinetic energy of one kind or another. Third, it may also be the case that a one-many TT-reduction may be decomposed into a number of one-one TT-type reductions by dividing the set of macro-objects into natural kinds. In terms of the standard example, if humans and computers can both have a mental property, it may well be that this property, as far as humans are concerned, can be reduced to a certain, say neuro-physiological, micro-type that may apply to humans, and as far as computers are concerned, to another, say electronic, micro-type that may apply to computers. This kind of 'partial' reduction is what Kim (1996, 2000) calls *local (TT-)reduction* or *species-restricted reduction.*

Of course, if, and only if, there is something like a psychological law holding for humans as well as for computers, that is, a 'common law', there is considerable reason to suppose that there is some similarity between the corresponding human and computer micro-types, which may lead to their unification and hence to the unification of the two local reductions.

As a matter of fact, and roughly speaking, temperature was first 'locally' reduced to mean kinetic energy of translation in gases, then to mean kinetic energy of vibration in solids, and finally to the mean kinetic energy for every degree of freedom of the molecules, leading to a complex account for the temperature of liquids.

In sum, multiple TT-reduction may well combine a number of successful local singular TT-reductions that may guide the search for common laws and anticipate a global singular TT-reduction. Hence, from the point of view of research strategies, the current reservations against multiple TT-reduction, e.g. Macdonald (1989), seem misplaced.

5.3.3 *Reductionistic, Holistic, and Mixed Strategies*

A further question to be discussed in this section is the relation between the presented epistemological analysis of reduction of laws and concepts and all kinds of metaphysical positions and methodological research strategies. The latter were already briefly indicated in Chapter 1. This subsection can be seen as an attempt to indicate the different use-values of the presented models.

In the discussion we will restrict ourselves to *vertical reduction*, that is, reduction of concepts in one of the explicated senses, and reduction of laws based on aggregation and/or identification. This leaves reduction of laws based on approximation, that is, corrective reduction, apart as a kind of *horizontal reduction*. From here on 'reduction' is to be read as 'vertical reduction', except when otherwise stated.

In Chapter 1 we formulated three methodological approaches or research strategies with respect to a certain macro-domain. Let us briefly reformulate them here. In the *radical reductionistic strategy* all attention is directed to the reduction of macro-laws and -concepts and the establishment of aggregate concepts and aggregated laws. In the *radical holistic strategy* all attention is directed to the formation of macro-concepts, to discovering macro-laws, and finally to the explanation of macro-laws by encompassing macro-theories (hence, without using aggregation or identification). Finally, there is the *mixed strategy*, according to which all these activities take place alternately, depending on all kinds of considerations, such as the stage of the research, what seems possible, and, last but not least, what can stimulate what, etc. To be sure, the mixed strategy favors reduction when possible. Roughly speaking, in the mixed strategy one describes the macro-phenomena and their possible relations in macro-terms, and tries to explain them in micro-terms as far as possible, and hence in macro-terms as far as necessary.

This brings us to the formulation of three current philosophical (ontological cum epistemological) positions with respect to a certain 'macro-domain'. They are formulated as general statements about the possibility of reduction of the concepts and laws of that domain. *Radical reductionism* is the belief that every macro-concept and macro-law can be reduced. *Radical holism* is the belief that all (interesting) concepts and laws of the domain cannot be reduced. Finally, *restricted reductionism* (and holism!) is the belief that some concepts and laws may be reducible, but others may not be.

It is important to note that the terminologically corresponding strategies and positions are not strictly coupled, except perhaps that radical philosophical holism leaves only room for the radical holistic strategy. The converse is not self-evident. There are excellent examples of research according to the radical holistic strategy, e.g. phenomenological thermodynamics and macro-economics, where it is seen as a compatible, but separate task to try to reduce the macro-concepts and -laws, i.e. to work according to the radical reductionistic strategy. To be sure, it has to be conceded that reductionistic strategies in general have been very successful in the history of science.

However, the radical reductionistic strategy often leads to impressive minute research, constituting perfect aggregate concepts and aggregated laws, which, however, nobody is waiting for. On the other hand, the radical holistic strategy frequently degenerates into hardly testable and transferable insights.

In line with these roughly formulated impressions, it is plausible to formulate the working hypothesis that the mixed strategy will in many cases be the best strategy. For, to reduce concepts and laws of a certain domain, they have first to be established. In its turn, it is frequently the case that the search for concepts and laws has been considerably stimulated by reductionistic questions.

We would like to conceive examples of the mixed strategy more specifically as important cases of interaction of research programs, as outlined in Chapter 1. In the case of reductive interaction the guide program is of course a program on the macro-level, whereas at least one supply program is supposed to deal with the micro-level. When the supply program is, like the guide program, in the internal phase, the reductive interaction is symmetric, when it is in the external phase it is asymmetric.

Zandvoort's paradigm example of a supply program in the natural sciences is the NMR(nuclear magnetic resonance)-program, engaged in asymmetric reductive co-operation with chemical and biological research programs. An important asymmetric example in the social sciences is the utility maximization or rational choice program. It has proved its strength in micro-economics and nowadays it co-operates with guide programs in macro-economics (Janssen, 1993) and macro-sociology (explanatory sociology). An historical example of the symmetric reductive type is the interaction between phenomenological thermodynamics and statistical mechanics.

CONCLUDING REMARKS

The type of interdisciplinary research just indicated fits well with the following view of the different types of explanation. Beside the suggested 'vertical' forms of reductive explanation, there are not only other vertical forms of non-reductive explanation (e.g. based on correlation), but also many types of 'horizontal' explanation (e.g. based on approximation). As far as the nomological explanation of laws is concerned all these types of explanation seem to fit well in our general decomposition model. However, as we have seen in the previous chapter, there have always been claims that there are types of explanation which are essentially different from nomological explanation. More specifically, we have argued that, e.g. intentional explanation and functional explanation (in biology) indeed constitute separate forms of explanation and that they fit into a general model of so-called *explanation by* (intentional, functional, etc.) *specification*, which even includes certain types of causal explanation. The relevance for the present context is the following. Explanation by specification may be of a horizontal or of a vertical nature. In particular, it may be the dominant way of horizontal explanation in a guide program. Consequently, instead of disqualifying explanation by specification, it may play a crucial role in reductive interdisciplinary mind-body research: it can open the way for interesting forms of reductive neurophysiological explanation, a way that would be indistinguishable in the overwhelming number of possibilities confronting the radical reductionistic strategy. Conversely, reductive neurophysiological explanation may very well suggest new ways for further development of psychological explanation, a possibility which is blocked by the radical holistic strategy.[20] In the next chapter we will illustrate several kinds of explanation and reduction by focussing on mind-body research.

CHAPTER 6

LEVELS, STYLES, AND MIND-BODY RESEARCH

INTRODUCTION

In Chapter 1 we already indicated several goals and types of symmetric and asymmetric co-operation between research programs, where the programs may or may not deal with different levels of aggregation or even belong to different disciplines. Moreover, they may or may not use the same or different styles of description and explanation. In this chapter we will elaborate these themes, with emphasis on mind-body-research.

In Section 6.1 we will set out a matrix of four kinds of so-called interfield research, starting from a survey of possible interlevel research between epistemological levels in general and the two major divisions of them, viz. in ontological levels and epistemological styles, in particular. In Section 6.2 we will deal with the mutual relations between the styles, with emphasis on interstyle monolevel research. In Section 6.3 we will focus on monostyle interlevel research, notably biophysical mind-body research. In Section 6.4 we will broaden the scope by discussing some models and examples of interlevel and interstyle interfield mind-body research. The main example will concern two types of juvenile delinquency. Finally, in Section 6.4 we will give some examples of interdisciplinary research which is nevertheless of a monostyle monolevel nature. This chapter will use several means developed in the previous chapters, either in a straightforward way or in a rather new way.

6.1 INTERLEVEL AND INTERFIELD RESEARCH

Introduction

In the previous chapters we have frequently spoken about conceptual or epistemological levels. Such levels may or may not be associated with ontological levels and/or epistemological styles. We will start with a global survey of epistemological interlevel research, following Burton (1993). Then we will differentiate the picture, on the one hand in terms of the matrix of ontological levels and epistemological styles and on the other the concept of so-called interfield research, introduced by Darden and Maull (1977). This will lead to a matrix of four general kinds of co-operation in interfield research. In the present as well as in later sections we will select a number of examples and specific topics, mainly related to mind-body research.

6.1.1 *Interlevel Research*

Burton has given a plausible survey of possibilities of interlevel research. In general he takes levels just in the epistemological sense, but his examples sometimes reflect more specifically an ontological levelling or a confrontation of epistemological styles. He focusses on cognitive science and his examples, some of which will be indicated, reflect this focus, but his four types of 'interlevel' methodology are completely general.

> '[Cognitive] scientists must select the level or levels of focus for their research. But if a phenomenon is not explainable at its own level and in terms of its own implicit theory, to which level or levels should the researcher turn? It is easy to see why the reductionist methodology of seeking the lower, more comprehensive levels has often been a good one. But this is not always the case. Sometimes it is good methodology to explain lower level phenomena in terms of the higher. A third methodology is the investigation of phenomena at an intermediate level constrained by both higher and lower levels. A fourth methodology is the investigation of phenomena using models and/or methods borrowed from other disciplines at the same level.' (Burton, 1993, p. 77)

We propose to call these methodologies, or methods, *downward*, *upward*, *intermediate*, and *lateral*, of which the last one is of course not interlevel but intra- or monolevel. We will indicate some of Burton's examples, some that were discussed before, and some new ones.

For the downward method, Burton mentions the explanation of the Purkinje effect, an observational law regarding color vision, in terms of a causal mechanism starting from differences between retinal rods and cones. Burton stresses the interdependence of the two (ontological) levels, and hence, in his terms, that such research cannot be radically reductive, although it certainly is of a reductive nature. It even seems plausible that the color vision example can be reconstructed as a case of micro-reduction. However this may be, all unambiguous examples of micro-reduction presented in Chapter 3 illustrate the downward method, notably those analyzed in detail, viz. the kinetic explanation of the ideal gas law and the utilistic explanation of the collective goods law, but also, for instance, the explanations that have resulted from the atomic theory program. As the non-reductive, but nevertheless downward explanation of the interbreeding laws by Mendel's theory (also presented in Chapter 3) illustrates, the downward method may also lead to impressive though non-reductive results. Hence we do not subscribe to Burton's implicit claim in the quotation above that the downward method is always reductive.

For the upward method Burton refers to, among others, explanations of demographic phenomena, such as birth rate changes, in terms of intentional changes in reproductive behavior under socioeconomic pressures. However, from our point of view, this does not exclude the fact that the resulting, obviously, utilistic explanation is of a micro-reductive nature. Crucial in this respect is the nature of the auxiliary specification hypotheses concerning beliefs and desires, notably utilities. On average, people may in general prefer to have fewer children than natural sexual behavior results in, and in particular under bad socioeconomic conditions. In the first assumption a general birth rate decrease can be explained as the aggregate effect of the availability of easy contraceptives; in the second, a specific birth rate decrease can be explained by the start of an economic depression, as was the case in the US in the 1930s. In the latter case, socioeconomic phenomena are taken into account in the specification hypotheses of a utilistic, micro-reductive, explanation of a demographic phenomenon. In sum, the upward method may well lead to reductive explanations, viz. micro-reductive explanations using higher level auxiliary hypotheses, which function as macroscopic constraints.

Burton also refers to the need of the upward method in dealing with 'the case of the world class weight lifter who, through incredible concentration, achieves the explosive yet highly coordinated muscular contraction and movement in a world record lift'. Although there is a neurophysiological story to tell, this will be incomplete without some account of the mental aspects, such as the disciplined concentration. However, in contrast to what Burton seems to suggest, this does not exclude that a biophysical account of the mental (and bodily, see Section 6.3) aspects can be given. In other words, although the upward method is required, this may turn out to be a necessary detour for a full biophysical account. Here 'necessary detour' is meant in two senses, first, we could never find a full biophysical account without the detour, and second, the detour will not be eliminated but be conserved by such an account in one way or another. In other words, there is no short-cut for this type of detour, but it is a detour arising from two epistemological faces, a mental (and bodily) face and a biophysical face, where the latter can only be discovered in co-operation with the former. In Section 6.4 we will come across another example of the upward method, viz. the explanation of juvenile delinquency.

Let us now turn to the third method, the intermediate method, according to which a level in between may be postulated that is constrained by both higher and lower levels. Let us start with our favorite examples. To begin with, one of Dalton's main innovations of the atomic theory of the antiques was the postulation of the intermediate level of molecules, called by him 'compound atoms'. His molecules and molecular reactions constitute the glue level between the macro-level of substances and the base-level of individual atoms. Another typical example is the level of chromosomes as the intermediate level between that of genes and cells. This example has been analyzed in great detail by Darden and Maull (1977) and later expanded by Darden (1991); more specifically, the theory of chromosomes is characterized as an 'interfield theory' (see below) between genetics and cytology.

We turn now to Burton's two examples. First he localizes 'connectionism' between the neurobiological level and the 'rules and representations' level characteristic for the traditional approach to AI. Similarly, Smolensky (1988) has extensively elaborated the claim that connectionism should be seen as an intermediate level of analysis, between the symbolic and the neural level. Of course, the corresponding theories are also intermediate: the connectionist theory of the mind/brain is an intermediate theory between the symbolist theory of the mind and the neurobiological theory of the brain. The connectionist theory borrows from the symbolic theory its serial grasp on the competence of the human mind in language processing and problem solving, and from the neurobiological theory its parallel, adaptive picture of neuronal networks, as a model for learning behavior and other cognitive performances. As a second example Burton mentions Paul Churchland's (1990) neurocomputational theory of prototype activation, as an intermediate between (an extreme neurobiological version of) connectionism and prototype theory, as it is developed by Rosch (1981) and Lakoff (1987), following the 'family resemblance' theory of concepts initiated by Wittgenstein.

Although Burton's two examples are evident cases of 'intermediating' theories, they are not theories postulating an ontological level in between, as in the case of molecules and chromosomes. The connectionist 'intermediating' theory may well be a theory that is more realistic than the symbolic, without being as detailed (and untractable) as a neurobiological theory could be. In other words, as suggested by Smolensky (1988) and elaborated by Rantala and Vadén (1995) (see also Rantala, 1997), but phrased in our favorite terms, the symbolic theory may well be an idealized, limiting case, and the connectionist theory a concretization of it, going in the direction of a further neurobiological concretization. Hence, a sequence of epistemological levels and theories of increasing sophistication, pertaining to the individual level, described in three successive vocabularies: macro-terms, some kind of intermediate terms, and neurobiological aggregate terms.[1]

The fourth, lateral method, viz. the investigation of phenomena using models and/or methods borrowed from other disciplines at the same level, may lead to formal or even substantial analogies. Burton's main example is the extrapolation of Chomsky's theory of linguistic competence in theorizing about sociocultural systems, such as religious ritual systems, by Lawson and McCauley (1990). In Section 6.5 we will meet some other examples of this monolevel method of co-operation.

6.1.2 *Interfield Research*

Let us now introduce the two main divisions of epistemological levels we have met before. On the one hand there are the epistemological levels corresponding with the ontological levels of aggregation,[2] on the other hand we have the three main epistemological styles of description and explanation. With each ontological level and each epistemological style still more epistemological levels may be associated as well, generated, for instance, by a distinction between observation and theoretical terms. From now on in this chapter, 'levels' and 'styles' refer to ontological levels and epistemological styles, respectively, except when otherwise stated. In Figure 6.1 levels and styles have been put together in a matrix, the LS-matrix, where we restrict the attention to those levels that underlie the social level, hence neglecting e.g. cosmological levels (for a similar scheme, see

aggregation level	styles of description and explanation causal	functional	intentional	take-off level of disciplines
social	X	X	X	social sciences
individual	X	X	X	psychology
organs	X	X		
cells	X	X		biology
substances	X			
molecules	X			chemistry
atoms	X			
el. particles	X			physics

Figure 6.1 LS-matrix of levels and styles underlying the social level.

(Franssen, 1998)). Moreover, we have neglected several kinds of intermediate ontological levels.

As indicated in it, the LS-matrix enables us to localize the take-off level of disciplines. This is of course not to say that lower levels are not taken into consideration by these disciplines. For instance, biologists frequently do chemical research in order to functionally classify molecules, but the cell seems to be the smallest unit of reference for that purpose and called alive (or dead). Note, moreover, that the take-off level of a discipline may change in the course of history. Before the general acceptance of Dalton's atomic theory, (amounts of pure and mixed) substances were the point of departure of chemistry.

Let us confront the LS-matrix with a perspective on co-operative interlevel (and lateral) research that is more specific than that of Burton. Darden and Maull (1977, p. 44) introduced the concepts of a 'field' of research and of an 'interfield theory'.

> 'A field can be specified by reference to a focal problem, a domain consisting of 'facts' related to that problem, explanatory goals providing expectations as to how the problem is to be solved, special methods and techniques, and sometimes, but not always, laws and theories.'

Hence, a field is more or less the sum-total of the aims and the findings up to now of one or more research programs regarding one or more related domains of phenomena.

In many cases it will be possible to localize a field on a knot in the LS-matrix, that is, a combination of level and style. Note that there will be more than one field on a knot. Hence, those fields are comparable in level and style. However, a field need not be restricted to one level and one style. It may well have successfully integrated two or more levels and/or two or even three styles.

Co-operation between fields is usually directed at the study of the relations between them, called interrelations or interfield relations, to be grasped by 'interfield theories'. Darden and Maull (1978, p. 49) distinguish the following four relations, relations we have met already before in one way or another.

(1) *Localization relations*: 'A field may provide *a specification of the physical location* of an entity or process postulated in another field.' A specific localization relation is a part-whole relation, such as the localization of genes as parts of chromosomes, and chromosomes as parts of cells
(2) *Identity relations*: 'A field may provide the *physical nature* of an entity or process postulated in another field.' We have indicated already several identities. For example, biochemistry, more specifically, the DNA-theory, provided the molecular nature and structure of chromosomes, and hence, of genes, and subatomic physics indentified the atomic number of chemical elements with the number of protons of an atom.
(3) *Structure-function relations*: 'A field may investigate the *structure* of entities or processes, the *function* of which is investigated in another field.' Whereas biochemistry primarily deals with the function of certain molecules, physical chemistry deals with their structure.
(4) *Causal relations*: 'Fields may be linked *causally*, the entities postulated in one field providing the causes of effects investigated in the other.' Causal relations in this sense cover causal correlations, in the sense of Chapter 5, but may also apply to reductive relations; e.g. the kinetic theory gives a kind of causal explanation of the ideal gas law. Moreover, many structure-function relations are at the same time of a causal nature; e.g. physico-chemical analysis may provide causal explanations of functional patterns.

Research directed at interfield theories and claiming to make such relations between fields explicit is called 'interfield research'. Such interfield research may originate in different ways. It may originate from (a research program in) one of the two fields, in which case the co-operation does not (yet) take place in a social sense, and we get 'boundary breaking' research. It may also originate from the active co-operation of research programs in the two fields, in which case we get 'boundary bridging' research.[3] Interfield research may concern fields of two disciplines, and hence it may break or bridge the boundaries between the two disciplines. It is plausible to reserve the term 'interdisciplinary' research for the latter, co-operative case, in agreement with the general models of hierarchical and symmetric interdisciplinary co-operating research programs introduced in Chapter 1.

Interfield research may concern fields of the same or different levels and of the same or different styles.[4] Hence, we get a matrix of four types of interfield research, called the Interfield matrix (IF-matrix): mono-(or intra-)level versus multi-(or inter-) level and mono-(or intra-)style versus multi-(or inter-)style interfield research. In the IF-matrix below we have indicated for each type the section that is primarily devoted to that type.

Interfield (IF-)matrix: matrix of (sections devoted to) types of interfield research, theories, and programs

	monostyle	interstyle
monolevel	(Section 6.5)	(Section 6.2)
interlevel	(Section 6.3)	(Section 6.4)

Although we would prefer the 'intra/inter' qualifications of levels and styles, we choose for the 'mono/inter' qualifications because they seem to be less confusing in reading. Moreover, we do not choose for the 'mono/multi' qualifications because e.g. 'multilevel' does not suggest interest in the relation between the levels. In other words, interlevel and intersyle research can by definition be supposed to refer to multilevel and multistyle *interfield* research. On the other hand, 'monolevel research' or 'monostyle research' only refer to interfield research if 'interfield' is explicitly added.

As defined, interfield research aims at an interfield theory, that is, a theory relating phenomena of two different fields. Hence, interfield theories may also be localized in the IF-matrix. Research programs of which the core theory is an interfield theory are of course called 'interfield research programs', and may also be classified in the IF-matrix.

In the rest of this chapter we will deal with the four options of the IF-matrix, with emphasis on mind-body research. In Section 6.2, of which the details may be skipped in a first reading, we will study the interstyle relations. We will reconsider and elaborate on the causal explication of the functional style and introduce the functional explication of the intentional style. In Section 6.3 we will focus on interlevel, biophysical, mind-body research. More specifically, we will investigate the prospects for interlevel and monostyle correlation and reduction in this interfield research. In Section 6.4 we will broaden the scope by discussing some examples and models of interlevel and interstyle interfield research. Finally, in Section 6.5 we will add to the above mentioned Chomskyan example of Burton of the lateral method, some other examples of monolevel and monostyle research of an interdisciplinary nature.

We conclude this section with a global and abstract survey of interfield research, in which 'research' always indicates interfield research. We will start from the IF-matrix and place the emphasis on research combining one level or style with more than one style or level. Hence, let us first consider *interlevel-monostyle* research. Assuming the causal style, interlevel research amounts to causal decomposition of a causal process at the (relative) macro-level in terms of the action and interaction of the relevant micro-entities. The result may be that macro-concepts and macro-laws, if any, are correlated or even reduced to micro-concepts and micro-theories.

Similarly, assuming the functional style, interlevel research amounts to functional decomposition of a functional process at the (relative) macro-level in terms of the functions and 'co-operation' of the relevant micro-entities. In brain research this particular kind of monostyle-interlevel research is frequently called 'homuncular analysis'. The result may again be that macro-concepts and macro-laws are correlated or even reduced to micro-concepts and micro-theories.

Finally, assuming the intentional style, interlevel research amounts to research in the interrelation of individual and social intentions, where social intentions may or may not be the intended effect of individual intentions. Again, correlation and even reduction of concepts and laws is possible.

In Sections 6.3 and 6.4 we will extensively deal with the theoretical nature, reductive prospects, and examples of such interlevel research, with emphasis on mono- and interstyle versions.

Let us now look at *monolevel-interstyle* research. As a matter of fact, we have conducted a causal analysis of the search for functions in biology in Chapter 4. More specifically, our meaning explication amounted to a causal interpretation of biofunctional descriptions and explanations of items, that is, an interpretation in terms of their causal role, backed by the theory of evolution or at least by reproduction and survival considerations. Hence, apart from (monostyle) functional decomposition (see above), biofunctional research is interstyle; viz. the function items are specified in terms of their causal role. Moreover, for the selection version of that explication there is a kind of *general* interfield theory, derived from the theory of evolution, originating with Ruth Millikan (1984, 1993), and called *proper function theory*. In Section 6.2 we will elaborate this causal interpretation of the functional style and deal with some more complicated cases than those in Chapter 4.

It is important now to note here that the causal interpretation of functional items should not be confused with the material realization of these items. That is, the causal role(s) of a functional (state of an) item is one thing, the way in which that (state of the) item is materially realized in such a way that it plays these roles is another. In other words, functional items do not coincide with their causal roles.

Although functional research has become inherently interstyle, it is precisely the search for the material realization of causal roles of functional items, that creates interesting new kinds of interstyle, viz. causal-functional, research, for which the evolutionary interfield theory is not directly useful, and new specific interfield theories, of a mono- or interlevel nature, may be required.

Intentional research, as analyzed in Chapter 4, on the other hand, is by definition more or less interstyle. On the contrary, the decomposition of intentional descriptions and explanations of actions and attitudes typically was in the same terms of actions, beliefs and desires, that is, actions and propositional attitudes. In other words, whereas functional specification is primarily of an interstyle nature, intentional specification is normally monostyle. In this respect, intentional specification resembles causal specification, for that is also primarily monostyle. This situation raises two interstyle questions in view of the foregoing remarks about the functional style. First, can a functional, and hence indirectly causal, interpretation be given of intentional items? Second, how are intentional items materially realized? In Section 6.2 we will discuss these questions in some detail and, regarding the first question, we will see that proper function theory is again a plausible general interfield theory, now providing a (selection version of a) functional, and hence causal, explication of the intentional style. However, this explication is not as straightforward as it is in the case of the causal interpretation of the functional style. The other question, about the material realization of actions and propositional attitudes, will bring us to the heart of interlevel mind-body research, to be dealt with in Sections 6.3 and 6.4.[5]

6.2 EXPLICATION OF THE RELATIONS BETWEEN THE STYLES

Introduction

In the previous chapter we announced already that Millikan (1984, 1993) has introduced a way to integrate the three main styles of thought. She has argued, among other things, that biological items, such as organs, states and behavioral traits are classified and explained in terms of their so-called proper function, and that (folk-)psychological categories, such as beliefs and desires, also have proper functions. Mackor (1995, 1997) and Guichard (1995, 1997) have elaborated Millikan's analysis. Mackor has argued that the proper function theory can be extended to analyze social, rule guided behavior, providing in this way a strong argument against the alleged autonomy of psychology and the social sciences. Guichard has argued that proper function theory provides an adequate interfield approach to stress research and makes possible substantial co-operation between certain biological and psychological stress research programs.

This section heavily leans on Millikan's explication of functional terminology in terms of reproduction and survival in 'the selection environment'. We leave the question open whether there is a meaningful translation of this approach in terms of reproduction and survival in 'the present environment'.

We will first address the functional style and show that our analysis of functional descriptions and explanations, amounts to a partly abbreviated version of Millikan's causal explication of the functional style. Moreover, we will show in detail how our analysis can be applied to innate or learned 'representational' behavior. Then we will address the intentional style and show how our (monostyle) analysis of intentional descriptions and explanations can be functionally explicated along the lines suggested by Millikan. The consequence is that mental affairs, at least some important examples, such as beliefs, desires and actions, can ultimately be causally explicated, as we will see, even in terms of specific causal explanations.

In both cases it would be wrong to conclude that the functional and intentional styles have been eliminated in favor of the causal style, for their specific nature, and hence their usefulness, remains intact. Hence, the explications amount to conservative explications of the styles, both leaving open the question of their material realization. This section will draw heavily upon Guichard's detailed causal reconstruction of some examples(Guichard, 1997, Chapter 12). Moreover, it profits from Mackor's conceptual analysis (Mackor, 1997, Chapter 3 and 4).

6.2.1 *Causal Explication of the Functional Style*

6.2.1.1 *Elementary Analysis and Cases*

Recall that in Section 4.2 we proposed the following strong (selection) explication of the specific functional statement $FH(x, y, z)$, according to which the (or one) function of trait y of x-organisms is its contribution to trait (process /phenomenon) z in the selection environment, was:

$$H(x, y) \text{ and } P_s(x, y, z) \text{ and } P_s(x, z, rs) \text{ and } CE_s(x, y, z)$$

It is important to recall that the non-trivial components either concern (positive) causal correlations, viz. $P_s(x, y, z)$ and $P_s(x, z, r\&s)$, or a causal-historical claim, viz. $CE_s(x, y, z)$. In this explication the causal-historical component, $CE_s(x, y, z)$, according to which $P_s(x, y, z)$ and $P_s(x, y, r\&s)$ were evolutionary effective for $H(x, y)$, amounts to an abbreviated version of the 'reproductive history' – conditions Millikan (1984, 1993) has formulated for what she calls 'proper functions' or 'teleofunctions' (substituting our symbols):

> '... for an item y to have a function z as a 'proper function', it is necessary (and close to sufficient) that one of these two conditions should hold. (1) y originated as a ''reproduction'' (to give one example, as a copy, or a copy of a copy) of some prior item or items that, *due* in part to possession of the properties reproduced, have actually performed z in the past, and y exists because (causally historically because) of this or these performances. (2) y originated as the product of some prior device that, given its circumstances, had performance of z as a proper function and that, under those circumstances, normally causes z to be performed by *means* of producing an item like y.' (Millikan, 1993, pp. 13–4)

Guichard and Mackor argue that this complicated 'formula' amounts to, using Guichard's version, assuming that we deal with x-organisms: (1) ancestors of x-organisms possessed y; (2) it is the property z of y that caused the proliferation of those ancestors; (3) and it is due to this proliferation that x now possesses y. It is in this decomposition that all aspects of $FH(x, y, z)$ are easily recognizable. Accordingly, the last component $(CE_s(x, y, z))$ and hence the second causal correlation $(P_s(x, z, rs))$ are instantiations/illustrations of the theory of evolution. Millikan's proper function theory refines the 'reproductive history' interfield theory of proper functions by adding an analysis in terms of normal conditions and normal performance. As we will see, in this and the next subsections, such conditions and qualifications can be identified with causal correlations in our meaning analysis. Our analysis puts the emphasis on the causal correlations composing a (connected) causal chain, implying that the item was causally positive for reproduction and survival in the selection environment. It is important to note that 'normal' in the sense of Millikan, as in 'normal conditions' and 'normal performance', does not mean 'usually', but 'appropriate' for the performance of the function. Hence, normal conditions in the present, proper function sense may well be rare, in contrast to the normal conditions in the regular sense as, for instance, referred to in the context of explanation by causal specification. Normal conditions in the regular sense are supposed to obtain as a rule.

As we have seen in Section 4.2, in its most simple form the chain may be reconstrued as consisting of two causal correlations in which case they are said to represent the proximate and the ultimate function of the behavior. For example, for the claim that the function of the fanning movement by sticklebacks is to supply the eggs with oxygen, it was enough to distinguish the proximate and ultimate functions of the fanning movement: the fanning movement causes an oxygen supply, at least normally, and the latter is causally positive for reproduction and survival, again at least normally. As suggested by the formulation of this example, both correlations are what Millikan would call normal conditions for the proper functioning of the fanning movement.

6.2.1.2 *Analysis of Some Complex Examples*

In cases in which more than two causal correlations can be distinguished it is plausible to speak of an intermediate or *distal function*. For instance, an internal system of chameleons has the proximate function of adapting the color of the skin to the color of the surface the chameleon is sitting on. The distal function of this system is to become less visible to predators, and its ultimate function is to benefit species reproduction and survival.

In the case of, innate or learned, representational behavior it seems that we have always to introduce a third causal correlation component, creating room for talking about the proximate, the distal and the ultimate function. We will first deal with an example of innate representational, but non-communicative, behavior, viz. magnetosome bacteria, then with an example of innate communicative (hence representational) behavior, viz. bee dances, and finally with a case of learned representational behavior, viz. soap-avoidance by rats.[6] We will concentrate on the causal correlations, leaving aside the trivial component and the causal-historical component, which is nevertheless always presupposed. We will also adopt Millikan's insight that representations as used by animals usually have an indicative and an imperative function at the same time. In human beings (and higher animals?) these two functions of representations have become separated into beliefs and desires.

Magnetosome bacteria have little magnets, called magnetosomes, that not only point in the direction of the geomagnetic north, but more importantly in the regions where they happen to live, north is also the direction away from oxygen rich water, a kind of water that is not good for them. A causal mechanism ensures that the bacteria swim in the direction of oxygen-poor water. Hence, we may say that the (proper) function of the magnetosomes is to cause the bacteria to swim in the direction of oxygen-poor water. The causal components can be arranged in the following causal chain: (1) the direction of oxygen-poor water is causally correlated with the direction of the geomagnetic field, (2) which causes the direction indicated by the magnetosomes; (3) which causes the bacteria to swim in that direction, and hence (3′) in the direction of oxygen-poor water, (4) which is causally positive for their reproduction and survival.

The following remarks link this causal analysis with Millikan's terminology. The correlations (1), (2) and (4) are called the *normal conditions* under which the magnetesomes can fulfil their proper function; (1) is a matter of an (organism-) *external* nature, whereas (2) and (4) are primarily a matter of an *internal* nature. Fulfillment of correlation (3) is called *normal performance*. Note that normal conditions and normal performance require in their turn normal conditions in the regular sense. The *proximate* (representational) function of the magnetosomes is twofold: an indicative one, indicating the direction of the magnetic field (2) and an imperative one, causing the bacteria to swim in that direction (3). The *distal* (representational) function of the magnetosomes is also twofold: indicative, viz. indicating the direction of oxygen-poor water (a consequence of (1) and (2)), and imperative, viz. causing the bacteria to swim in that direction (3′). Of course, the ultimate function of the magnetosomes remains (indirectly) the fact that they contribute to reproduction and survival (4).

A similar, even more simple, analysis can be given of the case of innate communicative behavior: the function of bee dances is to cause other bees to fly in the direction of nectar. The causal chain now amounts to: (1) the direction of nectar causes a certain dance pattern, (2) which causes spectator bees to fly in the indicated direction, hence (2′) the direction of nectar, (3) which is causally positive for reproduction and survival.

Now the correlations (2′) and (3) are the normal conditions, both of an external nature, and (1) and (2) express normal performance. The proximate (representational) function of the bee dance is twofold: an indicative one, indicating a certain direction, implied by (1), and an imperative one, causing the spectator bees to fly in that direction (2). Again there is also a twofold distal (representational) function of the bee dance, viz. indicating the direction of nectar (1) and causing the bees to fly in that direction (2′). The ultimate function of the bee dances is to contribute to reproduction and survival (3).

Let us now turn to a case of learned behavior, viz. soap-avoidance by rats, where we will refrain from spelling out the twofold character of the representational nature of the relevant rules. Here we may distinguish three behavioral rules, the distal rule (DR) to avoid eating poisonous substances, the proximate rule (PR) to avoid eating illness-generating substances, and the derived proximate rule (dPR), avoid eating substances tasting like soap. DR is evidently causally positive for reproduction and survival, and hence PR as well, since illness generating substances may well be poisonous. PR is an innate learning rule, in fact an instance of a general law called the law of effect, and dPR is learned behavior based on PR, that is, dPR is an instantiation or 'application' of PR for all rats that have once eaten soap and become ill afterwards (which may or may not have been caused by the soap). The core of the functional explanation of dPR is the following causal chain: soap consumption (probably) caused illness, which caused, by PR, soap avoidance. The functional explanation of PR is governed by the following causal chain: avoiding illness-generating substances is causally positive for avoiding poisonous substances, i.e. DR, which is causally positive for reproduction and survival. The important point of the example is that the learned behavior depends on contingencies. Hence, the relevant behavior is not shown by all rats, but only by those that happen to have become ill from soap consumption.

6.2.1.3 *General Considerations*

The examples above illustrate the causal explication of functional concepts as they occur in specific (bio-)functional statements and hence in functional descriptions and explanations. In general, the search for a functional explanation amounts to the specification of the causal correlations, providing the normal conditions, such that a connected causal chain arises. The number of causal links that can be distinguished is partly a matter of taste, but also depends on the nature of the specific case. As we have seen, when representations are involved there seem necessarily to be more than two relevant causal components. Increasing the number of causal components influences, of course, the number of instantiations of the conditions, and hence the number of 'positive instances' of the proper functioning. That is, the number of successfully pumping hearts, pointing magnetosomes, dancing bees, and soap-avoiding rats highly depends on the nature of these conditions. Recall that, although these conditions are called normal relative to the proper functioning, it does not mean that they frequently, let alone usually, apply.

Here two divisions can be made that determine to a large extent the number of function instantiations. First, whereas internal conditions always seem to play a substantive role, this is not the case for external conditions, e.g. in the case of the functioning of the heart as opposed to all other examples. Secondly, the conditions for proper functions may or may not be uniformly distributed across the species. This is particularly the case for external conditions that may or may not form part of the learning history of individuals, e.g. in the rat example. Only when it is part of the learning history can the relevant learning rule be triggered. But also internal conditions can be non-uniformly distributed within species. Think in the first place about differences between males and females. But also other conditions can vary among people, e.g. hormone levels, the quality of the bones, and, who knows, the neurophysiological quality of the brains. Such differences give rise to different kinds of function and/or different degrees and ways of function performing. In line with the philosophy of mind literature, Guichard calls functional properties depending on nonuniformly distributed external conditions 'relational' and all others 'monadic'. However, using the terminology of systems theory, notably input, state and output, we will call items depending on external non-uniform conditions, input-dependent, and items depending on internal non-uniform conditions, state-dependent. In this terminology we can say that the functional laws, that is, laws dealing with functional states, state transitions and outputs and depending on both types of conditions are system laws in the standard sense. That is, functional state transitions are causally determined by inputs and previous states, as is also the case for functional outputs, that is, behavior. In this terminology it is evident that the number of occurrences of certain functional states, state transitions and outputs may vary from occurring in all members of a species at all times (of their life), e.g. heart beat, to occurring once in the whole lifetime of one individual, e.g. avoidance of eating of a very rare substance. The latter possibility corresponds to a feature that frequently holds for intentional states as well; that is, they may be very rare. So let us see how the intentional style can be explicated in functional terms.

6.2.2 *Functional and Causal Explication of the Intentional Style*

Recall now that we proposed in Section 4.1 the following strong explication of the specific intentional statement $IP(x, y, z)$, according to which x performed action y with the intention of approaching goal z,

$$P(x, y) \text{ and } D(x, z) \text{ and } BU(x, y, z) \text{ and } CE(x, y, z)$$

Of the non-trivial components only the last one, $CE(x, y, z)$, according to which $D(x, z)$ and $BU(x, y, z)$ were causally effective for $P(x, y)$, is explicitly causal. The desire and belief component ($D(x, z)$ and $BU(x, y, z)$, respectively) are not explicitly functional, let alone causal. Hence, in contrast to the explicitly causal, hence interstyle, nature of the initial explication of specific functional statements, the initial explication of specific intentional statements chiefly remains within the intentional style. But this does not, of course, exclude the possibility of a functional or even causal explication of that style. For a causal explication of a specific intentional statement and hence of intentional description and explanation, it is at least necessary to explicate the belief and desire component in functional language. In the foregoing subsection we have seen that it then is, at least in principle, possible to reduce them to causal language. So we turn to their explication in functional language.

Let us first intuitively compare the soap example of Subsection 6.2.1.2 with the example of Jane opening the window in order to cool down the room. As Guichard notes, after Millikan: 'One cannot say that rats don't *wish* to eat some stuff because they *believe* it to be soap: with rats indicative and imperative representations aren't separated. With humans they are, and one does ascribe beliefs and desires.' Hence, for instance, one says of Jane that she opens the window because she believes that it will cool down the room and that is what she wants.

According to Millikan, for humans it is important that beliefs and desires can be distinguished, and corresponding belief- and desire-generating mechanisms. We should first realize that having separate beliefs and desires, and hence having separate mechanisms, can be functionally explained. Desires are, as a rule, such that their fulfillment is causally positive for reproduction and survival.[7] Hence, the proper function of desires is their own fulfillment, which fits well in our general scheme: a desire is causally positive for its fulfillment (proximate component) and the latter was already assumed to be, as a rule, causally positive for reproduction and survival (ultimate component). For beliefs the situation is somewhat more complex. The relevant mechanism generates beliefs, directly or indirectly via other beliefs, which do not have specific proper functions, but which are helpful for the proper function of desires; viz. they guide the realization of such desires by, as a rule, appropriate actions. For that purpose it will frequently be helpful that beliefs are true, but this need not always be the case (see below). If beliefs would be false most of the time they would not fulfil their action guiding function. Hence, the mechanism will be such that the beliefs it produces are frequently sufficiently correct, at least as far as their relevant consequences are concerned.

It should be noted, however, that for certain kinds of belief it is better that they are false, for others it does not matter. A well-known common sense example of the first kind is planning, in particular of non-routine work, for instance scientific research, including writing it down in order to publish it. Productive scholars are, as a rule, at least for themselves, if not for their PhD-students, much too optimistic in planning. If they had believed in advance that it would take so much more time, they would frequently not have started the realization of the plan at all. Such an awareness is in fact what frequently inhibits unproductive scholars: their realistic expectation of the amount of work involved deters them from starting in the first place. Tiger (1980) has called the phenomenon of exaggerated optimism the 'self-serving bias' and has given an evolutionary basis for it. The paradigm, but at least as controversial, example of the second kind is religious belief. Irrespective of whether such beliefs are true or false, they are for many people very stimulating in realizing desires. As a matter of fact, this view is the core of the evolutionary explanation of the development of more or less similar kinds of religious belief all over the world (Wenegrat, 1990; Crisler, 1994). Of course, the position does not exclude the recognition that religious belief may occasionally or perhaps even frequently be disfunctional for people.

In sum, according to Millikan, the proper function of beliefs is to be causally positive for the realization of desires. Depending on the kind of belief, their being true, or their being false, may be helpful for that purpose, or its truth-value may be irrelevant.

Finally, it is now easy to functionally analyze actions. The function of an action usually coincides with its intention; viz. they help bring about the realization of a desire.

In this way we have explicated the intentional language of beliefs, desires and actions into functional language and hence into causal language, in view of the causal explication of the functional style in the preceding subsection. The theory of evolution is, as in the previous case, the cornerstone of the relevant interfield theory. The latter is an adapted extension of the proper function theory to the intentional domain.

Not only an explication of the intentional style completely in causal language is now within our reach, it is even possible to make a straightforward link with *specific* causal statements and explanations as explicated in Chapter 4: usually an action is not only functional for the realization of a certain desire but the desire is even the 'abnormal factor' for causing the action (where 'abnormal' is meant in the regular sense of not being the case almost all the time). For instance, when someone entering Jane's room asks for a causal explanation of the open window, Jane or a witness may merely refer to her desire for a cool room as the 'intervening causal factor', taking the other relevant causal factors for granted. In sum, Jane's opening the window can be intentionally, functionally and causally explained in terms of Jane's desire to cool down the room. Note in addition that the cooling down of the room can also be explained by causal specification, viz. viewing Jane's action as the intervening event and taking the other relevant causal factors for granted. Hence, an action cannot only be intentionally, functionally and causally explained, but a causal explanation of its indirect causal effect can also be provided, where causal explanation refers in both cases to explanation by causal specification.

As at the end of the preceding subsection, it is again plausible to introduce some systems theory language. Intentional states, i.e. propositional attitudes such as beliefs and desires, may or may not be (specific) input-dependent and state-dependent. Intentional laws, that is, laws dealing with intentional states (beliefs and desires), state transitions and outputs (actions, including speech acts), are (quasi-) system laws. That is, intentional state transitions are causally determined by their inputs and previous states, as is also the case for intentional outputs. Even more than in the functional case it is evident from this perspective that the number of occurrences of certain intentional states, state transitions and outputs may be very high or very low, completely in agreement with our intuitive judgement.

On the level of individuals this results in the following relations between causal, functional and intentional terms. Individuals have causal properties, some (or most) of which are functional. In the latter case we may at least distinguish between proximate and ultimate functional properties, and sometimes intermediating distal functional properties. The corresponding causal properties may be named in the same way. As we have seen, intentional properties are, as a rule, directly or indirectly, functional. Hence, intentional properties are causal properties of sorts. However, not every instance of a functional property needs to be functional in any given situation, in view of the nature of the mechanisms generating desires, beliefs and actions, it is even less plausible that every ('activated') instance of an intentional property is functional in any given situation. However, even in that case it is nevertheless causally active in some way or other.

The last point brings us to the core of the next section. The causal explication of (bio-)functional statements and the related functional interpretation of intentional phenomena does not yet answer the following two questions. First, if the function of an item is known, it remains to be seen how the item performs that function, that is, how the function is materially realized by causal mechanisms in the body. For example, one may know that the heartbeat serves to circulate blood through the body, but not yet know how the heart performs that function, that is, what mechanisms inside and outside the heart cause the heart to beat. We will not study this question about causal explanation any further here. See (Wouters, 1999) for several detailed examples. The second question relates to intentional phenomena. Beliefs, desires and actions may be functional, at least as a rule, and conscious beliefs and desires may be assumed to be always causally active, but how are conscious mental states materially realized such that they are always causally active in one way or another, and occasionally functional? We will address the general presuppositions and prospects of biophysical (interlevel) research into this question in the next section.

In sum, we have argued in this section that the proper function theory provides a general interfield theory between the causal and (bio-)functional style, as well as between the intentional and the functional style. However, this does not yet provide a general, let alone specific, interlevel theory, relating properties of individuals to lower, or for that matter, higher levels of aggregation. As we will see, (bio-) physicalism provides such a theory by primarily, but not exclusively, relating the individual mental (and bodily!) properties of individuals to lower level entities and their interactions.

6.3 BIOPHYSICAL MIND-BODY INTERLEVEL RESEARCH

Introduction

In this section we will concentrate on interlevel mind-body research in the form of symmetric co-operation between research programs or even disciplines, hence boundary-bridging interlevel research. More specifically, we will deal with interlevel reduction and correlation of mind-body concepts and with interlevel explanation and reduction of mind-body laws. The guiding perspective is the material (more specifically, biophysical) realization of such concepts and laws in micro-states and -processes. First we will do so from a general perspective, then with emphasis on the functional style.

In what amounts to an interdisciplinary dream, we will apply the mixed strategy of moderate reductionism or moderate holism (see Chapter 1 and 5) to the non-eliminative reduction that might result from the co-operation between psychology and (neuro-)physiology.

Schwartz (1991) has rightly noted that the antireductionist arguments of many philosophers of psychology (he mentions Baker, Fodor, Davidson) are motivated by the worry that successful reduction would eliminate rather than conserve the mental realm, and points out in his specific terms, but in the same spirit as ours, that some paradigm natural science examples of reduction of concepts, such as water, genes and tables, are not at all eliminative. Indeed, 'Philosophy of psychology needs more detailed attention to issues in natural science which serve as analogies for reduction of the mental' (p. 203). Our examples of non-eliminative concept and law reduction in Chapter 3 and 5 may help in this respect.

Another general point is that philosophers seem to favor discussions involving the reduction of whole disciplines. Possibilities discussed include the reduction of sociology and economics to psychology, the reduction of psychology to (neuro-) physiology and biology, the reduction of biology to chemistry and physics and, finally, the reduction of chemistry to physics. One does not even hesitate to concentrate on the ultimate question of whether all sciences can be reduced to physics. Although providing very different answers, both Fodor (1975) and (Paul) Churchland (1990) champion such an approach. From our point of view in general, and the mixed strategy in particular, such discussions are rather pretentious. The really interesting question is whether some mental concepts or even mental laws can be reduced, straightforwardly or approximately, to neurophysiological concepts and laws.

6.3.1 *Interlevel Reduction and Correlation of Mind-Body Concepts and Laws*

To state the reductionist ambition more precisely, let us start from the common sense division of properties (events, states, dispositions, etc.) of human individuals into bodily and mental properties. That is, some individual properties are purely or largely bodily, e.g. weight, temperature, strength etc. Some others are typically mental, e.g. beliefs and desires. And there are many mixed properties and the like in between: actions, including speech acts, pains, etc. Note that mental states need not be conscious states, e.g. passive memory. In common sense terms all these properties are causally active (or at least causally reactive) and not only within their own sphere (mind/mind and body/body interactions) but also between the spheres (mind/body interactions). Many of these properties can be described in functional, and many of them in intentional terms, and these ways of description may well go together.

To describe and explain these common sense ideas, classical mind-body dualism postulated two kinds of substances, which can also interact. The modern, prevailing biophysical point of view is quite different. On lower levels of aggregation or organization everything is supposed to be of a biophysical (and perhaps functional) nature, such that all individual properties supervene as macro-properties of the individual conceived as an aggregate of a biophysical nature. In other words, the macro-properties are supposed to be (materially) realized by biophysical micro-states. Hence, and this is the core idea of our approach, *mental, bodily and mixed terms can be treated in the same way, and that way is such that their being causally interactive is unproblematic.*

It is important to note that actual biophysical states of human beings have (distributed) traces of earlier states and inputs of one kind or another. This fact is not only relevant for many bodily properties, e.g. specific muscular features of sportsmen, but also for many mental properties, e.g. what one believes and, very important, how that is realized. More specifically, it is plausible to assume that the specific learning history of an individual has left its biophysical traces in such a way that the biophysical realization of the same belief, may be different between different people. Moreover, the biophysical realization of the same belief of one and the same individual may change in time, with the consequence, for instance, that the strength of the belief changes. Long-term learning is nowadays conceived not only as a matter of electrical and chemical processes, within and between neurons, but also as a matter of morphological changes of the neurons (Gispen, 1993; Kandel, 2000). Of course, beliefs may be biophysically realized by a certain type of (biophysical) micro-states, and for a particular type of belief, e.g. classificatory, a subtype may be involved. But, there is no reason to suppose that the same belief has always to be realized in precisely the same way within one individual, let alone across individuals.

Hence, the reduction ambition can now be stated more clearly. We distinguish B(ody)-properties/-concepts/-terms, such as weight, M(ental)-properties, such as having a certain belief, and MB-properties, such as sexual arousal, without supposing that the division is always clear. The potential ambiguity will turn out to be of no importance. We will talk about B-laws (-regularities/ quasi-laws) and M-laws, depending on whether they relate only B-properties or only M-properties. However, it remains a matter of dispute whether (pure) M-laws exist at all. MB-laws are laws that relate M- and B-properties and/or MB-properties. Note that in our approach, MB-laws are as a rule or even always correlations, MB-correlations for short. That is, the question of whether there are MB-identities is irrelevant in our approach. In this terminology the really interesting question is whether some M- or MB-concepts or even M- or MB-laws can be reduced, straightforwardly or approximately, to biophysical concepts and theories. As long as reduction of, say, an M-concept is not successful, it may well be that we can find one or more MB-correlations and 'ontological' correlations, that is, correlations between the M-concept and micro-concepts. (Recall that correlations may be causal or ontological correlations, in the latter case they are incomplete, e.g. of a part-whole nature, such that they cannot be considered as ontological identities.) For instance, neuro-imaging techniques may show that certain areas and processes of the brain are specifically related to the M-concept. Such MB-correlations may well be a crucial step for an ultimate reduction of the M-concept. Moreover, electrophysiological, neuro-anatomical, and neurochemical experiments, for instance, may show that certain types of neurons or neurotransmitters are involved in M. Such ontological correlations may well be part of an ultimate reduction. Similarly, as long as the reduction of, say, an M-law is not successful, it may well be that we can find a micro-theory that enables the explanation, together with suitable concept correlations, of the M-law.

In the following analysis, we will essentially apply the findings of Chapters 3 and 5. In order to simplify the presentation we will restrict the attention to MB-concepts and MB-laws, relating MB-concepts, since the story is essentially the same for pure M-concepts and M-laws, if any exist, for B-concepts and B-laws, and for MB-laws relating pure M- and B-concepts. Let us concentrate for a while on the macro-concepts and let us assume a finite number of families of MB-types, i.e. sets of mutually exclusive and together exhaustive, monadic MB-predicates, each giving rise to an MB-representation function. The MB-predicates at the start are supposed to be conceived as types in the sense that their application is stable and strongly intersubjective and/or because they are lawfully correlated with other types.

Starting with some families of MB-predicates, it is always possible to define in terms of the available ones many other (families of) MB-predicates, by forming conjunctions and disjunctions of predicates. Defined MB-predicates themselves may or may not be considered as MB-types, for the same reasons as the

MB-predicates from the start. If, as a matter of empirical fact, a certain MB-type is correlated with another one belonging to another family, this implies that there are MB-laws.[8]

Let us now conceive individuals, humans and other animals, as aggregates, to be described in terms derived from base-terms. It is important to distinguish between the genuine base-concepts for the constituents, e.g. molecules or cells, etc. and their mutual relations, and the representation of aggregates of these constituents, e.g. substances and tissues. The latter may be done by construing set-theoretic micro-structures in terms of the (biophysical) base-concepts. The micro-structures will from now on to be indicated as CS-structures, because the micro- or base-entities are supposed to be *c*ells, in particular neurons, and (micro-pieces or -amounts of) *s*ubstances. The latter are included in order to be able to take non-biological micro-entities, e.g. in the environment, into account. Each CS-structure represents a conceptual possibility of a state of an aggregate. According to a first, very strong idealization, depicted in Figure 6.2, it will specify a base set of neurons (fixed for one person, to be concretized), an asymmetric connection relation (also fixed for one person, also to be concretized), and a time-related excitation or activation function of neurons (with only two values, also to be concretized).

$\langle N,T,c,a\rangle$ is a neuro-structure iff:

N: the set of neurons of an organism
T: a discrete time-interval (in milliseconds)
c: $\subseteq N{\times}N$: (directed) connection relation
$c(n,n')$: neuron n is (via an axon) connected with (dendrite of) neuron n'
a: $N{\times}T \dashrightarrow \{0,1\}$: activation function
$a(n,t)=1\,(0)$: neuron n is at t (not) activated

BASIC (OBSERVATIONAL) LAW:
if $a(n,t)$ then $a(n', t+1)$ for every n' such that $c(n,n')$

Figure 6.2 *A neuro-structure.*

In Figure 6.2 the basic observational law is also indicated in a super-idealized way. Such a structure is a token of as many (disjunctions of) CS-types as can be meaningfully defined as sets of structures, using aggregate concepts, i.e. concepts characterizing a certain aspect of the structure, e.g. the total number of neurons, the ratio of activated neurons in a certain layer of neurons, etc. Moreover, the idea is of course that many different, though probably in some way or other related

CS-structures, may realize the same MB-concept, e.g. the same belief or desire, called the realization-class of that MB-concept.

Let us now formulate some possible reduction results concerning some MB-families that would be considered as successes. For this purpose we have to assume (see Chapter 5) at least the Token-Identity Hypothesis; i.e. every state of an individual can be represented by an MB-type of each MB-family as well as by a CS-structure. Additionally the Realization Hypothesis has to be adopted; i.e. every CS-structure uniquely determines, as a matter of ontological fact, an MB-type for each MB-family. Recall that the Realization Hypothesis implies that each structure can be represented as a member of the realization-class of precisely one MB-property of each family. In other words, that structure is a token of the realization-types of those MB-properties.

From the general analysis in Chapter 5 it follows that we now can distinguish three degrees of concept reduction (see Table 5.1). A result of the first degree is the quasi-type-type reduction of some MB-type, say an MB-state. This result presupposes that it is possible to characterize a set of CS-structures as the realization-class of that MB-state. Assuming that the CS-structure representation can be experimentally established, the reduction enables the prediction of the MB-state on the basis of the CS-structure representation (quasi-reduction) and, conversely, being in an MB-state predicts that the CS-structure representation belongs to the realization-class of that MB-state. Of course, given that the realization-class is not exclusively defined in CS-terms, for it is by definition 'MB-induced', both kinds of prediction concern CS-structure representations that are (very) similar to those that were used to characterize the realization-class.

As soon as a realization-class can also be characterized independently in CS-terms, in the form of a disjunction of CS-(micro-)types, exemplifying a case of the Multiple Type Identity Hypothesis, we get the possibility of a one-many or multiple type-type reduction of an MB-state: the second degree of reduction. A second degree reduction enables the deduction of the realization-type and hence the MB-state when one knows that the CS-structure of the state belongs to the disjunction of the corresponding CS-(micro-)types. The special case that the realization-type corresponds to just one CS-type, exemplifying the (singular) Type Identity Hypothesis, is a third degree result: the one-one or singular type-type reduction of an MB-state.

The foregoing classification immediately leads to three degrees of reduction of laws. First, if two MB-types are lawfully connected and both can be quasi-reduced, it may be possible that the law can be reduced by a theory in a weaker sense than occurs in reduction by identification according to the general model in Chapter 3, viz. that, starting from one or more specific theory applications, the transformation step consists of replacing realization-type attributions by macro-type attributions. We call this reduction of a law of the first degree, or quasi-reduction. If two MB-types are one-many type-type reducible and lawfully connected, there must be a lawful connection of unions of CS-types. Formally, this must even be the case in such a way that the first can be reduced to the second, in the sense of reduction by identification, based on the appropriate Multiple Type Identity Hypotheses as presented in Chapter 3. We call this reduction of a law of the second degree, or multiple reduction of a law. In the special case that MB-types can be reduced to proper CS-types, the law reduction is a result of the third degree, called singular or strong reduction of a law. All discussed degrees of reduction of concepts and laws concern straightforward non-eliminative reduction. Of course, for all cases we may leave room for at least some approximation, apart from the cases of quasi-reduction, since approximation is conceptually impossible in these cases.[9]

According to the mixed strategy every reduction result counts. Although one may be skeptical about overall reduction, it seems defensible to be optimistic for successful cases. Successes of concept reduction of the first degree largely depend on our experimental techniques, and it seems already possible to reconstruct some results of present-day neurophysiological, neuro-anatomical and neurochemical techniques as successes of this kind, e.g. certain neurobiological findings about learning and memory (Gispen, 1993; Kandel, 2000). In 2000 Eric Kandel got (part of) the Nobel prize for physiology precisely for this kind of research. However, the possibility of successes of the second and third degree largely depends on the existence of stable MB-types. That some MB-types are stable is not a very optimistic assumption. For example, certain syndromes seem rather stable. In many cases, however, MB-types seem at most to be multiply realizable. For example, PET-scan experiments, reported by Pardo et al. (1993), suggest that sadness is realized in the brain metabolism in different people in rather different ways.[10] Hence, concept reduction of the third degree may remain an exception.[11] To be sure, in all three cases the core of the reduction is the establishment of the Token-Identity Hypothesis, the Realization Hypothesis, and, in the second and the third case, the relevant Type Identity Hypothesis. As we have argued in Section 5.1 for the singular case, following Causey, this will not be possible in a conclusive way, but it may be possible in a convincing way.

Successes of law reduction of one of the three degrees of course largely depend on the existence of MB-laws in terms of stable MB-types. Again, in particular, results of the lower degrees are very well conceivable. Think, e.g. of laws of (mental) aging and association. Neurophysiological brain models and connectionist mind-brain models are at least able to simulate something like aging and association of people with and without the Alzheimer disease. Moreover, the better results of certain standardized (talking) therapies compared with certain tailor-made treatments in psychotherapy strongly suggest lawlike features of mental disorders. They are reported by Nelson-Gray et al. (1989) in their studies of depression, by Schulte et al. (1992) in their research into phobic disorders, and by Emmelkamp et al. (1994) in their work about obsessive-compulsive disorders. Of course, this is not in conflict with the surprising fact that certain other tailor-made treatments, such as re-incarnation therapy, turn out to be even more effective than certain standardized therapies, at least for lots of people (Gomperts, 1996). The effectiveness of the former is likely to be due to non-specific aspects of such therapies.

The suggested forms of reduction of MB-concepts and MB-laws are of course ideals. In practice we will have to accept that we will not in most cases come further than (ontological) correlation of MB-concepts with aggregate CS-concepts, and the explanation of MB-laws by CS-theories and appropriate correlations. But such results are impressive in themselves, and they will remain the target of future reductive research.

As suggested before, the situation is essentially similar for B-concepts and B-laws, and M-concepts and M-laws. However, MB-concepts and MB-laws may well be conceived as paradigms, with purely mental or purely bodily concepts and laws as extreme cases. The examples of phobic and obsessive-compulsive disorders just mentioned typically illustrate MB-concepts and MB-laws. At least their quasi-reduction may well be possible.

6.3.2 *Functional Versions*

Recall that we have described in Chapter 5 the main forms of correlation and reduction of concepts (three types of each) and in Chapter 3 the main forms of explanation and reduction of laws by theories (five basic steps, among which three are called reductive in the literature). Correlation and reduction of concepts may or may not be (ontologically) interlevel, similarly for the identification and the correlation step. The aggregation step is by definition interlevel, while the remaining ones, viz. the application and the approximation step, are by definition monolevel.

Some of the interlevel examples we elaborated were straightforwardly in the causal style, i.e. of a causal or ontological nature, viz. the reductive explanation of the ideal gas law, the reduction of the thing-predicate 'water' and the reduction of the attribute-predicate 'temperature'. The correlative explanation of Mendel's crossing law also belongs to this category, notwithstanding the fact that the notion of gene or more precisely allele is functional. That is, although the causal bridge principles between alleles and outer characteristics form part of the functional explanation of alleles, the statistical effects expressed in crossing laws are just causal (macro-)effects of the internal and the bridge principles. Similarly, although the reductive explanation of Olson's law deals with the social level starting from the (quasi-)intentional style on the individual level, the social phenomenon expressed in Olson's law is just the causal macro-effect of the individual behavior. In this case, however, the causal correlation is not part of a functional explanation, let alone intentional explanation, of the individual behavior. From these examples it is clear that it may be difficult to indicate where the functional and intentional terms play a crucial role.

Hence, so far we have not dealt with a straightforward example of interlevel reduction (or correlation or explanation) within the functional style. We would like to elaborate this kind of reduction and correlation a bit, viz. of functions on a macro-level to functions on the relative micro-level: functional decomposition. We closely follow here the main lines of Bechtel's argument (1988a) for what he calls 'mechanistic explanation'. In the context of mind-body research interlevel theories relate behavior of animals or humans described in functional terms with functions performed by their subsystems. They are supposed to influence the macro-function, if not produce it. The point of departure of functional decomposition is the so-called *modularity assumption* expressed by Simon (1980), according to which nature is organized in a 'nearly decomposable' way. That is, organisms and organs can be decomposed in modules. On the level of organisms, functional decomposition starts with the question of where the 'locus of control', or the control organ, if any, is situated. In higher animals there is something we call the brain, which is (amongst other things) also called the organ of the mind, e.g. by Gerald Fischbach (1992). The next question is how the control functions are fulfilled by the control organ. Here more detailed strategies for functional decomposition are required. It is possible to distinguish at least two such strategies. The first is called 'direct localization' and the second 'complex localization'.

By *direct localization* we try, for example, to identify a subsystem of the brain that is crucially responsible for a certain task. In this case we start from the strong modularity assumption according to which one subsystem fulfills a crucial task. In order to successfully identify such a subsystem we have to look for enough causal correlations between the relevant task and the natural or artificial stimulation of a subsystem. The only conclusion which can be drawn from such research is that the subsystem indeed plays a crucial role in the task performance; that is, there is a correlation between the relevant micro- and the macro-function in question. *Complex localization* is based on the weak modularity assumption, according to which, e.g. an organism or an organ is nearly decomposable in terms of subsystems with subtasks allocated such that the macro-task in question results. Complex localization hence starts with functional analysis, which provides hypotheses about subsystems and their subtasks and requires interlevel theories, in the present context, relating the level of one or more subsystems and the level of the main system. Functional analysis is followed by direct localization of the postulated subsystems and the study of their actual performance, separately as well as interactively. In brain research complex localization is also called homuncular analysis, since the subtasks seem to be performed by little creatures. Note that the ideal result is a structure of subtask performing subsystems producing, by their interrelations, the macro-function in question. Hence, although the micro-phenomena and the resulting macro-phenomenon are both described in functional terms, the connecting interfield theory is essentially of an ontological nature, explaining how the macro-function results from the micro-functions. If this theory is complete in the sense that it shows that and how the macro-function is the 'causal-ontological result' of the micro-phenomena, we have fully grasped the 'material realization' of the macro-function by the micro-phenomena.[12] As long as it is incomplete, additional, not yet understood, bridge principles have to be assumed in order to be able to derive the macro-effect from the micro-phenomena. In the first case the functional decomposition is a genuine functional micro-reduction (of a function), otherwise it is a functional micro-correlation or micro-explanation.

Bechtel gives an interesting illustration. O'Keefe and Nadel (1978) studied spatial memory, in particular the capacity to make a mental map of some area. First they found out by direct localization that the hippocampus is the subsystem of the brain that is mainly responsible for this task. Secondly, by way of complex localization they found the neuro-morphological structure of the hippocampus, with many different subsystems, and the neurophysiological interactions between these subsystems. In this way they can explain the functional task of map making to a large extent in terms of the structure and working of the hippocampus. Note that in this example we have at least four levels of organization, the organism, the brain, the hippocampus, and its subsystems.

In general, functional decomposition is a kind of explanation by functional specification in the following sense. The primary question of functional specification is what the function of something is, whereas in the case of functional decomposition, the (or at least one) function is known, but the question is how does the something perform that function. If the postulated subsystems exist and perform the postulated functions they get a functional specification, viz. in terms of the function we started with. In other words, functional explanation of a subsystem starts with the question of what its function is, whereas functional decomposition of a system starts with the question of how a certain function is performed. In both cases the result is a functional explanation of some subsystem.

As Bechtel rightly stresses, besides in functional decomposition itself, the functional perspective is also very useful in guiding causal research, not in the least because there are important limits to functional decomposition. In particular, the connectionist research program conjectures the existence of these limits by assuming that many phenomena, such as representation and association, are in fact distributed. Hence, depending on the type of phenomena, there may be good reasons to assume that there is a level below which the quest for the material realization of functional items on that level has to be addressed in purely causal terms, that is, it asks for interlevel, interstyle research.

Let us finally and briefly consider interlevel research primarily within the intentional style. It amounts to research in the interrelation of individual and social (institutional) intentions. As is well-known, institutional intentions, such as intentions of organizations, may or may not be realized by individual actions, and if so, they may or may not be the intended effect of the individual actions. Hence, interlevel research remaining within the intentional style will only be fruitful in a limited number of cases. However this may be, within and across the intentional style, correlation and even reduction of concepts and laws from the social level to the individual level is possible (see Franssen (1997) for a broad survey, see also Pettit (1993), Searle (1995), Tuomela (1995) and Balzer and Tuomela (1997).

6.4 INTERLEVEL AND INTERSTYLE MIND-BODY RESEARCH

Introduction

Most mind-body research is of an interlevel and interstyle nature. We will present and illustrate a general model for such research, put together from previously presented elements and some new ones. But we will start with two related examples of typically interlevel and interstyle research.

6.4.1 *Juvenile Delinquent Behavior*

Moffitt (1993) has suggested that there are two rather different types, and corresponding explanations, of juvenile delinquent behavior, one of which is reconstructed by Mackor (1997) in the proper function terms of Millikan, in particular her specification in terms of normal conditions and performance. We will confine ourselves to a global sketch of both types in these terms. From this point of view, normal conditions are obviously crucial in functional and intentional laws governing 'normal behavior'. Moreover, abnormal conditions frequently provide the clue to (specific) causal explanation of abnormal behavior. They are suggested by the normal conditions featuring in intentional, functional or unspecific causal explanations of normal behavior.

Moffitt makes a distinction between two types of juvenile delinquents, those who start when their puberty starts, and those who start some years earlier. Members of the second group usually continue in delinquent behavior after their adolescence age, whereas members of the first group usually stop with it at that age. The former are called 'persistent delinquents' and the latter 'adolescence delinquents'. According to Moffitt, the behavior of the two types of delinquents should be explained in two radically different ways, a primarily 'social explanation' for the adolescence delinquents, and a primarily 'biophysical explanation' for the persistent delinquents. To begin with the latter, they happen to show the ADHD-pattern (attention deficit hyperactivity disorder). Hence, some causal explanation may be supposed for the apparent correlation between the ADHD-pattern and persistent delinquent behavior, typically an MB-correlation, for mental and bodily affairs are both crucially involved. As Mackor argues, it is plausible to hypothesize that this is caused by an abnormal neurophysiological architecture of the brain, which is not sufficiently compensated by medication or therapy. Such a brain defect may be due to genetic or other causes, such as oxygen shortage at one's birth. However this may be, it may be such that it prevents the normal performance of innate learning rules in the presence of normal inputs, which causes an abnormal functional development, including the replacement of normal 'intentional representational abilities' (derived intentional learning rules) by the ADHD-pattern. Under normal social conditions this may lead to persistent delinquent behavior. Of course, the ADHD-pattern, let alone the abnormal brain architecture, is no guarantee for persistent criminal behavior. These factors typically are INUS-conditions in the sense of Mackie (see Section 4.3). That is, the abnormal brain architecture is an Insufficient but Necessary part of a condition which is itself Unnecessary but Sufficient for the ADHD-pattern. In its turn, the ADHD-pattern is an INUS-condition for persistent criminal behavior.

In short, persistent delinquent behavior is explained by referring to abnormal biophysical conditions leading to abnormal functional development, which under normal social conditions may lead to persistent delinquent behavior. The really important point is that the abnormal biophysical, internal condition plays a crucial role here, it is the abnormal causal factor for the specific causal explanation of the delinquent behavior.

Such an abnormal biophysical condition does not play a role in the other type of delinquent behavior. Adolescence delinquents have a perfectly normal functional development, including the (functional) tendency to choose age-specific role models. However, in the absence of classical role models and the presence of other delinquents, of a persistent or adolescence nature, they join delinquent behavior, up to the age that other role models, specific for that age, become dominant. In sum, in the case of adolescence delinquents the external social factors are crucial; they provide the abnormal factors for the specific causal explanation of the delinquent behavior. Of course, like in the other case, in this case these factors are at most INUS-conditions, they do not guarantee criminal behavior.

In general, at least in the context of criminology, there are neither sufficient nor necessary conditions for criminal behavior, but only INUS-conditions. In this connection it is interesting to note that Raine et al. (1998) observed with brain scans of 41 murderers that they, more frequently than the members of the control group, had more or less severely damaged prefrontal cortex. Since the prefrontal cortex is supposed to be crucial for rational behavior, damage of it is in fact interpreted as an INUS-condition for murder. However, Raine c.s. also observed that the damage was most severe for the murderers grown up in a protected environment. Apparently, in such families the protecting social factors function as a kind of reverse INUS-condition, which needs to be compensated by more severe brain damage in order to become a murderer. Another interesting correlation suggests that being an unwanted child is an INUS-condition for criminal behavior in general. According to Levitt and Donohue (1999) the significant decrease of crimes in the US since the early 1990s is for this reason related to the legalization of abortion some 20 years before, leading to a severe decrease in the number of unwanted children. Indeed, the decrease in crime statistics starts earlier in those states where the legalization was earlier.

These examples illustrate the way in which mind-body behavior may be explained by taking biophysical as well as social factors into account, freely passing from one style to another. In the following we want to study this kind of interlevel and interstyle research with emphasis on the interlevel aspect. We will exploit the fact, discussed in Section 6.2, that the functional style essentially amounts to a constellation of specific, more or less complex, forms of application of the causal style, and that the same holds true for the intentional style, but in a still more complicated way. Moreover, we also use the fact that, as we have argued, a biophysical justification can in principle be given for the common sense talk about mental causation, viz. states of the mind and/or of the body may causally influence states of the body and/or of the mind.

6.4.2 *Interactionism*

Although the classical dualist mind-body substance ontology has generally been replaced by a monist, in particular, biophysical ontology,[13] a dualist aspect theory is still around. In a dualist aspect theory, two epistemological layers are distinguished, a (bio-)physical and a mental level, both of which are usually seen as being to some extent autonomous. Nevertheless, some kind of presupposition seems to be unavoidable: the mental 'needs' the physical. But in the same way, it is then also argued that the physical needs the mental, in some sense. The latter intuition seems typical of a classical interactionist nature. In some seminal articles, Engel (1977, 1980) has proposed to increase the number of layers to three: biophysical, mental and social, and has argued that, at least in biomedical research, these layers, and their relation should be taken into account, giving rise to the so-called *biopsychosocial (BPS-)model*. Panhuysen and Tuiten (1993, 1995) have transformed this model into a full-blown interactionist model for guiding research, with a core idea, which we would like to call the 'initial condition-principle' (*IC-principle*), for reasons that will become clear. Unfortunately, Panhuysen and Tuiten consider the interlevel interaction primarily as causal, which brings us back to a classical, interactionist ontology, though of a pluralist nature.

In this subsection we will develop an epistemologically, not causally, interactionist BPS-model of research, governed by the IC-principle. Hence, the model reflects a pluralist aspect theory based on a monist but stratified biophysical ontology.

The point of departure is the ontological level of one individual I and its environment E. We start with I. It has various kinds of lower ontological levels (see Figure 1). Recall from Chapter 5 that it is possible to distinguish two epistemological levels for I, a macro-level and a micro-level. On the macro-level, I is described and explained in terms of individual properties of a bodily, mental, and mixed nature, giving rise to the (generalized) macro-type representation of I, indicated by $R(I)$. The individual properties are supposed to be more or (usually) less directly observable. On the micro-level, I is described and explained in (proper or quasi) biophysical micro-types, starting from the structure representation of I. This procedure gives rise to a (partially proper and partially quasi) micro-type representation of I, indicated by $r(I)$. Recall that the micro-level, hence $r(I)$, has two (epistemological) sides, a (macro-induced) upper or macro-side and a (base-induced) lower or micro-side.

The individual will interact in various ways with the environment E, that is the social and the natural environment. In recent publications the crucial, if not constitutive, role of the environment for the mind is stressed more and more, see e.g. Clark (1997), Keijzer (2001), and Kim (2000). The environment E will also be conceptualized into two epistemological levels, giving rise to the macro-type representation $R(E)$ in appropriate macro-terms and the micro-type representation $r(E)$ in biophysical terms. In this way we get two main conceptual or epistemological portraits of I and E, a macro- and a micro- one. The model to be developed will be called the *stratified IE-model*.

We will present it by first critically exposing 'six plus one' principles that Panhuysen and Tuiten (1993) claim to derive from applying General Systems Theory to interlevel research in general.[14] Before presenting them, it should be noted that, although they speak of levels of organization, suggesting ontological levels, it is clear that they intend at least as much to refer to epistemological levels.

I: Reality is organized at different levels and each level has its own units, properties and laws.
II: The different levels constitute a hierarchy, each higher level presupposes all lower levels, and the lower levels constrain the possibilities at the higher level.
III: The properties, distinguishing a phenomenon at a higher level from phenomena at lower levels, cannot be explained by reducing the phenomenon to the properties and laws of those lower levels.
IV: Events at a higher level can have influences upon the events at lower levels.
V: Different lower level conditions can underlie the same phenomenon at a higher level
VI: The same lower level condition can underlie different phenomena at a higher level

According to a seventh principle, here called the initial condition (IC-)principle, mechanisms at different levels 'operate in part independently, they also gear into one another. *Some effects of a mechanism at the one level constitute the initial conditions of a mechanism at the other level, and this second mechanism in its turn changes the initial conditions of the first mechanism.*'

At first sight all these principles have some plausibility. They suggest that it makes sense to study a phenomenon at a certain level as soon as and as long as that phenomenon seems to be causally, functionally or intentionally (including conceptually) related with other phenomena on that same level of description and explanation. Of course, the determination of levels is essentially man-made, though governed by our findings, hence reality-guided. However, being related in one or more of the suggested ways is not a sufficient condition for being localized on the same level. The levelling is typically a matter of a pragmatic overall judgment, that is, a matter of a pragmatic holistic nature, if you want. Moreover, in one context it may be useful to distinguish a number of levels, including intermediate levels, in other contexts it may be useful, to contract some related levels and retain only a few global levels. In this respect, the traditional levelling within the natural sciences, viz. a physical, a chemical and a biological level, is highly comparable to the *prima facie* (pf-)levels, suggested by Engel and that we are presently discussing: a biological, a psychological and a sociological level.

The epistemological denomination of these pf-levels stresses the (partially) man-made nature of the resulting level-specific concepts and laws. To be sure, in order to be useful, concepts need not be of the ideal nature; that is, they need not be characterizable by necessary and sufficient conditions, e.g. in the form of existence and uniqueness conditions. They may well be of a 'family resemblance' or 'prototype' nature. Similarly, the laws to be distinguished need not be of a universal nature, the largest possible range, they may well be laws of the middle range or the small range, including the extreme case of a one-object-range of (intended) application. Moreover, the laws need not apply strictly or all the time, that is, it may well concern quasi-regularities. Note that, with these qualifications the attribution of a stable property, including a disposition, is lawlike. This may seem to be a matter of stretching the concept of law too far. However, a moment's reflection shows that all this is a relative matter, depending on our particular point of view. As soon as we talk about entities remaining the same though their space-time location changes, we are implying that such entities have some stable features.

An example illustrating the principles I–VI and the IC-principle applied to the mind-body case is that of eating and such related phenomena as digesting the food and enjoying the meal. Let us uncritically illustrate these principles in terms of this example. Ad. I: Several levels can be distinguished. Eating is a physical process, leading to at least temporarily weight increase; it is a (bio-)chemical process in which substances are transformed into other substances; it is a physiological process, partly with neuro-physiological aspects (think of the transport processes in the gullet and the guts); it is a mental process, not only of tasting and smelling, but also of pleasure or dislike, and of coordinated action; and, finally, it has small and large scale social, economic and cultural aspects. Ad. II: The levels constitute a hierarchy. Something cannot taste salty, unless it is a (pure or mixed) substance of a certain chemico-physically specifiable kind. Ad. III: Higher level phenomena do not seem to be reducible to lower level ones. To taste a salty taste, although having a neurophysiological substratum, does not seem to be reducible to a neurophysiological event. Ad. IV: Higher level phenomena may influence lower level ones. The fact that you love or hate the person who prepared the food may well influence the degree in which you find the meal delicious. Ad. V: Higher level phenomena can be multiply realized. There are different kinds of salt tasting substances. Ad. VI: The same lower level conditions may lead to different higher level phenomena. The same food may go together with quite different appreciations of the meal. Finally, the IC-principle, according to which effects at one level constitute initial conditions at other levels, is also easily illustrated by a combination of illustrations of the foregoing kinds. One special food, if not a drug, may cause you to feel better, which may improve the atmosphere at the table, and stimulate more enjoyment of the rest of the food.

In sum, in their intuitive form the seven principles are valid and very useful in stimulating interlevel mind-body research, as, for instance, Panhuysen and Tuiten illustrate in their research on anorexia nervosa and depression (Panhuysen and Tuiten, 1993, 1995).

However, in its present form the mind-body applications of the principles are of a classical interactionist, though pluralist, nature. It is a challenge to reformulate them in such a way that they reflect an essentially monist, more specifically biophysical, ontology, accounting for a pragmatic pluralist perspective of relatively separate epistemological levels.

We will do this by taking the following principles into consideration. Regarding the macro- and the micro-levels of an individual *I*, and the corresponding representations $R(I)$ and $r(I)$, we assume that, although we may like to localize an event primarily at a certain epistemological level, one event has always a representation, whether directly or indirectly interesting for our purposes or not, on the other level. This is the Token Identity Hypothesis. Moreover, according to the Realization Hypothesis, properties on the macro-level are supposed to be supervenient to the properties on the micro-level, i.e. the same conditions at this level guarantee the (non-)realization of the property in question. Finally, properties on the macro-level may be (single or multiply) type-type identical to properties on the micro-level.

The possibility of supervenient and type-type identical properties creates the possibility of various formulations of causal mechanisms, and functional and intentional coherence. For type-type identical properties this is obvious, sunshine raises the temperature and the mean kinetic energy at one and the same time. However, if a causal mechanism has an effect which has a certain supervenient property, it is tempting to neglect additional conditions, and to say that the mechanism caused the supervenient property. A certain drug may cause a certain change in brain state, which may, depending on the constitution and state of departure, lead to a brain state upon which a certain mood supervenes. In this case, we easily say that the drug consumption not only caused the change of brain state, but also the change of mood. Note, by the way, that this example fits well in our analysis of causal specification, it is the drug which is considered to be the 'abnormal' causal factor for both effects.

Such examples illustrate that we may, in the face of supervenient and type-identical properties, distinguish monolevel (macro- or micro-) and interlevel (i.e. between the macro- and the micro-level) formulations of causal mechanisms and the like. However, interlevel formulations are essentially abbreviated formulations, which can always be split in a monolevel genuine empirical relation and an interlevel conceptual cum ontological relation. Taking this into account, the seemingly interlevel interaction can be reduced to monolevel causal interaction and interlevel 'conceptual interaction'.

Essentially the same holds true for functional and intentional transitions. At first sight they may seem to be of an interlevel nature, but they can be reconstructed, by monolevel causal, functional or (only for the macro-level) intentional analysis and conceptual interlevel analysis. The demonstration of how this point follows from the causal explications of the functional and the intentional style given in Section 6.2 is left as an exercise.

We now turn to an individual's environment. The nature of the macro- and the (bio-)physical micro-portrait of the environment, $R(E)$ and $r(E)$, and of their mutual relation, is in many respects essentially similar to that of the individual itself. The environment can be described on both levels in causal and functional terms, and on the macro-level also in intentional terms as far as 'intentional subjects' are concerned. Again, interlevel transitions can be reconstructed in monolevel causal, functional or (on the macro-level) intentional interaction and interlevel 'conceptual interaction'. Whereas $R(I)$ and $r(I)$ are supposed to deal with one and the same individual, $R(E)$ and $r(E)$ bring the whole outside-world into the picture, as far as relevant, of course. Hence, $R(E)$ and $r(E)$ lack the special features of a biological organism. However this may be, an individual interacts with its environment, with traces and hence changes in both. The interactions of the individual and the environment may sometimes be directly described and explained in causal terms on the macro- or the micro-level of both, but frequently they can better be described and explained indirectly in functional terms on one of these levels, or in intentional terms on the macro-level. As we have seen in Section 6.2, the latter styles can be explicated in causal terms. However, there is no point in a direct interlevel characterization of the interactions, that is, between the macro-portrait of the individual and the micro-portrait of the environment, or the other way around.

The resulting model is called the stratified IE-model for mind-body research. It is depicted in Figure 6.3 In an explanation only some transitions or relations of influence, of a causal, functional or intentional nature, will be specified. For a relatively complete explanation only a connected chain is necessary.

The two types of delinquent behavior dealt with in the first subsection can now roughly be represented by Figure 6.4 and Figure 6.5, respectively. It should be stressed once more that the depicted influence relations are always of the INUS-type, they are neither necessary nor sufficient, but they form an necessary part of a sufficient condition.

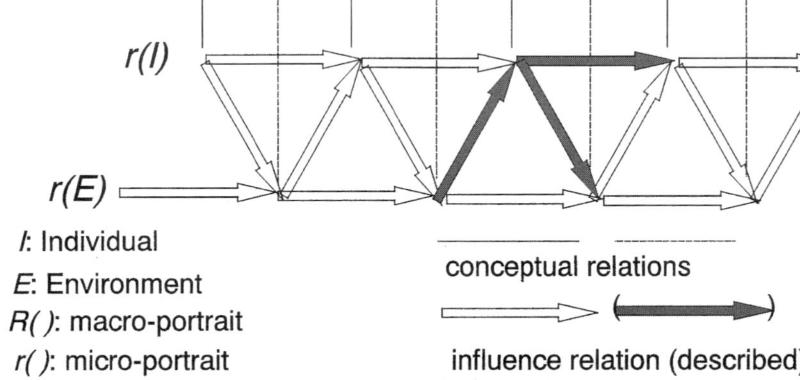

Figure 6.3 The stratified IE-model of mind-body research.

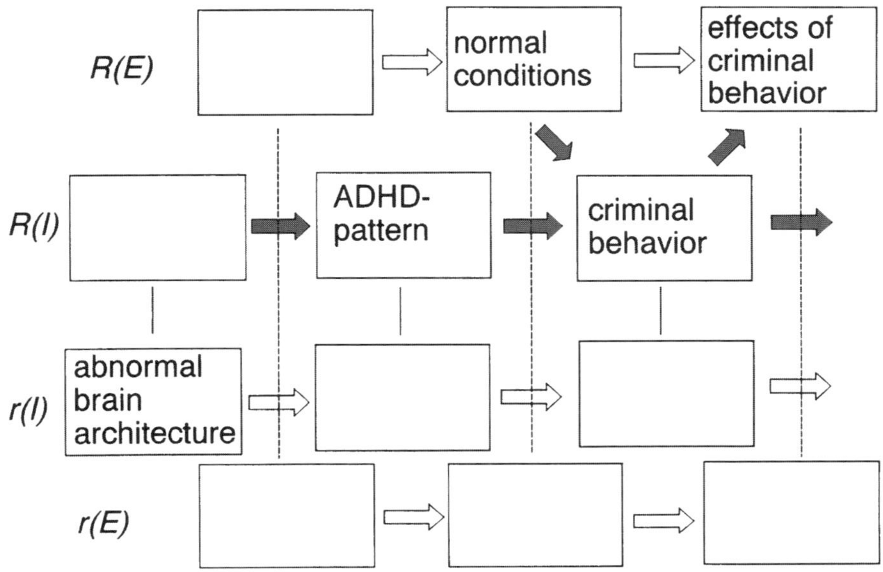

Figure 6.4 Persistent delinquent behavior.

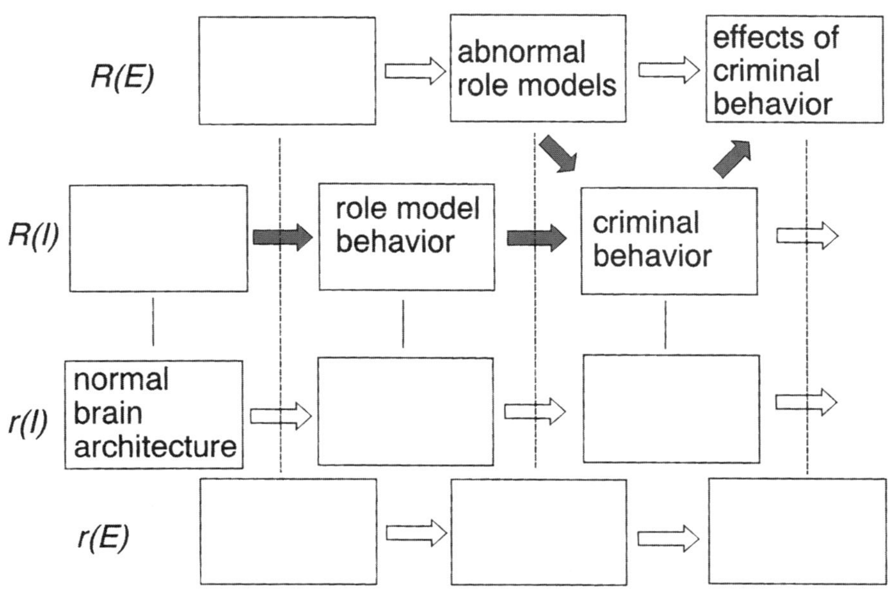

Figure 6.5 Adolescent delinquent behavior.

6.4.3 *Resulting Principles*

For the sake of completeness, let us finally look back at the seven principles of Panhuysen and Tuiten from the general perspective of Chapter 5 on a macro- and one or more micro-levels, all relating to one aggregate-level. In the mind-body context one may think of the macro- and micro-level as discussed above. Recall that there are similar levels for the environment. We will come back to these points later.

The IC-principle is highly compatible with the above approach, but the six basic principles need critical reconsideration. Principle I is reinforced by the fact that the division in levels is governed by the question of which properties are best able to form lawful connections with each other. The general analysis in Chapter 5, assuming the Token Identity Hypothesis, perfectly leaves room for this possibility, at least for 'ontologically induced' epistemological levels: realization-types corresponding to supervenient properties may not be type-type identical to micro-types. In this case the supervenient properties, and their lawlike relations, are particularly specific for the corresponding level. Principle II is more difficult. It requires a hierarchy of levels, such that the higher levels presuppose the lower ones, which are said to constrain the higher ones. Where does the hierarchy come from? Does the presupposing/ constraining relation always point in the same direction? Let us first consider how conceptual relations can be directed, where 'upward' is of course just a convention. Supervenience is obviously upward directed, ('upward supervenience'); a single type-type identity may be, but need not be, directed on ontological grounds; e.g. temperature and mean kinetic energy, where the upward direction goes from micro-objects, via aggregates of micro-objects, to macro-objects ('upward aggregation'), and a multiple type-type identity may be directed on the same grounds, and it is additionally upward directed on conceptual grounds, from a multiple of types on a lower level to a single type on the higher ('downward multiplicity'). In sum, the revised ordering Principle II includes three criteria: upward supervenience, upward aggregation and downward multiplicity. In this form it complies with our analysis.[15]

The antireductionist Principle III is more difficult to accept. First, by its general formulation it suggests that the irreducibility of higher level properties is responsible for the division of levels and their ordering. This is certainly not very attractive, for it might well be that the above reasons for forming levels and their ordering leave room for properties and even laws that can be reduced to properties and laws of lower levels. Hence, it seems wise to weaken III at least to a principle explicitly just leaving room for irreducible properties and laws. Although it may be true as a (moderate) philosophical position, that is, it is theoretically possible that there are such properties and laws, as a heuristic principle it will not be very fruitful. Hence, we will further plea for some kind of reduction principle. For that purpose, note first that it is not necessary to subscribe to III in order to make research on the relevant level useful. On the contrary, as we have stressed before, striving at non-eliminative reduction is a strong reason for preliminary and simultaneous monolevel research. Hence, strategically it is plausible to replace III by its moderate opposite, viz. the moderate reductionistic strategy distinguished before. It amounts to: all properties and laws at a certain level can be reduced, at least in the quasi-sense, to properties and laws of lower levels, not as an attempt to eliminate them, but to consolidate and reinforce them and hence the hierarchy of levels, either straightforwardly or in a corrected way. Of course, concepts and laws may occasionally have to be eliminated, in which case it may well be that they can be replaced by new concepts and laws on that level, fulfilling to some extent the same functions as those played by the eliminated ones.

Let us now turn to Principle IV. It postulates a 'downward' influence of events, which strongly suggests downward causation and hence also the possibility of upward causation. In our view, even if both possibilities are not in conflict with the first two principles of the theory of levels, which is unclear, they need not be assumed. An event at one level can also be described on another higher or lower level and on that level lateral causation may lead to some effect. In this way, the causing seems to be interlevel, but can be redescribed in an monolevel way, augmented with suitable conceptual relations. In this interpretation, IV essentially becomes equivalent to the IC-principle. Moreover, it can be extended to the functional style and, as far as the individual level is concerned, to the intentional style.

Principle V amounts to a statement of the so-called multiple realizability of phenomena at higher levels by a lower level. This is essentially captured by the idea of realization-classes ('downward token multiplicity'), e.g. the case of substances, pressure and temperature, and multiple type-type identities ('downward type multiplicity'), e.g. the case of substances (to aggregates of (molecules of) atoms of different isotopes) and the case of temperature (to kinds of molecular kinetic energy).

Finally, Principle VI was already remarked (Note 15) to suggest downward supervenience or upward multiplicity, which are from our point of departure conceptually impossible. However, it is perfectly possible to make sense of VI by taking the E-level into account: the same lower level condition can underlie different phenomena at a higher level, assuming that the higher level description takes other factors, e.g. from the past and present environment, into account, a process which may be called 'external enrichment'. Different reproductive and learning histories provide the paradigm backgrounds of *prima facie* upward multiplicity.

Let us now summarize the principles we have adopted for two or more levels related to one aggregate level, e.g. the *I*- and *E*-level:

> I^*: Reality is organized at different levels and each level has its own units, properties and laws. The same event 'occurs on all levels' although only some levels may have the conceptual means to describe it (Token Identity Hypothesis).
>
> II^*: The different levels constitute a hierarchy, each higher level presupposes all lower levels, and the lower levels constrain the possibilities at the higher level. The ordering derives from 'upward supervenience' or 'upward aggregation' or

'downward multiplicity' or a combination of these phenomena (ordering principle).

III*: The properties and laws can be reduced in one or another sense to the properties and laws of those lower levels, leading to the elimination or the consolidation of the former (reduction principle). This possibility implies at least that all higher level concepts supervene on lower level conditions, taking enough conditions into account (Realization Hypothesis).

IV*: Some effects of a (causal, functional or intentional) 'mechanism' on one level constitute the initial conditions of a mechanism on another level, and this second mechanism in its turn changes the initial conditions of the first mechanism (epistemic IC-principle).

V*: Different lower level conditions can underlie the same higher level phenomenon (principle of downward token multiplicity, sometimes downward type multiplicity).

VI*: The same lower level condition can underlie different higher level phenomena, by taking external aspects into account (principle of external enrichment).

No doubt these starred principles, although in practice similar, are fundamentally rather different from those of Panhuysen and Tuiten. First, the starred ones, except VI*, are only supposed to apply to (the epistemological macro- and micro-level relating to) one aggregate-level. In VI* the environment of the aggregate is taken into account. Second, even with this restriction, several of them differ essentially. Third, the above comparative evaluation of the first five principles does not apply to (the interaction with) the (macro- or micro-level of the) environment. Apart from noting that the IC-principle (hence IV and IV*) then seems unproblematic and claiming that interlevel interactions between an individual and its environment can again be reconstructed as monolevel empirical interactions and interlevel conceptual interactions, we will leave this evaluation as an exercise for the reader.

6.5 LATERAL INTERFIELD RESEARCH

Let us finally address the possibility of interfield research between fields of the same ontological level and epistemological style. We will indicate some examples of this kind, in particular cases in which the fields stem from different disciplines: interdisciplinary monolevel and monostyle interfield research. The examples are taken from Bechtel (1988a).

All kinds of descriptive and explanatory linking research between fields of the same level are possible. Assuming that in both fields some more or less stable phenomena have been established, usually figuring in some laws, the question arises whether there is some causal or ontological relation between them. If so, the question is whether it is possible that a theory in one field might explain (causally or reductively) a law of the other. In case of a causal explanation, it may even give rise to a functional explanation, or another kind of explanation by specification.

In an interdisciplinary context such ideas governing monolevel interfield research may relate the corresponding disciplines in a symmetric or an asymmetric way. Recall first the possibilities for globally symmetric or asymmetric co-operation between research programs in general (Section 1.2), and hence in the case of two programs on the same level stemming from different disciplines in particular.

Other types of symmetric and asymmetric relations between disciplines may arise from boundary-bridging and boundary-breaking research programs (Abrahamsen, 1987). To begin with boundary-breaking research, a research program originating from one discipline, may be used to (re-)direct the research in another discipline. Blumenthal (1987) has, interestingly enough, described two such cases in the opposite direction in the interfield studied by what nowadays is called psycholinguistics. First it was linguists at the end of the 19th century who wanted to redirect their research by applying the related, empirically and quantitatively oriented, psychological research programs of Hobart and Wundt. In the middle of this century, Chomsky introduced a linguistic research program that not only aimed at redirecting linguistic research, but also at redirecting psychological research. Two ideas constituted the core of the interfield research program: grammar, at least universal grammar, is innate, and since linguistics primarily deals with linguistic competence, starting from the idea of an innate universal grammar, psychology should deal with linguistic performance in relation to linguistic competence. Note that in both fields the main style is functional.

Similarly, a symmetric relation between disciplines may arise in the case of boundary-bridging research. Bechtel (1988a) indicates several examples of psycholinguistic research.[16] Here we restrict ourselves to their general pattern. All this kind of research originates from the fact that language is something very different for linguists and psychologists. For linguists it is a cultural product with an abstract structure which can be analyzed. For psychologists language is something which is understood and produced in real-time processes taking place in human beings (in interaction with their human and non-human environment). To study language as a process one needs a description of language as a product. This description should primarily be delivered by linguists, but it should be reconceptualized in order to make it useful for studying the process. This may lead to a causal interfield theory between the structure and the processing, which may in its turn have severe repercussions on the study of language as a product. Again the primary style in both fields is functional.

CONCLUDING REMARKS

In this chapter the emphasis was on mind-body research, where we have elaborated to some extent the way in which the moderate reductionist style can be applied in this type of research. Unfortunately, other possibilities for such reduction cannot be treated in this book. For an application to biology in general and ecology in particular, see Looijen (1998/2000); for an application to economy, see Janssen (1993); finally, for an application to social science in general, see Franssen (1997).

In the first three parts of the book we have concentrated on (units of) description and explanation of various kinds, without bothering about details of testing and evaluation of the relevant hypotheses. This will be the subject of the next part of the book.

PART IV

CONFIRMATION AND EMPIRICAL PROGRESS

INTRODUCTION

We will start this part with a brief exposition of the main ideas behind confirmation and falsification by the so-called HD-method, followed by a sketch of the 'landscape of qualitative and quantitative confirmation' as we prefer to reconstrue it. Confirmation of a hypothesis, however, has the connotation that the hypothesis has not yet been falsified. Whatever the truth claim associated with a hypothesis, as soon as it has been falsified, the plausibility (or probability) that it is true becomes and remains nihil. In this part we will elaborate how theories can nevertheless be evaluated after falsification.

In the rest of Chapter 7 the attention is directed at the more sophisticated qualitative HD-evaluation of the merits of theories, in terms of successes and counter-examples, obtained by testing the test implications of theories. The resulting evaluation report leads to three interesting models of separate HD-evaluation of theories. Special attention is paid to the many factors that complicate HD-evaluation and, roughly for the same reasons, HD-testing.

In Chapter 8 it will be pointed out that the evaluation report resulting from separate evaluation naturally leads to the comparative evaluation of theories, using the crucial notion of 'more successfulness', which in its turn suggests 'the rule of success' and which indicates empirical progress. It will be argued that this 'instrumentalist' or 'evaluation methodology', by denying a dramatic role for falsification, and even leaving room for some dogmatism, is methodologically superior to the 'falsificationist methodology', which assigns a theory eliminative role to falsification. Moreover, the former methodology will be argued to have better perspectives for being functional for truth approximation than the latter.

As a matter of fact, this part, together with Chapter 9, gives a brief impression of *From instrumentalism to constructive realism* (ICR). That book gives a detailed analysis of the various kinds of confirmation, of separate and comparative evaluation of theories, and of (basic and refined) truth approximation. A recurring theme in that book, is the localization and comparison of the main standard epistemological positions, as they have been presented in Subsection 2.4.3 (and will briefly be summarized in the introduction of Chapter 7), viz. instrumentalism, constructive empiricism, referential realism, and constructive (theory) realism.

To make the present book a self-contained volume, the present part has been taken over almost *verbatim* from the methodologically most important chapters of the other, viz. on separate and comparative evaluation. To be precise, Chapter 7 (Introduction and Sections 7.2 and 7.3) corresponds to Chapter 5 of ICR and Chapter 8 to Chapter 6 of ICR. Moreover, Subsection 7.1.1 of Chapter 7 is based on the Introduction to Chapter 2 of ICR, that is, the first of the three chapters on confirmation in ICR. Finally, Subsection 7.1.2 of Chapter 7 presents some of the main points of those three chapters.

CHAPTER 7
TESTING AND FURTHER SEPARATE EVALUATION OF THEORIES

INTRODUCTION

We will start this chapter with a brief exposition of HD(hypothetico-deductive)-testing, that is, the HD-method of testing hypotheses, and indicate the related qualitative explication of confirmation as well as quantitative extensions. HD-testing attempts to give an answer to one of the questions that one may be interested in, *the truth question*, which may be qualified according to the relevant epistemological position.[1] However, *the (theory) realist*, for instance, is not only interested in the truth question, but also in some other questions. To begin with, there is the more refined question of which (individual or general) facts[2] the hypothesis explains (its explanatory successes) and with which facts it is in conflict (its failures); *the success question* for short. We will show in this chapter that the HD-method can also be used in such a way that it is functional for (partially) answering this question. This method is called HD-evaluation, and uses HD-testing. Since the realist ultimately aims to approach the strongest true hypothesis, if any, i.e. *the (theoretical-cum-observational) truth* about the subject matter, the plausible third aim of the HD-method is to help answer the question of how far a hypothesis is from the truth, *the truth approximation question*. Here the truth will be taken in a relatively modest sense, viz. relative to a given domain and conceptual frame. In the next chapter we will make plausible that HD-evaluation is also functional for answering the truth approximation question.

As we will indicate in a moment, the other epistemological positions are guided by two related, but more modest success and truth approximation questions, and we will show later that the HD-method is also functional for answering these related questions. But first we will articulate the realist viewpoint in some more detail. For the realist, a hypothesis is a statement that may be true or false, and it may explain a number of facts. A theory will here be conceived of as a hypothesis of a general nature claiming that it is the strongest true hypothesis, i.e. the truth, about the chosen domain (subject matter) within the chosen conceptual frame (generated by a vocabulary). This claim implies, of course, that a theory claims to explain all relevant facts. Hence, a theory may not only be true or false, it may also explain more or fewer facts, and it may even be more or less near the truth. To be sure, presenting the realist notion of a theory as the indicated special kind of hypothesis is to some extent a matter of choice, an arbitrariness which will turn out to be very useful. The same holds for adapted versions for the other epistemological positions.

Let us briefly look at the relevant questions from the other main epistemological viewpoints, repeating the relevant version of the truth question. Regarding the four perspectives for each position indicated in Subsection 2.4.3, that is, true-false versus truth approximation and 'the actual world' versus 'the nomic world', we only (continue to) address the first issue. The reason to (continue to) suppress the second issue, is that whether or not one conceives the actual or the nomic world as the primary target of theorizing does not make very much methodological difference. However, this issue will certainly play a role in the next part. The *constructive empiricist* is interested in the question of whether the theory is empirically adequate or observationally true; i.e. whether the observational theory implied by the full theory is true. This type of researcher is interested in the refined success question about what its true observational consequences and its observational failures are, and in the question of how far the implied observational theory is from the strongest true observational hypothesis, *the observational truth*. The *referential realist* is, in addition, interested in the truth of the reference claims of the theory and how far it is from the strongest true reference claim, *the referential truth*. The *instrumentalist* phrases the first question of the empiricist more liberally: for what (sub-)domain is it observationally true? He retains the success question of the empiricist. Finally, he will reformulate the third question: to what extent is it the best (and hence the most widely applicable) derivation instrument?

The method of HD-evaluation will turn out, in this part, to be a direct way to answer the success question and, as will only be indicated in this book (but see ICR), an indirect way to answer the truth approximation question, in both cases for all four epistemological positions. This part will primarily be presented in relatively neutral terminology, with specific remarks relating to the various positions. The success question will be presented in terms of successes and counter-examples:[3] what are the potential successes and counter-examples of the theory?

In sum, two related ways of applying the HD-method to theories can be distinguished. The first one is *HD-testing*, which aims to answer the truth question. However, as soon as the theory is falsified, the realist of a falsificationist nature, i.e. advocating exclusively the method of HD-testing, sees this as a disqualification of an explanatory success. The reason is that genuine explanation is supposed to presuppose the truth of the theory. Hence, from the realist-falsificationist point of view a falsified theory has to be given up and one has to look for a new one.

However, the second method to be distinguished, *HD-evaluation*, keeps taking falsified theories seriously. It tries to answer the success question, the evaluation of a theory in terms of its successes and counter-examples (problems) (Laudan, 1977). For the (non-falsificationist) realist, successes are explanatory successes and, when evaluating a theory, they are counted as such, even if the theory is known to be false. It is important to note that the term '(HD-)evaluation' refers to the evaluation in terms of successes and counter-examples, and not in terms of truth approximation, despite the fact that the method of HD-evaluation will nevertheless turn out to be functional for truth approximation. Hence, the method of HD-evaluation can be used meaningfully without any explicit interest in truth approximation and without even any substantial commitment to a particular epistemological position stronger than instrumentalism.[4]

In Section 7.1 we will see that and how the HD-method can be used for testing. Roughly speaking, the HD-method prescribes to derive test implications and to test them. In each particular case, this may either lead to confirmation or to falsification. Whereas the 'language of falsification' is relatively clear, the 'language of confirmation' is a much disputed one. However, the particular choice of confirmation language is not very important, provided one is clear about one's choice. For this reason we will sketch the non-standard view developed in ICR. Then we will turn to the question of the relation between the HD-method used for testing and for evaluation.

In Section 7.2, it is shown that a decomposition of the HD-method applied to theories naturally leads to an explication of the method of separate HD-evaluation, using HD-testing, even in terms of three models. Among other things, it will turn out that HD-evaluation is effective and efficient in answering the success question.

In Section 7.3, so-called falsifying general facts will first be analyzed. Then the decomposition of the HD-method will be adapted for statistical test implications. Finally, it is shown that the decomposition suggests a systematic presentation of the different factors that complicate the straightforward application of the HD-methods of testing and evaluation.

In the next chapter we will use the separate HD-evaluation of theories for the comparative HD-evaluation of them. Strictly speaking, only Sections 1 and 2 of this chapter are required for that purpose.

7.1 FALSIFICATION AND CONFIRMATION BY THE HD-METHOD

7.1.1 *Basics*

According to the leading expositions of the hypothetico-deductive (HD-)method by Hempel (1966), Popper (1934/1959) and De Groot (1961/1969), the aim of the HD-method is to determine whether a hypothesis is true or false, that is; it is a method of testing. On closer inspection, this formulation of the aim of the HD-method is not only laden with the epistemological assumption of theory realism, according to which it generally makes sense to aim at true hypotheses, but it also mentions only one of the realist aims, i.e. answering the 'truth question', called the aim of HD-testing.

For the moment, however, we will restrict attention to the HD-method as a method of testing hypotheses. Though the realist has a clear aim in undertaking HD-testing, this does not mean that HD-testing is only useful from that epistemological point of view. Let us briefly review the other main epistemological positions as far as the truth question is concerned. Hypotheses may or may not use so-called 'theoretical terms', in addition to so-called 'observation terms'. What is observational is not taken in some absolute, theory-free sense, but depends greatly on the level of theoretical sophistication. Theoretical terms intended to refer to something in the actual or nomic world may or may not in fact successfully refer to something. For the *(constructive) empiricist* the aim of HD-testing is to find out whether the hypothesis is observationally true, i.e. has only true observational consequences, or is observationally or empirically adequate, to use Van Fraassen's favorite expression. For the *instrumentalist* the aim of HD-testing is still more liberal: is the hypothesis observationally true for all intended applications? The *referential realist*, on the other hand, adds to the aim of the empiricist to find out whether the hypothesis is referentially true, i.e. whether its referential claims are correct. In contrast to the *theory realist*, he is not interested in the question whether the theoretical claims, i.e. the claims using theoretical terms, are true as well. Recall that claims may pertain to the actual world or to the nomic world (of physical possibilities).

Methodologies are ways of answering epistemological questions. It will turn out that the method of HD-testing, the test methodology, is functional for answering the truth question of all four epistemological positions. For this reason, we will present the test methodology in fairly neutral terms, viz. plausibility, confirmation and falsification.

The expression 'the plausibility of a hypothesis' abbreviates the informal qualification 'the plausibility, in the light of the background beliefs and the evidence, that the hypothesis is true', where 'true' may be specified in one of the four main senses: (1) observationally as far as the intended applications are concerned; (2) observationally, in all possible respects; (3) and, moreover, referentially; (4) and, even, theoretically. Admittedly, despite these possible qualifications, the notion of 'plausibility' remains necessarily vague, but that is what most scientists would be willing to subscribe to.[5] When talking about 'the plausibility of certain evidence', we mean, of course, 'the prior plausibility of the (observational!) hypothesis that the test will result in the reported outcome'. Hence, here 'observationally true' and 'true' coincide by definition of what can be considered as evidential statements.

Regarding the notions of 'confirmation' and 'falsification' the situation is rather asymmetric. 'Falsification' of a hypothesis simply means that the evidence entails that the hypothesis is observationally false, and hence also false in the stronger senses. However, what 'confirmation' of a hypothesis precisely means is not so clear. The explication of the notion of 'confirmation' of a hypothesis by certain evidence in terms of plausibility is the main goal of Part I in ICR. It is approached from the success perspective on confirmation, equating confirmation with an increase of the plausibility of the evidence on the basis of the hypothesis, and implying that the plausibility of the hypothesis is increased by the evidence.

The variety of empirical hypotheses is large. To stress this we mention a number of examples:

- Mozart was poisoned by Salieri,
- Dutch is more similar to English than to German,
- When people have bought something they have selective attention for information justifying their choice,
- People tend to choose that action which maximizes their expected utility,

- The function of lungs is to supply oxygen to the organism,
- The average rainfall per year gradually increased in the 20th century,
- The universe originated from the big bang,
- Action is minus reaction,
- Dalton's theory of the atom.

The last example will be used several times for illustrative purposes in this part of the book.

A general characterization of an empirical hypothesis is the following: an empirical hypothesis is a tentative statement which is, in some way or other, about the actual or nomic world and which is in addition testable. Statements about the world claim to tell us something about it, how it is or was (not), how it will (not) be, how it can (not) be. They are supposed to have, or may acquire a sufficiently clear meaning such that they are true or false in a sense which has to be specified, preferably such that a false statement may, nevertheless, be approximately true in some plausible sense. The above characterization of an empirical hypothesis still leaves room for hypotheses with nonsensical or otherwise redundant additions. Instead of trying to exclude such additions, which is not easy, it will become clear that such additions are relatively harmless.

A test for a hypothesis may be experimental or natural. That is, a test may be an experiment, an active intervention in nature or culture, but it may also concern the passive registration of what is or was the case, or what happens or has happened. In the latter case of a so-called natural test, the registration may be a more or less complicated intervention, but is nevertheless supposed to have no serious effect on the course of events of interest.

According to the HD-method a hypothesis H is tested by deriving test implications from it, and checking, if possible, whether they are true or false. Each test implication has to be formulated in terms that are considered to be observation terms. A test implication may or may not be of a general nature. Usually, a test implication is of a conditional nature, if C then F $(C \rightarrow F)$. Here C denotes one or more initial conditions which can be, or have been naturally or artificially realized, i.e. by nature or by experiment. F denotes a potential fact (event or state of affairs) predicted by H and C. If C and F are of an individual nature, F is called an individual test implication, and $C \rightarrow F$ a conditional one. When C is artificially realized, it is an experimental test, otherwise it is a natural test.

As is well-known, the basic logic of HD-testing can be represented by some (valid) applications of Modus (Ponendo) Ponens (MP), where '$\models$' indicates logical entailment:

$H \models I$	$H \models (C \rightarrow F)$
H	H, C
———	———
I	F
Scheme 7.1.1	*Scheme 7.1.2*

It should be stressed that $H \models I$ and $H \models (C \rightarrow F)$ are supposed to be deductive claims, i.e. claims of a logico-mathematical nature.

The remaining logic of hypothesis testing concerns the application of Modus (Tollendo) Tollens (MT). Neglecting complications that may arise, if the test implication is false, the hypothesis must be false, and therefore has been falsified, for the following arguments are deductively valid:

$H \models I$	$H \models (C \rightarrow F)$
$\neg I$	$C, \neg F$
———	———————
$\neg H$	$\neg H$
Scheme 7.2.1	*Scheme 7.2.2*

When the test implication turns out to be true, the hypothesis has of course not been (conclusively) verified, for the following arguments are invalid, indicated by '–/–/–':

$H \models I$	$H \models (C \rightarrow F)$
I	C, F
–/–/–	–/–/–/–/–/–
H?	H?
Scheme 7.3.1	*Scheme 7.3.2*

Since the evidence (I or $C\&F$) is compatible with H, H may still be true. However, we can say more than that. Usually it is said that H has been confirmed. It is important to note that such confirmation by the HD-method is in the strong sense that H has obtained a *success* of a (conditional) deductive nature; by entailing the evidence, H makes the evidence as plausible as possible. This will be called *the success perspective* on ((conditional) deductive) *confirmation*.

Falsification and confirmation have many complications, e.g. due to auxiliary hypotheses. We will deal with several complications, related to general and individual test implications, at the end of this chapter. There is however a great difference between falsification and confirmation. Whereas the 'logical grammar' of falsification is not very problematic, the grammar of confirmation, i.e. the explication of the concept of confirmation, has been a subject of much dispute.

7.1.2 *Ordering the Landscape of Confirmation*

Introduction

The first thematic part of ICR is in many respects a systematic exposition of well-known ideas about deductive, structural, and inductive confirmation. However, these ideas are presented in a non-standard way and refine and revise several standard solutions of problems associated with these ideas. That part consists of three chapters (2–4), the first one dealing with qualitative (deductive) confirmation that results from applying the HD-method, and the second one dealing with quantitative (deductive and non-deductive) confirmation and its qualitative consequences. The third one explicates the idea of (quantitative) *inductive* confirmation and gives a brief survey of the main systems of inductive confirmation in the Carnap–Hintikka tradition of so-called inductive logic. The main non-standard aspect is the approach of confirmation from the 'success perspective', according to which confirmation is equated with evidential success, more specifically, with an increase of the plausibility of the *evidence* on the basis of the hypothesis. Hence, in contrast to standard expositions, confirmation is not equated with an increase of the plausibility of the *hypothesis* by the evidence. This is merely an additional aspect of confirmation under appropriate conditions and epistemological assumptions.

Here we will only give a sketch of the main lines of these three chapters on confirmation. It is important to note that, although the role of falsification and confirmation will be relativized in many respects in the rest of this part, it will also become clear that they remain very important for particular types of hypotheses, notably, for general observational (conditional) hypotheses, and for several kinds of (testable) comparative hypotheses, e.g. hypotheses claiming that one theory is more successful or (observationally, referentially or theoretically) even more truthlike than another.

The aim of the following survey is to present a coherent set of qualitative and quantitative notions of confirmation, partly by synthesizing the work of others, partly by taking distance to others. We start with qualitative deductive confirmation, then turn to quantitative, more specifically, probabilistic confirmation, and introduce the crucial distinction between structural and inductive confirmation. Finally, we address the question of a quantitative degree of confirmation.

7.1.2.1 *Types of Confirmation*

Deductive Confirmation

Contrary to many critics, and partly in line with Gemes (1990), we believe that the notion of *deductive (d-)confirmation* makes perfectly, good sense, provided the classificatory definition is supplemented with some comparative principles. More specifically, '(contingent) evidence E d-confirms (consistent) hypothesis H' is defined by the clause: H (logically) entails E, and further obeys:

Comparative principles:

P1: if H entails E and E entails E^* (and not vice versa) then E d-confirms H more than E^*.

P2: if H and H^* both entail E then E d-confirms H and H^* equally.

To be sure, this definition-with-comparative-supplement only makes sense as a *partial* explication of the intuitive notion of confirmation; it leaves room for non-deductive, in particular, probabilistic extensions, as we will see below. However, let us first look more closely at the comparative principles. They are very reasonable in the light of the fact that the deductive definition can be conceived as a (deductive) *success* definition of confirmation: if H entails E, E clearly is a success of H, if not a predictive success, then at least a kind of explanatory success. From this perspective, P1 says that a stronger (deductive) success confirms more than a weaker one, and P2 says that two hypotheses should equally be praised for the same success. In particular P2 runs against standard conceptions. However, in Chapter 2 of ICR we deal extensively with the possible objections and show, moreover, that the present analysis can handle the confirmation paradoxes discussed by Hempel and Goodman.

Probabilistic Confirmation

Probabilistic confirmation presupposes, by definition, a probability function, indicated by p, that is, a real-valued function obeying the standard axioms of probability, which may nevertheless be of one kind or another (see below). But first we will briefly deal with the general question of a probabilistic criterion of confirmation. The *standard* (or *forward*) *criterion* for probabilistic confirmation is that the *posterior* probability $p(H/E)$ exceeds the (relative to the background knowledge) *prior* probability $p(H)$, that is, $p(H/E) > p(H)$. However, this criterion is rather inadequate for 'p-zero' hypotheses. For example, if $p(H) = 0$ and E d-confirms H, this confirmation cannot be seen as an extreme case of probabilistic confirmation, since $p(H/E) = p(H) = 0$. However, for p-non-zero hypotheses and assuming $0 < p(E) < 1$, the standard criterion is equivalent to the *backward* or *success* criterion, according to which the so-called *likelihood* $p(E/H)$ exceeds the initial probability $p(E)$ of E: $p(E/H) > p(E)$.[6] Now it is easy to check that any probability function respects d-confirmation according to this criterion, since $p(E/H) = 1$ when H entails E, and hence exceeds $p(E)$, even if $p(H) = 0$. More generally, the success criterion can apply in all p-zero cases in which $p(E/H)$ can nevertheless be meaningfully interpreted.

Structural Confirmation

We now turn to a discussion of the kinds of probability functions and corresponding kinds of probabilistic confirmation. We start with *structural confirmation*, which has an objective and a logical version. Consider first an example dealing with a fair die. Let E indicate the even (elementary) outcomes 2, 4, 6, and H the 'high' outcomes 4, 5, 6. Then (the evidence of) an even outcome confirms the hypothesis of a high outcome according to both criteria, since $p(E/H) = p(H/E) = 2/3 > 1/2 = p(H) = p(E)$.

We define structural confirmation as (backward) confirmation based on a probability function assigning equal and constant probabilities to the elementary outcomes. Such a probability function may either represent an *objective* probability process, such as a fair die, or it may concern the so-called *logical* probability or logical measure function (Kemeny, 1953), indicated by m. Kemeny's m-function assigns probabilities on the basis of ((the limit of) the ratio of) the number of structures making a proposition true, that is, its number of models (cf. the random-world or labeled method in Grove, Halpern and Koller (1996)). These logical probabilities may or may not correspond to the objective probabilities of an underlying process, as is the case with a fair die. Hence, for structural confirmation, we may restrict the attention to (generalizations) of Kemeny's m-function.

Structural confirmation is a straightforward generalization of d-confirmation. For suppose that H entails E. Then $m(E/H) = (\lim)|Mod(E\&H)|/|Mod(H)| = 1 > (\lim)$ $|Mod(E)|/|Mod(Tautology)| = m(E)$, where e.g. '$|Mod(H)|$' indicates the number of models of H. Moreover, it is a probabilistic explication of Salmon's (1969) idea of confirmation by 'partial entailment', according to which an even outcome typically is partially implied by a high outcome.

It is important to note that the m-function leads in many cases to 'm-zero' hypotheses (cf. Compton, 1988). For instance, every universal generalization 'for all x Fx' gets zero m-value for an infinite universe. As we may conclude from the general exposition, such hypotheses may well be structurally confirmed by certain evidence, by definition, according to the success criterion, but not according to the standard criterion. For example, a black raven structurally confirms 'all ravens are black' according to the success criterion, even if the universe is supposed to be infinite. Typical for the m-function is that it lacks the property which is characteristic for inductive probability functions.

Inductive Confirmation

Inductive confirmation is (pace Popper and Miller (1983)) explicated in terms of confirmation based on an inductive probability function, i.e. a probability function p having the general feature of 'positive relevance', 'inductive confirmation' or, as we like to call it,

instantial confirmation: $p(Fa/E\&Fb) > p(Fa/E)$

where 'a' and 'b' represent distinct individuals, 'F' an arbitrary monadic property and 'E' any kind of contingent evidence. Note that this definition is easy to generalize to n-tuples and n-ary properties, but we will restrict the attention to monadic ones. Since the m-function satisfies the condition $m(Fa/Fb\&E) = m(Fa/E)$, we get for any inductive probability function p:

$$p(Fa\&Fb/E) = p(Fa/E) \cdot p(Fb/E\&Fa) > m(Fa\&Fb/E)$$

Inductive (probability) functions can be obtained in two ways, which may also be combined:

- 'inductive priors', i.e. positive prior p-values $p(H)$ for m-zero hypotheses

and/or

- 'inductive likelihoods', i.e. likelihood functions $p(E/H)$ having the property of instantial confirmation

Note first that forward confirmation of m-zero hypotheses requires inductive priors, whereas backward confirmation of such hypotheses is always possible, assuming that $p(E/H)$ can be interpreted. Below we will give a general definition of inductive confirmation in terms of degrees of confirmation.

In terms of the two origins of inductive probability functions we can characterize the four main theories of confirmation in philosophy of science:

	inductive priors	inductive likelihoods
Popper	no	no
Carnap	no	yes
Bayes	yes	no
Hintikka	yes	yes

Popper rejected both kinds of inductive confirmation, roughly, for three reasons: two problematic ones and a defensible one. The first problematic one (Popper, 1934/1959) is that he tried to argue, not convincingly (see e.g. Earman, 1992, Howson and Urbach, 1989, Kuipers, 1978), that $p(H)$ could not be positive. The second one is that any probability function has the property '$p(E \rightarrow H/E) < p(E \rightarrow H)$' (Popper and Miller (1983)). Although the claimed property is undisputed, the argument that a proper inductive probability function should have the reverse property, since '$E \rightarrow H$' is the 'inductive conjunct' in the equivalence '$H \Longleftrightarrow (E \vee H) \& (E \rightarrow H)$', is not convincing. The indicated reverse property may well be conceived as an unlucky first attempt to explicate the core of (probabilistic) inductive intuitions, which should be replaced by the property of inductive confirmation. The defensible reason is that the latter property merely reflects a subjective attitude and, usually, not an objective feature of the underlying probability process, if there is such a process at all.

Carnap, following Laplace, favored inductive likelihoods, although he did not reject inductive priors. The so-called Bayesian approach in philosophy of science reflects inductive priors (but Bayesian statistics uses inductive likelihoods as well, see Festa, 1993). Finally, Hintikka introduced 'double inductive' probability functions, by combining the Carnapian and the Bayesian approach. For details, see ICR (Chapter 4).

7.1.2.2 *Degrees of Inductive Confirmation*

We now turn to the problem of defining a *degree of inductive confirmation* such that it entails a general definition of *inductive confirmation*. The present approach is not in the letter but in the spirit of Mura (1990) (see also e.g. Schlesinger (1995)) and Milne (1996) and Festa (1999). The idea is to specify a measure for the degree of inductive influence by comparing the relevant 'p-expressions' with the corresponding (structural) 'm-expressions' in an appropriate way. We proceed in three stages.

Stage 1. In the first stage we propose, instead of the standard difference measure $p(H/E)/p(H)$, the non-standard ratio measure $p(E/H)/p(E)$ as the degree (or rate) of (backward) confirmation (according to p), indicated by $c_p(H,E)$. This ratio has the following properties. For p-non-zero hypotheses, it is equal to the standard ratio measure $p(H/E)/p(H)$, and hence is symmetric ($c_p(H,E) = c_p(E,H)$), for p-non-zero hypotheses, but it leaves room for confirmation (amounting to: $c_p(H,E) > 1$) of p-zero-hypotheses. Moreover, it satisfies the comparative principles of deductive (d-)confirmation P1 and P2. Note first that $c_p(H,E)$ is equal to $1/p(E)$ when H entails E, for $p(E/H) = 1$ in that case. This immediately implies P2: if H and H^* both entail E then $c_p(H,E) = c_p(H^*,E)$. Moreover, if H entails E and E^*, and E entails E^* (and not vice versa) then $c_p(H,E) > c_p(H,E^*)$ as soon as we may not only assume that $p(E)p(E^*)$ (which follows from the assumption that E entails E^*), but even that $p(E) < p(E^*)$. This condition amounts to (a slightly weakened version) P1. In agreement with P2 we obtain, for example, that an even outcome with a fair die equally d-confirms the hypotheses {6}, {4,6}, and {2,4,6}, with degree of (structural and deductive) confirmation 2. This result expresses the fact that the outcome probabilities are multiplied by 2, raising them from 1/6 to 1/3, from 1/3 to 2/3, and from 1/2 to 1, respectively. Note also that in the paradigm example of structural non-deductive confirmation, that is, an even outcome confirms a high outcome, the corresponding degree is $(2/3)/(1/2) = 4/3$.

As suggested, there are a number of other degrees of confirmation. Fitelson (1999) evaluates four of them, among which the logarithmic forward version of our backward ratio measure, in the light of seven arguments or conditions of adequacy as they occur in the literature. The ratio measure fails in five cases. Three of them are directly related to the 'pure' character of r, that is, its satisfaction of P2.[7] In Chapter 2 of ICR P2 is defended extensively. However, we also argue there, in Chapter 3, that as soon as one uses the probability calculus, it does not matter very much which 'confirmation language' one chooses, for that calculus provides the crucial means for updating the plausibility of a hypothesis in the light of evidence. Hence, the only important point which then remains is to always make clear which confirmation language one has chosen.

Stage 2. In the second stage we define, as announced, the degree of inductive influence in this degree of confirmation, or simply the degree of inductive (backward) confirmation (according to p), as the ratio:

$$r_p(H,E) = \frac{c_p(H,E)}{c_m(H,E)} = \frac{p(E/H)/p(E)}{m(E/H)/m(E)}$$

A direct consequence of this definition is that the degree of confirmation equals the product of the degree of structural confirmation and the degree of inductive confirmation.

Stage 3. In the third and final stage we generally define *inductive confirmation*, that is, E inductively confirms H, of course, by the condition: $r_p(H,E) > 1$. This definition leads to four interesting possibilities for confirmation according to p. Assume that $c_p(H,E) > 1$. The *first* possibility is *purely structural* confirmation, that is, $r_p(H,E) = 1$, in which case the confirmation has *no inductive features*. This trivially holds in general for structural confirmation, but it may occasionally apply to cases of confirmation according to some p different from m. The *second* possibility is that of *purely inductive* confirmation, that is, $c_m(H,E) = 1$, and hence $r_p(H,E) = c_p(H,E)$. This condition typically applies in the case of instantial confirmation, since, e.g. $m(Fa/Fb\&E)/m(Fa/E) = 1$. The *third* possibility is that of a combination of structural and inductive confirmation: $c_m(H,E)$ and $c_p(H,E)$ both exceed 1, but the second more than the first. This type of *combined* confirmation typically occurs when a Carnapian inductive probability function is assigned e.g. in the case of a die-like object of which it may not be assumed that it is fair. Starting from equal prior probabilities for the six sides such a function gradually approaches the observed relative frequencies. If among the even outcomes a high outcome has been observed more than expected on the basis of equal probability then (only) knowing in addition that the next throw has resulted in an even outcome confirms the hypothesis that it is a high outcome in two ways: structurally (see above) and inductively.

> Let n be the total number of throws so far, let n_i indicate the number of throws that have resulted in outcome i $(1, \ldots, 6)$. Then the Carnapian probability that the next throw results in i is $(n_i + \lambda/6)/(n + \lambda)$, for some fixed finite positive value of the parameter λ. Hence, the probability that the next throw results in an even outcome is $(n_2 + n_4 + n_6 + \lambda/2)/(n + \lambda)$, and the probability that it is 'even-and-high' is $(n_4 + n_6 + \lambda/3)/(n + \lambda)$. The ratio of the latter to the former is the posterior probability of a high next outcome given that it is even and given the previous outcomes. It is now easy to check that in order to get a degree of confirmation larger than the structural degree, which is 4/3 as we have noted before, this posterior probability should be larger than the corresponding logical probability, which is 2/3. This is the case as soon as $2n_2 < n_4 + n_6$, that is, when the average occurrence of '4' and '6' exceeds that of '2'.

Let us finally turn to the *fourth* and perhaps most surprising possibility: confirmation combined with the 'opposite' of inductive confirmation, that is, $r_p(H,E) < 1$, to be called *counter-inductive* confirmation. Typical examples arise in the case of deductive confirmation. In this case $r_p(H,E)$ reduces to $m(E)/p(E)$, which may well be smaller than 1. A specific example is the following: let E be $Fa\&Fb$ and let p be inductive then E d-confirms 'for all x Fx' in a counter-inductive way. On second thought, the possibility of, in particular, deductive counter-inductive confirmation should not be surprising. Inductive probability functions borrow, as it were, the possibility of inductive confirmation by reducing the available 'amount' of deductive confirmation. Further research will determine whether deductive and inductive confirmation can ever go together in a meaningful way.

This last analysis completes the treatment of HD-testing of a theory in terms of confirmation and falsification. We now turn to HD-evaluation, which will leave room for continued interest in theories after their falsification.

7.2 SEPARATE HD-EVALUATION OF A THEORY

Introduction

The core of the HD-method for the evaluation of theories amounts to deriving from the theory in question, say *X*, General Test Implication (GTI's) and subsequently (HD-)testing them. For every GTI *I* holds that testing leads sooner or later either to a counter-example of *I*, and hence a counter-example of *X*, or to the (revocable) acceptance of *I*: a *success* of *X*. A counter-example implies, of course, the falsification of *I* and *X*. A success minimally means a 'derivational success'; it depends on the circumstances whether it is a predictive success and it depends on one's epistemological beliefs whether or not one speaks of an explanatory success.

However this may be, from the point of view of evaluation falsification is, although an interesting fact, no reason to stop the evaluation of the theory. One will derive and test new test implications. The result of such a systematic application of the HD-method is a (time relative) *evaluation report* of *X*, consisting of registered counter-examples and successes.

Now, it turns out to be very clarifying to write out in detail what is implicitly well-known from Hempel's and Popper's work, viz. that the HD-method applied to theories is essentially a stratified, two-step method, based on a macro- and a micro-argument, with much room for complications. In the already indicated macro-step, one derives GTI's from the theory. In their turn, such GTI's are tested by deriving from them, in the micro-step, with the help of suitable initial conditions, testable individual statements, called Individual Test Implications (ITI's).

In this section we will deal with the macro-argument, the micro-argument, and their combination into three models of (separate) evaluation. In the second section special attention will be paid to so-called falsifying general hypotheses, to statistical test implications, and to complications of testing and evaluation.

7.2.1 *The macro HD-argument.*

As indicated, a theory *X* is tested and evaluated by deriving *General Test Implications* (GTI's) from it, and testing these separately. Each GTI deals with a certain domain specified in observation terms. A GTI is general in the sense that the domain is not related to a specific (object or) system *and* place *and* time. In at least one of these respects, the domain is supposed to be general, but not necessarily in the universal sense. That is, the domain may be restricted within some boundaries, e.g. to all systems of a certain type, to all places in a region, to all times in an interval.[8] Moreover, within these boundaries, it may concern all actual cases or it may concern all possibilities in reality, all nomic possibilities. If the GTI is true, one speaks in the first case about an accidental general fact, and in the second case about a lawlike or nomic general fact, or simply a law.

Example (Cf. Subsection 1.1.7): A famous example of a GTI is Dalton's derivation of what became known as the law of multiple proportions from his theory of the atom. The internal and bridge principles of Dalton's theory were the following, in abbreviated form: IP1: *atoms* are indivisible, unchangeable, hence undestroyable, small material particles. IP2: atoms are grouped into *molecules*, and they may regroup into other molecules. BP1: pure substances consist of one type of molecules. BP2: chemical reactions come down to systematic regrouping of the molecules of a substance.
The suggested GTI, i.e. the law of multiple proportions, says that when two different elements unite into two different compounds, the different proportions bear a simple numerical relation to one another. Note that Proust's law of definite proportions, stating that compounds always decompose into components with constant weight ratios, can also be reconstructed as a GTI of Dalton's theory. However, in fact this law was not only one of Dalton's starting points for his theory, but, as we will see, is also useful in testing the multiple proportions GTI.

A GTI is assumed to be testable in principle and in practice. It is testable *in principle* if it is formulated in observation terms. To be testable *in practice* too, several specific conditions, depending on the context, will have to be satisfied.

A GTI is an implication or conditional statement in two senses. First, in the sense that it claims that for all cases in its domain, satisfying certain Initial Conditions (IC), a certain other individual fact applies. The conditionally claimed fact can be, like the initial conditions, of a (simple or compound) deterministic or statistical nature.

In sum, a GTI is formally of the form:

$$I: \text{for all } x \text{ in } D \text{ [if } C(x) \text{ then } F(x)]$$

that is, for all x in the domain D, satisfying the initial conditions $C(x)$, the fact $F(x)$ is 'predicted'.

The second sense in which a GTI is an implication directly relates to the HD-method. It is assumed to be an implication of some theory, i.c. X, i.e. to follow from this X (and auxiliary hypotheses, see below) by some logico-mathematical derivation. Scheme 7.4. represents this, what we call, *macro HD-argument*. (LMC indicates the suggested *Logico-Mathematical Claim* and MP indicates Modus Ponens.) Of course, it is a subtype of the MP-argument scheme (Scheme 7.1.1) given in Section 7.1. Though not indicated in the scheme, it is assumed that X is necessary for the derivation, i.e. I is not a logico-mathematically provable truth.

theory: X
LMC: if X then I
____________________ MP
GTI: I

Scheme 7.4 The macro HD-argument.

7.2.2 *Individual Problems and General Successes*

Let us now concentrate on the results of testing GTIs. When testing a GTI of a theory, we are interested in its truth-value, hence we use the test terminology.

Successive testing of a particular GTI I will lead to two mutually exclusive results. The one possibility is that sooner or later we get falsification of I by coming across a falsifying instance or counter-example of I, i.e. some x_0 in D such that $C(x_0)$ and $not\text{-}F(x_0)$, where the latter conjunction may be called a falsifying combined (individual) fact.

Assuming that LMC is correct, a counter-example of I is, strictly speaking, also a counter-example of X, falsifying X, for not only can not-I be derived from the falsifying combined fact, but also not-X by Modus Tollens. Hence, from the point of view of testing, it is plausible to speak also of falsification of the theory. However, it will frequently be useful in this chapter, and perhaps more in line with the evaluation terminology, to call the counter-example less dramatically a *negative instance*, and further to speak of a *negative (combined) individual fact*, or simply an *individual problem* of X.

The alternative possibility is that, despite variations in members of D and ways in which C can be satisfied, all our attempts to falsify I fail, i.e. lead to the predicted results. The conclusion attached to repeated success of I is of course that I is established as true, i.e. as a general (reproducible) fact.

Now one usually calls the acceptance of I as true at the same time a confirmation or corroboration of X, and the realist will want to add that I has been explained by X and hence that it is an explanatory success of X. However, this terminology is laden with the assumption that X has not yet been falsified, via another general test implication, or even that it is accepted as true. To block these connotations, it will be useful for evaluative purposes to call I a *positive general fact* or simply a *(general) success* of X.

It may well be that certain GTIs of X have already been tested long before X was taken into consideration. The corresponding individual problems and general successes have to be included in the evaluation report of X (see below).

Registered problems and successes are (partial) answers to the success question: what are the potential successes and problems of the theory? Hence, testing GTIs derived according to the macro HD-argument is effective in answering this question. Moreover, it is efficient, for it will never lead to irrelevant, neutral results.

Neutral results only come into the picture when we take the comparative evaluation of two or more theories into consideration (see the next chapter). In sum, evaluating a theory along the lines of the macro HD-argument is effective and efficient for answering the success question, whatever the epistemological beliefs of the scientist.

We call the list of partial answers to the success question, which are available at a certain moment t, the *evaluation report* of X at t, consisting of the following two components:

the set of individual problems,
i.e. established counter-examples of GTIs of X,
the set of general successes,
i.e. the established GTIs of X, that is, general facts derivable from X.

Hence, the goal of separate theory evaluation can be explicated as aiming at such an evaluation report.

Note that the two components of the report concern matters of a different nature. An individual problem specifies (a combined statement concerning) an individual item of the domain. A general success is a general statement considered to be true.

Note also that the derivation of a general test implication from a theory is essentially a prediction of a certain general fact. If this predicted general fact turns out to be true, this general fact is called a (general) predictive success. If the general fact was established before it was derived from the theory, it is sometimes called a postdictive success.

Of course, the realist will call a predictive as well as a postdictive (individual or general) success also an explanatory success and refer to the famous symmetry between explanation and prediction.[9] The only difference between explanation and prediction concerns the question of whether the relevant fact was established as a fact before or after the derivation. This relative non-distinction allows the realist to interchange explanation and prediction, depending merely on whether the supposed fact has, or has not yet, been established as such.

7.2.3 *The Micro HD-Argument*

Let us have a closer look at the testing of a general test implication, the micro-step of the HD-method. To study the testing of GTIs in detail, it is plausible to widen the perspective to the evaluative point of view on GTIs and to neglect their derivability from X.

Let us call a statement satisfying all conditions for being a GTI, except its derivability from some given theory, a *General Testable Conditional* (GTC). Let G be such a statement, which of course remains of the same form as a GTI:

$$G : \text{for all } x \text{ in } D \ [\text{if } C(x) \text{ then } F(x)]$$

To evaluate G we derive from G for some x_0 in D and suitable *initial conditions*

(IC), viz. $C(x_0)$, the predicted individual fact or individual prediction $F(x_0)$, i.e. an *Individual Test Implication* (ITI). It is an *individual* prediction in the sense that it concerns a specific statement about an individual item in the domain, as do the relevant initial conditions. It is a *prediction* in the sense that the outcome is assumed not to be known beforehand. Hence, talking about a prediction does not imply that the fact itself should occur later than the prediction, only that the establishment of the fact has to occur later (leaving room for so-called retrodictions).

What is predicted, i.e. $F(x_0)$, is usually called an (individual) effect or event. Both are, in general, misleading. It may concern a retrodiction (even relative to the initial conditions), and hence it may be a cause. And it may concern a state of affairs instead of an event. For these reasons we have chosen the neutral term (predicted) individual 'fact'.

In Scheme 7.5 the micro-reasoning of the HD-method, the *micro HD-argument*, is represented, where UI indicates Universal Instantiation.

$$\begin{array}{cl} G\text{: for all } x \text{ in } D \text{ [if } C(x) \text{ then } F(x)] & \\ x_0 \text{ in } D & \\ \hline \text{if } C(x_0) \text{ then } F(x_0) & \text{UI} \\ \text{IC: } C(x_0) & \\ \hline \text{ITI: } F(x_0) & \text{MP} \end{array}$$

Scheme 7.5 The Micro HD-Argument.

We suggest speaking in general of *strictly individual* applications of the micro HD-argument when the items concern a specific object, time and place.

Example (Cf. Subsection 1.1.7.): Proust's law of definite proportions is in fact a GTC. Assuming that we have determined the weight ratio of two elements in one particular sample of a given compound (initial condition), we can derive the prediction (ITI) that the weight ratio in another particular sample is the same.

With respect to Dalton's proper GTI of his theory of the atom, that is, the one concerning multiple proportions, individual experiments can be reconstructed as starting from a given weight proportion of the elements in a sample of one of the compounds and predicting a 'simple' multiple, or inverse multiple, proportion in a sample of the other. In this case the testing of the multiple proportions hypothesis is strictly individual.

It is clear that in the present example, knowledge of relevant cases of Proust's law of definite proportions, one of Dalton's starting points, is sufficient to check whether there is a multiple proportion relation between the weight ratios. More precisely, given the definite proportion of the elements in one compound (initial condition), the GTI predicts the ITI that the definite proportion in the other bears a simple relation to the former. In this case IC and ITI are not related to a specific time and place, and hence this way of testing the multiple proportions hypothesis is not strictly individual.

7.2.4 *Individual Problems and Individual Successes*

The specific prediction posed by the individual test implication can come true or be proven to be false. If the specific prediction turns out to be false, then, assuming that the initial conditions were indeed satisfied, the hypothesis G has been falsified. The combined individual fact '$C(x_0)$ and *not*-$F(x_0)$', 'C_0 and *not*-F_0' for short, may be called a falsifying individual fact and x_0 a falsifying instance or counter-example of G. It will again be useful for evaluative purposes to speak of a *negative instance*, and a *negative (combined) individual fact* or simply an *individual problem* of G.

If the specific prediction posed by the individual test implication turns out to be true, we get the combined individual fact 'C_0 and F_0', which is not only compatible with (the truth of) G, but is, not in full, but *partially derivable* from G in the following sense. One of its conjuncts can be derived from G, given the other. Again, one may be inclined to talk about confirmation[10] or even about explanation. However, given that we do not want to exclude that G has already been falsified, we prefer again the neutral (evaluation) terminology: x_0 is called a *positive instance* and the combined individual fact 'C_0 and F_0' a *positive (combined) individual fact* or simply an *individual success* of G.

It is easy to verify that the same story can be told about '*not*-C_0 and *not*-F_0' for some x_0 in D, by replacing the role of 'C_0' by that of '*not*-F_0', the new initial conditions, and the role of 'F_0' by that of '*not*-C_0', the new individual test implication. The crucial point is that 'if C_0 then F_0' is logically equivalent to 'if *not*-F_0 then *not*-C_0'. By consequence, x_0 is a positive instance satisfying '*not*-C_0 and *not*-F_0', being a positive individual fact or an individual success of G.

The remaining question concerns how to evaluate the fourth and last combined individual fact '*not*-C_0 and F_0' concerning some x_0 in D. Of course, this fact is compatible with G, but none of the components is derivable from G and the other component. Hence, the fourth combined fact cannot be partially derived from G. Or, to put it differently, none of its components, taken as initial condition, can lead to a negative instance, whereas this is the case for (precisely) one of the components in the two cases of partially derivable facts. Hence the terms *neutral instance* and *neutral (combined) individual fact* or *neutral result* are the proper qualifications.

Consequently, the evaluation report of GTCs has, like the evaluation reports of theories, two sides; one for problems and the other for successes. Again, they form partial answers to the success question now raised by the GTC. However, here the two sides list entities of the same kind: negative or positive instances or individual facts, that is, individual problems and individual successes, respectively.

It is again clear that the micro HD-argument for a GTC G is effective and efficient for making its evaluation report: each test of G either leads to a positive instance, and hence to an increase of G's individual successes, or it leads to a negative instance, and hence to an increase of G's individual problems. It does not result in neutral instances. Note that it is crucial for this analysis that GTC's have, by definition, a conditional character.

What we have described above is the micro HD-argument for *evaluating* a GTC. When we restrict attention to establishing its truth-value, and hence stop with the first counter-example, it is the (micro) HD-argument for *testing* the GTC.

7.2.5 *Models of HD-Evaluation and HD-Testing of a Theory*

Applications of the macro- or micro-argument, or a combination of both, will be called applications of the HD-method. In this subsection, we construct three interesting models of (separate) HD-evaluation of a theory.

Concatenation of the macro and micro HD-argument gives the full argument for theory evaluation in terms of individual combined facts: individual initial conditions and individual conclusions, leading to individual problems and individual successes.

Instead of the two-step concatenated account, theory evaluation can also be presented completely in terms of contracted HD-evaluation, without the intermediate GTIs, directly leading to individual problems and individual successes. In the contracted argument, the premises are the theory X and the initial conditions $C(x_0)$, and the conclusion, i.e. ITI, is $F(x_0)$. The transition is based on logico-mathematical argumentation. It is easy to check that it is always possible to (re)introduce the intermediate level of a GTI.

Any application of the HD-method (concatenated or contracted) leading to an evaluation report with individual problems and individual successes will be called an application of the *micro-model* of HD-evaluation. It is clear that application of the micro-model is possible for all kinds of general hypotheses, from GTCs to theories with proper theoretical terms.

However, as far as theories which are not just GTCs, are concerned, the macro-step also suggests the model of *asymmetric* HD-evaluation of a theory, leading to an evaluation report with individual problems and general successes. In that case, GTIs are derived in the macro-step, and only tested, not evaluated, in the micro-step. Hence, individual successes not fitting into an already established general derivable fact, are not registered as successes. Note that it directly follows from the contracted argument that an individual problem can always be reconstructed as a counter-example of a GTI of the theory.

In the micro-model of HD-evaluation of theories, in particular when contraction is used, the intermediate general successes of theories, i.e. positive general facts, may disappear from the picture. However, in scientific practice, these intermediate results frequently play an important role. The individual successes of theories are summarized, as far as possible, in positive general facts. These general successes relativize the dramatic role of falsification via other general test implications. As we will see in the next chapter, they form a natural unit of merit for theory comparison, together with conflicting individual facts, as the unit of problems.

In the next chapter, the model of asymmetric HD-evaluation will play a dominant role. The results it reports will then be called counter-examples and (general) successes.

In the next section, we study the possibility that individual problems can be summarized in general problems, that is, in negative general facts. Hence, there is also room for a *macro-model* of HD-evaluation, where the evaluation report lists, besides general successes, general problems as well. In this case, all individual successes and individual problems are left out of the picture, as long as they do not fit into an established general fact derivable from or in conflict with the theory.

Note that there is also the possibility of a fourth model of HD-evaluation of an asymmetric nature, with individual successes and general problems, but as far as we can see, it does not play a role in scientific practice.

The three interesting models of HD-evaluation of theories can be ordered by increasing refinement: the macro-model, the asymmetric model and the micro-model. Of course, in models where all individual problems are taken into consideration, they may or may not be summarized as far as possible in general problems. The same holds for individual successes.

Finally, it is plausible to characterize models of *HD-testing of theories* as the application of one of the three models up to the point of registering the first individual or general problem, that is, a counter-example or a falsifying general fact. It is clear that the corresponding *test* reports now provide partial answers, not only to the truth question, but also to the success question, up to falsification. For the realist this implies, among other things, a restriction to successes that may be genuine explanatory merits. Moreover, it will be clear that the models of HD-testing produce the test reports in an effective and efficient way for the same reasons as the models of HD-evaluation do: HD-testing leads to successes or problems, and not to neutral results. As already suggested, the exclusive interest in HD-testing of theories will be called the (naive) falsificationist perspective or method.

Table 7.1 summarizes the four methodologically relevant categories and their terminological variants. It is easy to read off the four possible models of HD-evaluation and HD-testing.

Table 7.1 Methodological categories of theory evaluation and testing.

	problems	successes
individual	individual problem negative instance negative individual fact counter-example	individual success positive instance positive individual fact
general	general problem negative general fact falsifying general fact	general success positive general fact

7.3 FALSIFYING GENERAL HYPOTHESES, STATISTICAL TEST IMPLICATIONS, AND COMPLICATING FACTORS

Introduction

In this section we will first deal with so-called falsifying general hypotheses, that is, general problems, summarizing individual problems, i.e. counter-examples. Then we will show that the main lines of the analysis of testing and evaluation also apply when the test implications are of a statistical nature. Finally, we will deal with all kinds of complications of testing and evaluation, giving occasion to dogmatic strategies and suggesting a refined scheme of HD-argumentation.

7.3.1 *Falsifying General Hypotheses and General Problems*

In this subsection we pay attention to a methodological issue that plays an important role in the methodology of Popper and others: so-called falsifying general hypotheses. In contrast to their dramatic role in HD-testing of theories, in HD-evaluation they play the more modest role of a general problem which *summarizes* individual problems on the minus side of the evaluation report of a theory.

Let us return to the evaluation or testing of a general testable conditional (GTC) G. Finding a partially implied instance of G does, of course, not imply that G is true. Repetitions of the individual tests, varying the different ways in which the initial conditions can be realized, are necessary to make it plausible that G is true. To conclude that it is true implies making an inductive jump, at least in the case that G has some infinite aspect. But as soon as one makes this jump, one has established a new general fact, in this case, G. If G is in fact a GTI of X, it becomes a positive general fact of X, i.e. a general success of X, dealt with before.

If we are only interested in the truth-value of G, its conclusive falsification stops the relevance of its further testing. From this test perspective, repeating similar individual tests a number of times, after one falsification, only serves the purpose of making sure that the initial conditions and the falsifying outcome have been realized properly at least once.

However, further evaluation of a conclusively falsified GTC makes sense for several other reasons, in particular, when it is a test implication of some theory. The results of repeated tests may suggest an interesting division of counter-examples and individual successes of G, and hence one or more alternative general hypotheses. This may lead to the establishment of other general facts which are also relevant for theory comparison. In particular, it may turn out that an alternative GTC, with the same domain:

$$G^{*}\text{: for all } x \text{ in } D \text{ [if } C^{*}(x) \text{ then } F^{*}(x)]$$

can be established, with C^* implying C and F^* implying *not-F*, such that each partially derivable individual fact of G^* of type 'C^*&F^*' is a falsifying instance of G. In that case, G^* may be called a *falsifying* or *negative general fact* for G. When G is a GTI of X, G^* may also be called a lower level falsifying (general) hypothesis, to use Popper's phrase, now contradicting not only G but also X. However, more in line with the terminology of evaluation, we call it a *negative general fact* or *general problem* of X.

> *Example* (Cf. Subsection 1.1.7): An example is the law of combining volumes (Gay-Lussac). Apart from by Dalton himself, it was generally considered as a problem for Dalton's version of the atomic theory. It can easily be reconstructed as a general problem in the technical sense defined above. As is well-known, Avogadro turned the tables on the atomic theory by a fundamental change in order to cope with this problem.

A technical alternative to the structure of a falsifying general fact which was described above is the following. A G^* may have been established for which *not-*C^* implies C and *not-*F^* implies *not-F*. Each partially derivable individual fact of this G^* of type '*not-*C^* *and not-*F^*' is a falsifying instance of G. Hence, this G^* would also be a falsifying general fact for G, and a falsifying general fact, or simply a general problem for X, if G is a GTI of X.

Note that in both described cases G^* and G form a contradiction as soon as C^*, respectively *not-*F^*, can be satisfied in the domain.

Of course, when positive and negative instances of G occur, repeated tests may lead not only to negative general facts, but also to the establishment of weaker variants of G. In particular, consider the GTC

$$G'\text{: for all } x \text{ in } D \text{ [if } C'(x) \text{ then } F'(x)]$$

with C' implying C and F implying F', then G implies G' and every negative instance of G' is a negative instance of G. When G' becomes established one may call it a general fact derivable from G, and hence a general success of X if G is a GTI of X.

As soon as, and as long as, all negative individual facts of a theory can be summarized in negative general facts, the individual problems in the evaluation report can be replaced by the corresponding general problems. In this way we get on both sides of the record the same kind of conceptual entities, viz. general facts, forming the ingredients of the macro-model of HD-evaluation of theories. Hence, the foregoing exposition concerning general facts amounts to an additional illustration of the fact that general hypotheses and the like can play an important intermediate role in the evaluation of theories. Skipping the intermediate notions of general successes and general problems would hide the fact that they make the evaluation of theories very efficient, in theory and practice. Instead of confronting theories with all previously or subsequently established combined individual facts, it is possible to restrict the confrontation as much as possible to a confrontation with old or new summarizing general facts.

7.3.2 *Statistical Test Implications*

The presentation thus far may have suggested that the analysis only applies to hypotheses and their test implications as far as they are of a deterministic and non-comparative nature. In this subsection we will present the adapted main lines for statistical test implications, first of a non-comparative and then of a comparative nature. In the literature concerning statistical testing, one can find all kinds of variants and details.[11] Here it is only necessary to show that statistical general and individual test implications can essentially be tested in a similar way to non-statistical ones.

In the non-comparative form, a typical case is that the theory, e.g. Mendel's theory, entails a probabilistic GTI of the following abstract form: in domain D the probability of feature F on the condition C satisfies a certain probability distribution p (e.g. binomial or normal). The sample version, the proper Statistical GTI, and the corresponding IC and ITI are then respectively of the following form:

> GTI: for all $\alpha (0 < \alpha < 1)$ and for all random samples s of (sufficiently large, determined by the distribution) size n of individuals from domain D satisfying condition C there are a and b $(0 < a < b < 1)$ such that the probability that the ratio of individuals satisfying F is in the region $[0, a]$ does not exceed $\alpha/2$ and similar for the region $[b, 1]$[12]

> IC: sample s is a random sample of (sufficiently large) size n of individuals from D satisfying C

> ITI: the probability that the ratio in s of individuals satisfying F is in $R(\alpha, n) =_{\text{def}} [0, a] \cup [b, 1]$ does not exceed α

If α is small enough, according to our taste, we may, by a final non-deductive jump, hypothesize the non-probabilistic ITI:

> ITI_{np}: the ratio in s of individuals satisfying F will be outside $R(\alpha, n)$

In classical statistics a test of this kind is called a *significance test* with *significance level* α (standard values of α are 0.005 and 0.001) and *critical region* $R(\alpha, n)$. Moreover, the abstract GTI is called the *null hypothesis* and the classical decision rule prescribes to reject it when ITI_{np} turns out to be false and not to reject it when ITI_{np} is true. However, from our perspective it is plausible to categorize the first test result merely as a negative sample result or a (statistical) 'counter-sample' and the second as a positive one. Of course, strictly speaking, a counter-sample does not falsify the GTI, let alone the theory. Moreover, from repeated positive results one may inductively jump to the conclusion of a general statistical success GTI.

Statistical test implications are frequently of a comparative nature. This holds in particular when they derive from causal hypotheses. The reason is that such hypotheses are essentially double hypotheses, one for the case that the relevant (supposedly) causal factor is present and another for the case that it is absent. Typical examples are the causal hypotheses governing drug testing. An adapted standard significance test for a supposedly normally distributed feature F on condition C in domain $D1$ and $D2$, with the same variance, focusses on the (null) hypothesis that their expectation values are the same. It is called the 't-test' or 'Student-test' (after W.S. Gossett who wrote under the name 'Student') and the resulting proper Statistical GTI, and corresponding IC, ITI and ITI_{np} are now respectively of the following form:

> GTI(comp): for all $\alpha (0 < \alpha < 1)$ and for all sufficiently large (random) samples $s1$ and $s2$ of size $n1$ and $n2$ of individuals from domain $D1$ and $D2$, respectively, satisfying condition C there are a and b $(0 < a < b < 1)$ such that the probability that a certain function (the t-statistic) of the difference between the respective ratios of individuals satisfying F is in the region $[0, a]$ does not exceed $\alpha/2$ and similar for the region $[b, 1]$[13]

> IC(comp): sample $s1$ and $s2$ are random samples of (sufficiently large) sizes $n1$ and $n2$ of individuals from $D1$ and $D2$, respectively, satisfying C

> ITI(comp): the probability that the value of the t-statistic is in $R(\alpha, n) = [0, a] \cup [b, 1]$ does not exceed α

Again, if α is small enough, according to our taste, we may, by a final non-deductive jump, hypothesize the non-probabilistic ITI:

> ITI_{np}(comp): the value of the t-statistic will be outside $R(\alpha, n)$

This can again be described in classical terms of significance level and rejection conditions. However, in our perspective it is again more plausible to use the more cautious terminology of positive and negative test results for the null hypothesis, which are, of course, negative and positive test results for the primary (e.g. causal) hypothesis at stake.

7.3.3 *Factors Complicating HD-Testing and Evaluation*

According to the idealized versions of HD-testing and evaluation presented so far there are only cases of evident success or failure. However, as is well-known, several factors complicate the application of the HD-method. Let us approach them first from the falsificationist perspective. Given the fact that scientists usually believe that their theory is (approximately) true, they have, on the basis of these factors, developed strategies to avoid the conclusion of falsification. The important point of these dogmatic or conservative strategies is that they may rightly save the theory from falsification, because the relevant factor may really be the cause of the seeming falsification. Although the recognition of a problem for a theory is more dramatic from the falsificationist perspective, when evaluating a theory one may also have good reasons for trying to avoid a problem. Wolpert (1992) briefly discusses several examples.

We distinguish five complicating factors, each leading to a standard saving strategy. They show in detail, among other things, how Lakatos' methodology of research programs (Lakatos, 1970/1978), saving the hard core, can be defended and effected. Though perhaps less frequently, the same factors may also be used, rightly or wrongly, as point of impact for contesting some success. In this case, there is one additional factor. All six factors concern suppositions in the concatenated macro and micro HD-argument. We do not claim originality with these factors as such; most of them have been mentioned by Lakatos and have been anticipated by Hempel, Popper and others. However, the following systematic survey and localization of them is made possible by the decomposition of the macro and micro HD-argument. It is left to the reader to identify examples of the factors, e.g. in the history of the theory of the atom.

7.3.3.1 *Dogmatic Strategies*

(1) The derivation of a general test implication from the theory is usually impossible without invoking explicitly or implicitly one or more *auxiliary hypotheses*, i.e. hypotheses which do not form a substantial part of the theory under test, but are nevertheless required to derive the test implication. An important type of such auxiliary hypotheses are *specification hypotheses*, that is, hypotheses that specify particular constraints for the values of certain (theoretical or non-theoretical) quantities in the specific kind of cases concerned. Hence, to avoid falsification, one can challenge an auxiliary hypothesis. A particular problem of auxiliary hypotheses may be that they are too idealized; they may need concretization.

(2) The derivation of the general test implication presupposes that the *logico-mathematical claim* can convincingly be proven. One may challenge this affirmation. A successful challenge may, of course, result in the formulation of implicit auxiliary hypotheses, i.e. new instances of the first factor, which can be questioned as indicated under the first point.

(3) Test implications have to be formulated in observation terms. However, at present, there is almost general agreement that pure observation terms do not exist. All observation terms are laden with hypotheses and theories. New, higher order, observation terms are frequently defined on the basis of more elementary ones and certain additional presuppositions, which provide, e.g. the relevant existence and

uniqueness conditions. Such presuppositions form a bridge between more and less theory-laden observation terms. These presuppositions belong to the so-called background knowledge; they form the underlying hypotheses and theories that are taken for granted. We call them the *observation presuppositions*. The theory to be evaluated should itself not be an observation presupposition, i.e. the observation terms relative to the theory should not be laden with that theory itself, only with other ones. The relevant strategy now is to challenge an observation presupposition. Ultimately, any challenge may bring us back to the observation language of a lay-person who is instructed by the experimenter to 'materially realize' (Radder, 1988, 1996) an experiment.

(4) A general test implication specifies *initial (test) conditions*. They have actually to be realized in order to conclude that the individual test implication must be(come) true. One may challenge the claim that these initial conditions were actually fulfilled. One important reason for repeating an experiment a number of times is to make it sufficiently sure that these conditions have at least once been realized. But if another outcome than the one predicted systematically occurs, one may defend the idea that there are structural causes preventing the fulfillment of the intended initial conditions in the way one is trying to install them.

(5) Whether an outcome is or is not in agreement with the predicted outcome is usually not a straightforward matter. This is particularly the case when the observation terms have vague boundaries or are quantitative, or when the predicted effects and the observation data are of a statistical nature. In all such cases, the question is whether the actual outcome is approximately equal to the predicted outcome. To decide this we need a (previously chosen) *decision criterion*. In the case of statistical individual test implications, several *statistical* decision criteria have been standardized. Although a statistical decision criterion also concerns an approximation question, we propose to reserve the term *approximation* decision criterion for the case of vague or quantitative observation concepts. In general, so the strategy goes, a decision criterion may or may not be adequate in a particular case, i.e. it may or may not lead to the correct decision in that case.

(6) Finally, in order to conclude that the theory has acquired a new general success, the relevant general test implication has first to be established on the basis of repeated tests. As a rule, establishing a GTI requires an inductive jump or *inductive generalization*, which may, of course, always be contested as unjustified. Note that a similar step is involved in the establishment of a general problem, but we will not discuss it here.

To be sure, the localization of these factors need not always be as unambiguous as suggested by the previous exposition, although we claim to have identified the main occurrences. However this may be, the consequence of the first five factors (auxiliary hypotheses, logico-mathematical claims, observation presuppositions, initial conditions, and decision criteria) is that a negative outcome of a test of a theory only points unambiguously in the direction of falsification when it may be assumed that the auxiliary hypotheses and the observation presuppositions are (approximately) true, that the logico-mathematical claim is valid, that the initial conditions were indeed realized and that the used decision criteria were adequate in the particular case. Hence, a beloved theory can be protected from threatening falsification by challenging one or more of these suppositions.

7.3.3.2 *Refined HD-Scheme*

In the subjoined, refined schematization of the concatenated HD-test arguments (Scheme 7.6) the five plus one vulnerable factors or weak spots in the argument have been made explicit and emphasized by question marks. As in the foregoing exposition, the indication in the scheme of the different types of weak spots is restricted to their main occurrences in the argument. SEE indicates Substitution of presupposed Empirically Equivalent terms (in particular, C by C^* and F by F^*). We contract the application of Universal Instantiation (UI) followed by Modus Ponens (MP).

Theory under test: X Auxiliary Hypotheses: A ?1?
LMC: if X and A then GTI ?2?
______________________________ MP
GTI: for all x in D [if $C(x)$ then $F(x)$]
Observation presuppositions: C/C^*, F/F^* ?3?
______________________________ SEE
GTI*: for all x in D [if $C^*(x)$ then $F^*(x)$]
x_0 in D
Initial conditions: $C^*(x_0)$?4?
________________________ UI + MP
ITI*: $F^*(x_0)$
Data from repeated tests
____________ Decision Criteria ?5?

either sooner or later a counter-example of GTI*, leading to the conclusion not-GTI*	*or* only positive instances of GTI*, suggesting inference of GTI* by Inductive Generalization ?6?

Scheme 7.6 The refined macro + micro HD-argument.

Ignoring the complicating factors, the left-hand case in Scheme 7.6 at the bottom results in falsification of GTI* and hence of GTI, leading to falsification of X. Under the same idealizations, the right-hand case results first in (the implied part ITI* of) individual successes of GTI* (and indirectly of GTI), and then, after the suggested inductive generalization, in the general success GTI* (and hence GTI) of X.

CONCLUDING REMARKS

If the truth question regarding a certain theory is the guiding question, most results of this chapter, e.g. the decomposition of the HD-method, the evaluation report and the survey of complications, are only interesting as long as the theory has not been falsified. However, if one is also, or primarily, interested in the success question the results remain interesting after falsification. In the next chapter we will show how this kind of separate HD-evaluation can be put to work in comparing the success of theories. Among other things, this application will explain and even justify non-falsificationist behavior, including certain kinds of dogmatic behavior.

CHAPTER 8
EMPIRICAL PROGRESS AND PSEUDOSCIENCE

INTRODUCTION

In this chapter we will extend the analysis of the previous chapter to the comparison of theories, giving rise to a definition of empirical progress and a sophisticated distinction between scientific and pseudoscientific behavior.

In Section 8.1 we will first describe the main line of theory comparison that forms the natural extension of separate HD-evaluation to comparative HD-evaluation. Moreover, we will introduce the rule of success for theory selection suggested by comparative HD-evaluation, leading to an encompassing evaluation methodology of instrumentalist flavor. This methodology can be seen as the core method for the assessment of claims to empirical progress.

In Section 8.2 we will compare the evaluation methodology with the three methods distinguished by Lakatos (1970/1978): the naive and sophisticated falsificationist method and the method of research programs, favored by Lakatos. We will show how the evaluation methodology resembles the sophisticated falsificationist methodology the most and how it may well be more efficient for truth approximation than the naive falsificationist method.

In Section 8.3 we will argue that the, in some way dogmatic, method of research programs may be a responsible way of truth approximation, as opposed to pseudoscientific dogmatic behavior.

8.1 COMPARATIVE HD-EVALUATION OF THEORIES

Introduction

The presented analysis of separate HD-evaluation has important consequences for theory comparison and theory selection. The momentary evaluation report of a theory generated by the macro-step immediately suggests a plausible way of comparing the success of theories, of further testing the comparative hypothesis that a more successful theory will remain more successful and, finally, the rule of theory selection, prescribing to adopt it, for the time being, if it has so far proven to be more successful.

The suggested comparison and rule of selection will be based on the asymmetric model of evaluation in terms of general successes and individual problems. However, it will also be shown that the symmetric approach, in terms of either individual or general successes and problems, leads to an illuminating symmetric evaluation matrix, with corresponding rules of selection.

8.1.1 *Theory Comparison*

A central question for methodology is what makes a new theory better than an old one. The intuitive answer for the new theory being as good as the old is plausible enough. The new theory has at least to save the established strengths of the old one and not to add new weaknesses on the basis of the former tests. In principle, we can choose any combination of individual or general successes and problems to measure strengths and weaknesses. However, the combination of general successes and individual problems, i.e. the two results of the asymmetric model of (separate) HD-evaluation, is the most attractive. First, this combination seems the closest to actual practice and, second, it turns out to be the most suitable one for a direct link with questions of truth approximation. For these reasons we will first deal with this alternative and come back to the two symmetric alternatives in a separate subsection (8.1.4).

Given the present choice, the following definition is the obvious formal interpretation of the idea of (*prima facie*) progress, i.e. *increasing success*:

> Theory Y is (at time t) *at least as successful as* (*more successful than* or *better than*) theory X iff (at t)

- the set of individual problems of Y forms a subset of that of X
- the set of general successes of X forms a subset of that of Y
- (in at least one case the relevant subset is a proper subset)[1, 2]

The definition presupposes, of course, that for every registered (individual) problem of one theory, it has been ascertained if it is also a problem for the other, and similarly if the (general) successes of one are also successes of the other. The first clause might be called the 'internal' clause and the second the 'external' clause, for 'truth approximation' reasons given in Part 3 of ICR and hinted at in the next chapter. However, the name 'instantial' clause for the first one is more appealing and relatively neutral. From the realist perspective it is plausible to call the second clause the 'explanatory' clause. From other epistemological perspectives one may choose another, perhaps more neutral name, such as, the general success clause.

It is also obvious how one should define, in similar terms to those above, the general notion of 'the most successful theory thus far among the available alternatives' or, simply, 'the best (available) theory'.

It should be stressed that the diagnosis that Y is more successful than X does not guarantee that this remains the case. It is a *prima facie* diagnosis based only on thus far established facts, and new evidence may change the comparative judgment. But, assuming that established facts are not called into question, it is easy to check that the judgement cannot have to be reversed, i.e. that X becomes more successful than Y in the light of old and new evidence. For, whatever happens, X has extra individual problems or Y has extra general successes.

It should be conceded that it will frequently not be possible to establish the comparative claim, let alone that one theory is more successful than all its available alternatives. The reason is that these definitions do not guarantee a constant linear ordering, but only an evidence-dependent partial ordering of the relevant theories. Of course, one may interpret this as a challenge for refinements, e.g. by introducing different concepts of 'relatively maximal' successful theories or by a quantitative approach.

8.1.2 *Testing the Comparative Success Hypothesis*

Be this as it may, we have defined enough notions to introduce our explication of the core of HD-evaluation in the following heuristic principles. The first principle will be self-evident: as long as there is no best theory, one may continue the separate HD-evaluation of all available theories in order to explore the domain further in terms of general facts to be accounted for and individual problems to be overcome by an overall better theory. We will concentrate on the second principle, applicable in the case that one theory is more successful than another one, and hence in the case that one theory is the best.

Suppose theory *Y* is at t more successful than theory *X*. This condition is not yet a sufficient reason to prefer *Y* in some substantival sense. That would be a case of 'instant rationality'. However, when *Y* is at a certain moment more successful than *X*, this situation suggests the following comparative success hypothesis:

CSH: *Y* (is and) will remain more successful than *X*

CSH is an interesting hypothesis, even if *Y* is already falsified. Given that *Y* is known to be more successful than *X* at *t*, CSH amounts at *t* to two components, one about problems, the other about successes:

CSH-P: all individual problems of *Y* are individual problems of *X*
CSH-S: all general successes of *X* are general successes of *Y*

where 'all' is to be read as 'all past and future'.

Although there may occasionally be restrictions of a fundamental or practical nature, these two components will usually concern testable generalizations. Hence, testing CSH requires application of the micro HD-argument. Following CSH-P, we may derive a GTI from *Y* that does not follow from *X*, and test it. When we get a counter-example of this GTI, and hence an individual problem of *Y*, it may be ascertained if the problem is shared by *X*. If it is not, we have falsified CSH-P.

Alternatively, following CSH-S, we may derive a GTI from *X* which cannot be derived from *Y*, and test it. If it becomes accepted, its acceptance means falsification of CSH-S. Of course, in both cases, the opposite test result confirms the corresponding comparative sub-hypothesis, and hence CSH, and hence increases the registered success difference. In the following we will, for obvious reasons, call (these two ways of) testing CSH *comparative HD-evaluation*.[3]

8.1.3 *The Rule of Success*

The plausible rule of theory selection is now the following:

> *Rule of success (RS)*
> When *Y* has so far proven to be more successful than *X*, i.e. when CSH has been 'sufficiently confirmed' to be accepted as true, eliminate *X* in favor of *Y*, at least for the time being

RS does not speak of 'remaining more successful', for that would imply the presupposition that the CSH could be completely verified (when true). Hence we speak of 'so far proven to be more successful' in the sense that CSH has been 'sufficiently confirmed' to be accepted as true; that is, CSH is accepted as a (twofold) inductive generalization. The point at which CSH is 'sufficiently confirmed' may be a matter of dispute. However this may be, the acceptance of CHS and consequent application of RS is the core idea of *empirical progress*, a new theory that is better than an old one. RS may even be considered as the (fallible) criterion and hallmark of scientific rationality, acceptable for the empiricist as well as for the realist.[4]

As soon as CSH is (supposed to be) true, the relevance of further comparative HD-evaluation is diminished. Applying RS, i.e. selecting the more successful theory, then means the following, whether or not that theory already has individual problems. One may concentrate on the further separate HD-evaluation of the selected theory, or one may concentrate on the attempt to invent new interesting competitors, that is, competitors that are at least as successful as the selected one.

Given the tension between increasing the domain of a theory and increasing its (general observational) success, it is not an easy task to find such interesting competitors. The search for such competitors cannot, of course, be guided by prescriptive rules, like RS, but there certainly are heuristic principles of which it is easy to see that they stimulate new applications of RS. Let us start by explicitly stating the two suggested principles leading to RS. First, there is the *principle of separate HD-evaluation* (PSE): 'Aim via general test implications at establishing new laws which can be derived from your theory (general successes) or, equivalently, aim at new negative instances (individual problems) of your theory'. Secondly, the *principle of comparative HD-evaluation* (PCE) 'Aim at HD-testing of the comparative success hypothesis, when that hypothesis has not yet been convincingly falsified'.[5]

As already suggested, RS presupposes previous application of PSE and PCE. But also some additional heuristic principles, though not necessary, may promote the application of RS. To begin with, the *principle of content* (PC) may do so: 'Aim at success preserving, strengthening *or*, pace Popper, weakening of your theory'. A stronger theory is likely to introduce new individual problems but gain new general successes. If the latter arise and the former do not materialize, RS can be applied. Something similar applies to a weaker theory. It may solve problems without sacrificing successes. We would also like to mention the *principle of dialectics*

(PD) for two theories that escape RS because of divided success: 'Aim at a success preserving synthesis of two RS-escaping theories'. In ICR (Section 8.3), we explicate a number of dialectical notions in this direction. Of course, there may come a point at which further attempts to improve a theory and hence to discover new applications of RS are abandoned.

In sum, the asymmetric model of HD-evaluation of theories naturally suggests the definition of 'more successful', the comparative success hypothesis, the testing of such a hypothesis, i.e. comparative HD-evaluation, and the rule of success (RS) as the cornerstone of empirical progress. Separate and comparative HD-evaluation provide the right ingredients for applying first the definition of 'more successful' and, after sufficient tests, that of RS, respectively. In short, separate and comparative HD-evaluation are functional for RS, and HD-testing evidently is functional for both types of HD-evaluation. The method of HD-evaluation of theories combined with RS and the principles stimulating the application of RS might well be called the instrumentalist methodology. In particular, it may be seen as a free interpretation or explication of Laudan's problem solving model (Laudan, 1977), which is generally conceived as a paradigm specification of the idea of an instrumentalist methodology. However, it will be called, more neutrally, the *evaluation methodology*. It will be said that this methodology is governed by RS. The claim is that this methodology governs the short-term dynamics of science, more specifically, the internal and competitive development of research programs.

Note that the evaluation methodology demonstrates continued interest in a falsified theory. The reasons behind it are easy to conceive. First, it is perfectly possible that the theory nevertheless passes other general test implications, leading to the establishment of new general successes. Second, even new tests leading to new individual problems are very useful, because they have to be overcome by a new theory. Hence, at least as long as no better theory has been invented, it remains useful to evaluate the old theory further in order to reach a better understanding of its strengths and weaknesses.

8.1.4 *Symmetric Theory Comparison*

The symmetric models of separate HD-evaluation, i.e. the micro- and the macro-models, suggest a somewhat different approach to theory comparison. Although these approaches do not seem to be in use to the extent of the asymmetric one and can only indirectly be related to truth approximation, they lead to a very illuminating (comparative) evaluation matrix.

Let us first examine in more detail precisely what we want to list in the three types of evaluation reports corresponding to the three models. From the present perspective of theory comparison, such specification is very important. A better theory has to be at least as successful as the old one, and this fact suggests general conditions of adequacy for the definitions of a success, of a problem and of a neutral result. A success is typically something to be retained by a better theory; a better theory is supposed not to introduce a new problem, and; finally, neutral results should remain neutral or become successes. Note that the definition of 'at least as successful' in Section 8.1.1 just summarizes the first two of these conditions in terms of individual problems and general successes, and leaves neutral results out of the picture (see below). We will now see that the general conditions of adequacy are very instructive for the specification of what precisely can be listed as successes and problems.

The notions of general successes and general problems are not problematic. A better theory retains general successes as general test implications, and does not give rise to new general test implications of which testing leads to the establishment of new general problems. Furthermore, the notion of a neutral general fact does not create problems. The notions of individual successes, individual problems and neutral individual facts are also not problematic as long as we list them in terms of positive, negative and neutral instances, respectively. A better theory keeps the positive instances as such; it does not lead to new negative instances, and neutral instances may remain neutral or become positive. However, if we want to list individual successes and/or individual problems in terms of statements, the situation becomes more complicated. The fact and the way in which that is possible is presented below, but the reader might prefer to go immediately to the evaluation matrix, which may be read in both ways.

8.1.4.1 *Individual Successes and Problems in Statement Form*

Recall that 'C_0 and F_0' was called a 'positive (combined) individual fact' of the GTC 'if C then F' and hence of theory X if this GTC is a GTI of X. Recall also that the conjunction 'C_0 and F_0' was only partially derivable from G or X: F_0 was derivable from G and hence of X, given C_0. But do we want to retain that F_0 is, given C_0, derivable from a supposedly better theory Y? As long as we have not established this G, or another GTI for which 'C_0 and F_0' is also a positive individual fact, the answer is: no! For it may well be that there is no better theory that can bridge the gap between C_0 and F_0. How then should we specify in propositional terms what we want to retain? In the light of the foregoing analysis, the plausible answer is the generalization of the idea of partial derivability: F_0 is a positive individual fact for X iff there are initial conditions C_0^*, not necessarily equivalent to C_0, such that X and C_0^* together (and not separately) imply F_0. Of course, only the strongest positive individual facts need to be listed and the whole story can be repeated for positive combined individual facts of the form '*not*-F_0 and *not*-C_0'.

Let us now look at 'C_0 and *not*-F_0', a 'negative (combined) individual fact' of the GTC 'if C then F' and hence of theory X if this GTC is a GTI of X. The following definition is now suggested by the analysis of positive individual facts. *Not*-F_0 is a negative individual fact for X iff there are initial conditions C_0^*, not necessarily equivalent to C_0, such that X and C_0^* together (and not separately) imply F_0. Note first that this definition automatically implies that the relevant C_0^* is also a negative individual fact for *X*. Negative individual facts typically come in pairs, and a new theory should not introduce such new pairs.

What happens if a new theory is better in the sense that it loses some individual facts as negative facts. Let *not*-F_0 and C_0 be a pair of negative individual facts for *X*, and let *not*-F_0 be not a negative individual fact for *Y*. This situation does not imply that C_0 is not a negative individual fact for *Y*, for it might come into conflict with *Y* and some other individual fact than *not*-F_0. Hence, although negative individual facts come in pairs, they do not need to lose that status together with regard to some other theory.

That *not*-F_0 is not a negative individual fact for *Y* also does not imply that it is a positive individual fact for *Y* in the sense defined above. This possibility suggests the plausible definition of a neutral individual fact for *Y*: a fact which is neither positive nor negative for *Y*. Note that if F_0 is a positive individual fact for *Y*, due to C_0, then this information alone suggests that C_0 is a neutral fact for *Y*. But it may well be that other facts than F_0 and/or more information about *Y*'s consequences lead to another status of C_0.

8.1.4.2 *The Evaluation Matrix*

Let us now look more specifically at the symmetric micro-model, counting in terms of individual problems, successes and neutral results, that is, negative, positive and neutral instances or (statements of) individual facts. Hence, in total, the two theories produce a matrix (Table 8.1) of nine combinations of possible instances or individual facts. In order to make the matrix also useful for the macro-model, we present it in terms of facts. For the moment, these facts are to be interpreted as individual facts. The entries represent the status of a fact with respect to the indicated theories.

From the perspective of *Y* the boxes B1/B2/B3 represent unfavorable facts (indicated by '−'), B4/B5/B6 (comparatively neutral or) indifferent facts (0), and B7/B8/B9 favorable facts (+). The numbering of the boxes, anticipating a possible quantitative use, was determined by three considerations: increasing number for increasingly favorable results for *Y*, symmetry with respect to the \-diagonal, and increasing number for increasingly positive indifferent facts.[6]

It is now highly plausible to define the idea that *Y* is *more successful* than *X* in the light of the available facts as follows: there are no unfavorable facts and there are

Table 8.1 The (comparative) evaluation matrix.

		X		
		negative	neutral	positive
Y	negative	B4: 0	B2: −	B1: −
	neutral	B8: +	B5: 0	B3: −
	positive	B9: +	B7: +	B6: 0

some favorable facts, that is, B1/2/3 should be empty, and at least one of B7/8/9 non-empty. This state of affairs immediately suggests adapted versions of the comparative success hypothesis and the rule of success.

It is also clear that we obtain macro-versions of the matrix, the notion of comparative success, the comparative success hypothesis and the rule of success by replacing individual facts by general facts. A general fact may be a general success, a general problem or a neutral general fact for a theory. Note that combinations with individual and general facts are also possible.[7]

In all these variants, the situation of being more successful will again be rare. However, it is certainly not excluded. In ICR (Chapter 11) we argue, for instance, that the theories of the atom developed by Rutherford, Bohr and Sommerfeld can be ordered in terms of general facts according to the symmetric definition. Another set of examples of this kind is provided in Table 8.2 (from: Panofsky and Phillips (1962^2), p. 282) below, representing the records in the face of 13 general experimental facts of the special theory of relativity (STR) and six alternative electrodynamic theories, viz. three versions of the ether theory and three emission theories.

As is easy to verify, STR is more successful than any of the other ones; in fact it is maximally successful as far as the 13 experimental facts are concerned. Moreover, Lorentz's contraction version of the (stationary) ether theory is more successful than the contractionless version. Similarly, the ballistic version of the emission theory is more successful than the other two. However, it also clear that many combinations lead to divided results. For instance, Lorentz's theory is more

Table 8.2 Comparison of experimental record of seven electrodynamic theories. Legend: A: agreement, D: disagreement, N: not applicable.

Theory		Light propagation experiments							Experiments from other fields					
		Aberration	Fizeau convection coefficient	Michelson-Morley	Kennedy-Thorndike	Moving sources and mirrors	De Sitter spectroscopic binaries	Michelson-Morley, using sunlight	Variation of mass with velocity	General mass-energy equivalence	Radiation from moving charges	Meson decay at high velocity	Trouton-Noble	Unipolar induction, using permanent magnet
Ether theories	Stationary ether, no contraction	A	A	D	D	A	A	D	D	N	A	N	D	D
	Stationary ether, Lorentz contraction	A	A	A	D	A	A	A	A	N	A	N	A	D
	Ether attached to ponderable bodies	D	D	A	A	A	A	A	D	N	N	N	A	N
Emission theories	Original source	A	A	A	A	A	D	D	N	N	D	N	N	N
	Ballistic	A	N	A	A	D	D	D	N	N	D	N	N	N
	New source	A	N	A	A	D	D	A	N	N	D	N	N	N
Special theory of relativity		A	A	A	A	A	A	A	A	A	A	A	A	A

successful in certain respects (e.g. De Sitter's spectroscopic binaries) than the ballistic theory, but less successful in other respects (e.g. the Kennedy–Thorndike experiments).

In the present approach it is plausible to define, in general, one type of divided success as a liberal version of more successfulness. Y is *almost more successful* than X if there are, besides some favorable facts and (possibly) some indifferent facts, some unfavorable facts, but only of the B3-type, provided there are (favorable) B8- or B9-facts or the number of B3-facts is (much) smaller than that of their antipodes, that is, B7-facts. The provision clause guarantees that it remains an asymmetric relation. Crucial is the special treatment of B3-facts. They correspond to what is called *Kuhn-loss*: the new theory seems no longer to retain a success demonstrated by the old one. The idea behind the suggested relatively undramatic nature is the belief that further investigation may show that how B3-fact turns out to be a success after all, perhaps by adding an additional (non-problematic) hypothesis. In this case it becomes an (indifferent) B6-fact. Hence, the presence of B3-facts is first of all an invitation to further research. If this is unsuccessful, such a B3-fact becomes a case of recognized Kuhn-loss. To be sure, when it concerns a general fact of a nomic nature it is more impressive than when it concerns some general or individual fact that may be conceived as 'accidental'. Unfortunately, Table 8.2 does not contain an example of an almost more successful theory.

Cases of divided success may also be approached by some (quasi-)quantitative weighing of facts. Something like the following quantitative evaluation matrix (Table 8.3) is directly suggested by the same considerations that governed the number ordering of the boxes.

Table 8.3 The quantitative (comparative) evaluation matrix.

		X		
		negative	neutral	positive
	negative	B4: $-1/-1$	B2: $-3/+3$	B1: $-4/+4$
Y	neutral	B8: $+3/-3$	B5: $0/0$	B3: $-2/+2$
	positive	B9: $+4/-4$	B7: $+2/-2$	B6: $+1/+1$

It is easy to calculate that all qualitative (i.e. Table 8.1 induced) success orderings of electrodynamic theories to which Table 8.2 gave rise, remain on the basis of Table 8.3 in tact (which is not automatically the case). Moreover, we then get of course a linear ordering, with Lorentz's theory in the second position after STR and far ahead of the other alternatives. Of course, one may further refine such orderings by assigning different basic weights to the different facts, to be multiplied by the relative weights specified in the matrix of Table 8.3.

Note that the qualitative and the quantitative versions of the evaluation matrix can be seen as explications of some core aspects of Laudan's (1977) problem-solving model of scientific progress, at least as far as empirical problems and their solutions are concerned.

Let us briefly consider the possible role of simplicity. In Section 11.2 we will show that the examples of theory comparison presented by Thagard (1992) to evaluate the computer program ECHO can well be treated by the comparative evaluation matrix (CEM), that is, ECHO and CEM are equally successful with respect to these examples. However, ECHO is in many respects more complicated than CEM. One respect is the fact that CEM only uses success considerations, whereas ECHO applies simultaneously success and simplicity considerations, even in such a way that, in theory, success may be sacrificed to simplicity. Of course, as long as two theories are equally successful, we may add as a supplement to the rule of success, that simplicity considerations provide good, pragmatic, reasons to prefer the more simple one,[8] e.g. on the meta-level CEM is to be preferable to ECHO.[9]

This applies, by definition, always in case of observationally equivalent theories. Similar to the case of acceptance (ICR, Section 2.3), simplicity considerations and, more generally, our background beliefs, may determine which of two (perhaps already falsified) theories is more plausible (in the sense that it is closer to the theoretical truth, the theory realist will add). Background beliefs may determine in this way our preference among observationally equivalent (falsified) theories; of course, these beliefs do no longer serve this purpose if they diagnose the two theories as equally plausible.

Be this as it may, returning to just momentarily equally successful theories, we do not see good reasons to apply simplicity considerations as long as success criteria lead to an ordering of two theories. Lack of good reasons seems to hold at least for deterministic theories. In particular, in the case of such theories there does not seem to be any link between simplicity and truth approximation. However, recent publications (see Sober (1998) for an overview) indicate that such a link between simplicity and (observational) truth approximation exists. If so, simplicity may well be counted as a kind of success.

8.2 EVALUATION AND FALSIFICATION IN THE LIGHT OF TRUTH APPROXIMATION

Introduction

Although the method of HD-testing, HD-evaluation, and hence the evaluation methodology have a falsificationist flavor, each with its own aim, they are certainly not naive in the sense in which Popper's methodology has sometimes been construed. Naive falsificationism in the sense described by Lakatos (1970/1978) roughly amounts to applying HD-testing for purposes of theory evaluation and elimination. Its core feature then becomes to further discard (convincingly) falsified theories. Lakatos has also construed a sophisticated version of falsificationism such that, when comparing theories, he takes their 'unrefuted content' into account, a practice which allows falsified theories to remain in the game. Moreover, Lakatos has proposed a 'methodology of research programs', which operates in a sophisticated falsificationist way. However, it works in such a way that it postpones the recognition of falsifications of the 'hard core theory' as long as it is possible to roll off the causes of falsification dogmatically onto auxiliary hypotheses or background theories.

It will be argued that HD-evaluation can be seen as an explication of sophisticated falsificationism, leaving room for a dogmatic research program specification. For easy reference, we will simply use the terms 'falsificationist methodology' and 'falsificationism' in the naive sense, except when otherwise stated.

Now it can already be made plausible by a suggestive picture (1) that the falsificationist and the evaluation methodology may be functional for truth approximation; and (2) that the latter non-falsificationist methodology, ironically enough, is much more efficient for that purpose. The basic proof of both claims, in terms of the structuralist explication of truthlikeness, is given in ICR (Chapter 7).

We like to make these claims about truth approximation here, for it enables us to sketch the main methodological consequence, which may stimulate the reader to 'consume' ICR. This consequence is that a new explanation, even justification, can be given for the observation of Kuhn, Lakatos and others that there is quite a discrepancy between falsificationist (methodological) theory and non-falsificationist practice. This view will be elaborated in the next section.

8.2.1 *The Falsificationist Methodology*

As suggested, we call a methodology a falsificationist methodology when it includes the claim that a conclusively falsified theory has become essentially useless, and rules of theory selection are hence essentially restricted to not-yet-falsified theories. For example, in all standard probabilistic methodologies, including the Bayesian, all falsified theories are thrown on the scrap-heap of 'theories with zero posterior probability', and theory selection is something between 'theories with non-zero posterior probability'. Realist epistemologists also distance themselves from falsified theories, in so far as they consider the truth question to be the main question. Although empiricist methodology is usually not presented as falsificationist, both methodologies prescribe similarly negative conclusions concerning falsification: a falsified theory has passed the fundamental border of observational adequacy for which scientists are supposed to strive. Of course, for all approaches it is possible to contest the supposition that the falsification is genuine because of all kinds of complications dealt with in Subsection 7.3.3.

We get the technical core of the falsificationist methodology, based on the asymmetric model of separate HD-testing of theories introduced in Subsection 7.2.5, by restricting the interesting hypothesis (CSH) and the rule (RS) to not yet falsified theories, indicated by rCSH and rRS, respectively, and by adding the

Rule of Elimination (RE)
When a theory has been convincingly falsified, elimination should follow, and one should look for a new theory

Hence, whereas the evaluation methodology is governed by the unrestricted RS, the falsificationist methodology is governed by the rRS and RE (rRS&RE).

Let us first make some further comparative remarks of a methodological nature. Of course, according to both methodologies, the application of their respective rules (RS versus rRS&RE) has to a primary goal. As already suggested, RS presupposes the unrestricted application of the separate and comparative principles of HD-evaluation (PSE and PCE). Moreover, we suggested some other specific heuristic principles that might stimulate the application of RS, viz. the (unrestricted) application of the principle of content (PC) and the principle of dialectics (PD). The ideal of an unrestricted application of RS and hence of PCE, PC, PD and perhaps some additional principles can be summarized by the general heuristic principle:

Principle of Improvement (of theories) (PI)
Aim at a more successful theory, and successive application of RS

The term 'improvement' may be somewhat misleading, since the more successful theory may be either an improvement of an older theory, e.g. within the same research program, or really a new theory.

On the falsificationist side, rRS and RE both only presuppose the application of all mentioned principles restricted to not yet falsified theories, for as soon as we have obtained in this way a (convincingly) falsified theory, it is put out of the game by RE. In other words, the falsificationist methodology, governed by rRS&RE, presupposes the restricted application of PSE, indicated by rPSE, and the restricted version of PI, indicated by rPI. If one does not yet have an unfalsified theory at one's disposal, to apply rPSE, one has to invent such a theory.

It is also important to note that the application of RE and of PSE, whether the latter is restricted or not, presupposes that the relevant theory is testable, falsifiable or confirmable:

Principle of Testability (PT)
Aim at theories that can be tested, and hence evaluated, in the sense that test implications can be derived which can be tested for their truth-value by way of observation

Hence, the relativization of the methodological role of falsification, inherent in the evaluation methodology, should not be construed as a plea to drop falsifiability as a criterion of an empirical theory. On the contrary, empirical theories are supposed to be able to score successes or, to be precise, general successes. This requirement evidently presupposes falsifiability. However, we prefer the neutral term 'testability',[10] for falsifiability and confirmability obviously are two sides of the same coin. This observation is seriously in conflict with Popper's critique of the goal of confirmation. Note also that Popper's plea to give priority to testing the most unexpected test implications can equally well be conceived as aiming at as high confirmation as possible, for surviving such tests gives, of course, the highest success value.

In sum, the evaluation methodology, governed by RS, can now be summarized by PI, presupposing PSE and PT, whereas the falsificationist methodology, governed by rRS&RE, amounts to rPI&RE, both presupposing rPSE and PT.

Though Popper primarily argued that theories should be falsifiable, i.e. testable (PT), and that they should be tested, i.e. evaluated in our sense (PSE) and improved in our sense (PI), it is frequently suggested that he promoted the more inclusive falsification principle, including RE, and hence restricted PSE and PI to rPSE and rPI. However, the number of times that Popper seems to plea for RE, or for the combination rPI&RE, is negligible compared to the number of times that he pleads for the unrestricted PI. In this respect, it is important to stress that Popper uses the expression '(principle or rule of) elimination', almost always in the sense of the 'elimination of error', which is precisely what PI amounts to. Hence, as Lakatos has suggested, Popper is most of the time a sophisticated falsificationist. However this may be, it is at least evident that rPI&RE, due to RE, does not use all opportunities for empirical progress, whereas PI does. Hence, RE is not useful for empirical progress in the sense of RS (and, as we will see, not for truth approximation either). The only justification of RE that remains is a pragmatic one. If one wants to use a theory to design a certain product or process and if it is important in that context to avoid risks, it is plausible to apply RE whenever it is possible.

But besides retarding empirical progress in the sense suggested, it is also plausible to think that RE affects the prospects for truth approximation. A striking feature of PI in this respect is that the question of whether the more successful theory is false or not does not play a role at all. That is, the more successful theory may well be false, provided all its counter-examples are also counter-examples of the old theory. In ICR (Chapter 7) it is proved that RS, and hence PI, are not only functional (to say 'effective' would have here too strong connotations) for approaching the truth in a precise sense, whatever the truth precisely is, but that they are even efficient in doing so. On the other hand, it will be shown that rRS&RE, and hence rPI&RE, is also functional for truth approximation, due to rRS, but very inefficient, due to PE. The reason is that RE prescribes that when a theory encounters a counter-example, one always has to look for a new theory that is compatible with the data thus far. A short cut to the truth of a theory with many (types of) counter-examples, via theories with fewer ones, is excluded. To be sure, the falsificationist methodology, including the comparative part, is functional *and* efficient in searching for an answer to the principle question concerning the testing of a theory, viz. its truth-value.

To put it somewhat generally and dramatically, something like the cunning of reason is operative in scientific research. The evaluation and the falsificationist methodologies, though no guarantee for truth approximation, are both functional for truth approximation in a weaker but precise sense. Hence, realists may claim that they approach the truth by using the falsificationist method, at least as a rule. However, more surprisingly, if one applies the evaluation methodology, one comes, as a rule, closer to the truth, whether one likes it or not.

Last but not least, the irony is that the cunning of reason works more efficiently when the evaluation methodology is applied than when the falsificationist methodology is applied. The reason is that the falsificationist allows himself, as it were, to be distracted by something which turns out to be irrelevant for approaching the truth, viz. that the theory is false.

The proof of these claims starts from the asymmetric model of HD-evaluation and is based on the structuralist theory of truthlikeness and plausible interpretations of individual problems and general successes. The main theorem, called the success theorem, presented in ICR (Chapter 7), states that being the closest to the truth among the available theories guarantees that it will be the most successful theory. To extend the proof to the two symmetric models merely requires some transformation of individual successes and general problems. By way of an easy lemma, it follows that the result of a crucial experiment is always functional, though again, is no guarantee, for truth approximation.

8.2.2 *A Suggestive Picture*

It is easy to convince the reader that the claims of functionality and efficiency for truth approximation must be right. The set of statements (more precisely, the set of classes of equivalent statements, also called (the set of) propositions) that can be formulated within a certain vocabulary can be partially ordered according to logical strength, see Figure 8.1.

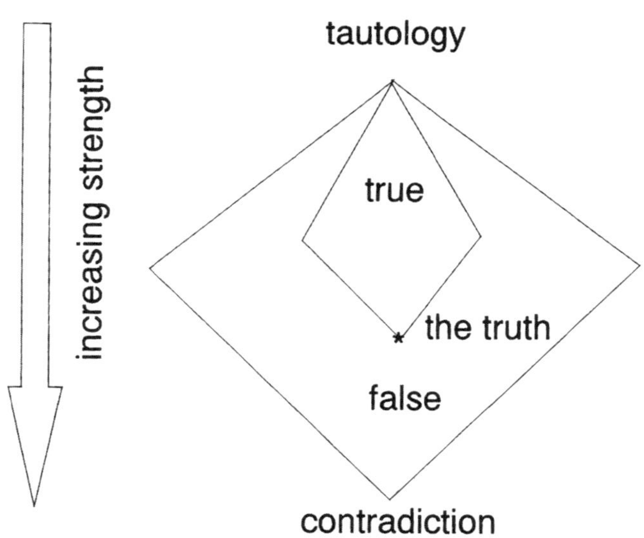

Figure 8.1 The landscape of the truth.

Tautology and contradiction form two unique extremes of this hierarchy. Between these extremes there is for every strength more than one (non-equivalent) statement. The number first increases and then decreases. A diamond-shape is formed. For an arbitrary statement it holds that the set of statements following from it, when ordered by strength, has roughly the same diamond structure. Naturally, this also applies to the set of statements following from the truth (recall, the strongest true statement), which is, of course, precisely the set of true statements. This is the case when the set of true statements are taken in isolation. However, the closed diamond of true statements in Figure 8.1 only arises after sufficient reshuffling of the true and false statements of the same strength (i.e. horizontally). But the points that will be made do not depend on this graphic simplification.

The following possibilities are easily read off from Figure 8.1, which may be said to represent *the landscape of the truth*. A false theory may well be (much) closer to the truth than a true one. And, although the question of whether a theory is true or false is relevant for the question of whether the theory coincides with the truth, the first question is irrelevant in determining its distance to the truth as long as the theory in question does not exactly coincide with the truth. Finally, it follows immediately from the landscape of the truth that it is possible to use a theory development strategy, such as idealization and concretization, to follow a path composed of a whole chain of false theories but leading to the truth.

It is clear that all these possibilities exist due to the plausible explication of 'the truth' as the strongest true statement. However, it is also important to note that the three indicated possibilities do not presuppose that the truth can be recognized as such, nor that the truth is verifiable, let alone that it can be established with certainty. What is needed is only that the truth gives recognizable signals, without making their source derivable from them.

These remarks are easy to combine with a literally geographical analogy for truth approximation in general and the possibility of the irony of the cunning of reason in particular. To find the most south-east spot in The Netherlands at the same level as N.A.P. (Normaal Amsterdam Peil(= level)) and assuming some very plausible arrangements, there must be precisely one such spot; there is no reason to try with spasmodic efforts to start and remain in areas not below N.A.P.

8.3 SCIENTIFIC AND PSEUDOSCIENTIFIC DOGMATISM

Introduction

As announced above, we have made these claims about truth approximation plausible in order to sketch the main methodological consequence of them. They enable a new explanation, even justification, of the observation of Kuhn, Lakatos and others that there is quite a discrepancy between falsificationist (methodological) theory and non-falsificationist practice. Straightforward truth approximation may be seen as the primary, conscious or unconscious, motive for non-falsificationist behavior. Dogmatic behavior, in the sense of working within a research program, is only a secondary motive for non-falsificationist behavior.

Whatever the main motive, as long as such behavior is directed at theory improvement within the program, it can be distinguished from pseudoscientific behavior.

8.3.1 *Falsificationist Theory and Non-Falsificationist Practice*

'Scientists do not practice what they, and philosophers of science, preach' is an observation constituting one of the most important findings in empirical studies of science. Popper, or at least his naive interpreters and followers, still felt themselves to have good reasons to declare that scientists should behave non-dogmatically. *Falsificationism* was proposed as a specific methodology: theories have to be tested deductively, and in case of convincing falsification, elimination of the theory should follow.

However, Kuhn and Lakatos showed that in practice it is seldom that one observes consistent falsificationist behavior. Appealing to the so-called *Duhem–Quine thesis*, Lakatos has argued convincingly that dogmatic behavior can be justified to a certain extent. As we have illustrated in detail in Subsection 7.3.3, deriving test implications from a theory usually requires all kinds of additional assumptions which can be blamed for the falsification. This enables one to retain the 'hard core' of the theory, giving rise to the methodology of (scientific) research programs. Opposed to this methodological justification of dogmatic behavior, there is Kuhn's social explanation of non-falsificationist behavior. He relativizes the importance of any method whatever. Science is a social process and the so-called scientific method is more or less irrelevant dressing. The similarity between Kuhn and Lakatos is that both call attention to the open discrepancy between falsificationist (methodological) theory and non-falsificationist practice.

There seems to be no reason to assume that there is just one general explanation for all cases of non-falsificationist behavior. The different explanations of Kuhn and Lakatos can be very plausible in concrete cases, and there is equally no reason to exclude the possibility that in still other cases other explanations have to be preferred.

However, it is possible to indicate a hierarchy of explanations. When a philosophical-methodological explanation is possible, it has priority over a purely social-psychological explanation. Moreover, one philosophical-methodological explanation may be more general than another.

A more general explanation than the one provided by Lakatos can be found in what is perhaps the most striking case of non-falsificationist behavior: the theory development strategy elaborated by Nowak and others called 'idealization and concretization'. According to this strategy, one starts with a theory that neglects some factors which one knows to be relevant. Hence, one knows beforehand that the theory suggested by strong idealization is false, even without having done any experimental testing. The same ('to be born refuted') holds for all concretizations of the theory that one obtains from successive accounting for what had previously been neglected factors. Only when one assumes that all factors that are supposed to be relevant have in fact been taken into account does it make sense not to exclude the possibility that the theory is true. One may say, of course, that the hypothesis that a certain factor is negligible is an *auxiliary* hypothesis that can be blamed for the falsification. But the problem is that the falsification is not at all a surprising result for which one has to find a cause. In other words, there need not be an inclination to dogmatism about to a certain theory that one might want to save.

Lakatos' and Nowak's strategies of theory development are in fact special cases of the evaluation methodology which can, but need not, go together. Hence, if we can give a justification for the evaluation methodology, this covers the rational explanation of the non-falsificationist aspect of both strategies.

In our view the evaluation methodology, including its non-falsificationist feature, is practiced because, as we have already indicated and will further do so in the next chapter, it happens to be functional and efficient for truth approximation. That is, scientists behave as if they are aiming at the truth concerning some subject matter, and as if they know that straightforward elimination of falsified theories is not the best way to achieve that goal.

8.3.2 *Responsible Dogmatic Behavior*

As mentioned, the evaluation methodology can be seen as an explication of Lakatos' sophisticated falsificationism. The basic common feature is that theory evaluation is primarily a matter of comparative evaluation, leaving falsified theories in the game as long as there are no better alternatives. Our analysis adds to this the justification that this procedure, though no guarantee for truth approximation, is functional for truth approximation, without detours. There are also some important differences. They concern the fact that Lakatos imposes two extra conditions on theory elimination, and hence progress. First, testing the comparative success hypothesis should not only show that the more successful theory remains more successful in the light of these facts, but also that it has to score new extra successes ('novel facts'). In other words, empirical progress not only requires additional explanatory or derivational success, but even additional predictive or retrodictive success.[11] Secondly, Lakatos ultimately proposes to use sophisticated falsificationist means in a 'methodology of (scientific) research programs'. That is, he considers something like RS primarily acceptable for theory transitions within a research program, where a research program not only presupposes some vocabulary, but also some hard core principles that should remain untouched, that is, one should follow the dogmatic strategy.

From our analysis it will follow that these extra conditions are not necessarily functional for truth approximation. However, concerning the first extra requirement, it can be justified on logical grounds that a theory closer to the truth than another will almost always be able to score 'new extra successes'. This property not only explains why Popper and Lakatos could show that in the history of science important theory transitions usually resulted from new predictive success, but also why they have made the success criterion a necessary feature.

Concerning the second of Lakatos' extra requirements, namely, sticking to the hard core of a research program, we have already suggested that this requisite is indeed an important specific, but not obligatory, theory development strategy. As a matter of fact, the evaluation methodology can easily be integrated in a 'methodology of research programs'. Theories within a research program are separately evaluated by the HD-method, if necessary, with recourse to dogmatic strategies, and theory transitions can be made by applying RS on the basis of a comparative HD-evaluation.

Our methodological portrait also leaves room for a well-conceived transition of one research program to another. It is possible that a new research program turns out to 'have a theory' of which the corresponding observational theory is more successful than that corresponding to the best known theory of an old program. Of course, this is always a momentary judgement, because, as Lakatos has stressed, the old program can later turn out to leave room for a theory that surpasses the success of the best known theory belonging to the new program. However this may be, Lakatos has argued that something like a principle of improvement of research programs is plausible: if it is difficult to make empirical progress within a program, aim at a more successful program, whether by repairing the old one or by replacing the old one by a new one.

There remains the justification of dogmatic behavior in the sense of a research program. Recall that, according to Lakatos, and we like to agree with him, scientists develop a research program on the basis of some main idea, in the form of one or more core hypotheses. They do so by continually thinking of new auxiliary hypotheses that form together with the core hypothesis a better total theory. By successfully doing so, there arises ideally a succession of ever better theories with the same hard core. Hence, in the development of a research program, the principle of improvement (PI) is applied while saving the central dogmas. For this reason, such research frequently has the character of presupposing or applying these dogmas, instead of evaluating them. Such research is nevertheless governed by PI, and hence functional for empirical progress and even truth approximation. Accordingly, dogmatic behavior may be responsible scientific behavior when it is combined with PI.

Although it is difficult to find a specific statement to this effect, it seems that Lakatos implicitly assumed that the hard core of a program can only be sustained with success during a long period of time when it is true. For otherwise it is difficult to explain why Lakatos did not account for the regularly occurring phenomenon that the hard core does not appear to be so hard as it seemed, and that it is frequently adapted. However, in the light of the foregoing, it appears that improving the 'hard' core, while sticking to the vocabulary, may be equally good for empirical progress and truth approximation. When this also does not work, it is time to look for completely different dogmas.[12] Whether one appreciates it or not, young researchers are frequently quicker off the mark in search of new dogmas than established scientists tend to be.

In sum, we may say that scientific dogmatic behavior satisfies a concretized version of PI:

> *Principle of improvement guided by research programs* (PIRP)
> One should primarily aim at progress within a research program, i.e. aim at a better theory while keeping the hard core of the program in tact. If, and only if, this strategy does not work, try to adapt the hard core, while leaving the vocabulary in tact. If, and only if, this second strategy is also unsuccessful, look for another program with better perspectives on progress

Note first that PIRP is so formulated that it contains something like a principle of improvement of programs, where the latter is in the first instance guided by a vocabulary. Since PIRP is a special version of PI, it is also functional for truth approximation. On the basis of the work published by Kuhn and Lakatos, it may be concluded that PIRP has been very successful in the history of science.

Descriptive research programs frequently do not have a proper hard core, at least not a specific one, but only a core vocabulary. If one wants to talk about a hard core in this case, it is the general idea that the core vocabulary generates, by itself or in extended versions, an interesting restriction of what is possible in reality. The existence of such terminological restraints suggests that there is, between the general PI and the special version PIRP, a broader special version in between, one which deals with the dogmatic adherence to 'core vocabularies':

> *Principle of improvement guided by core vocabularies* (PICV)
> One should primarily aim at progress within a core vocabulary, i.e. aim at a better theory while keeping the core vocabulary in tact; if, and only if, this does not work, look for another vocabulary, which may or may not be very different

Again, PICV is so formulated that it includes something like a principle of improvement of vocabularies. Moreover, since PICV is a special version of PI it is also functional for truth approximation. PICV has been very successful in the history of science, in particular where descriptive research is concerned. The history of 'descriptive (or inductive, see below) thermodynamics', dealing with the search for relations between volume, pressure and temperature, and the history of 'descriptive chemistry', dealing with chemical reactions, provide cases in point.

8.3.3 *Pseudoscience*

From the foregoing it follows that dogmatically dealing with theories has to be qualified as *un*scientific when one apparently does not aim at applying PIRP or PICV. Moreover, it is plausible to characterize *pseudoscience* as the combination of scientific pretensions and the neglect of PI, in particular its dogmatic versions PIRP and PICV. This characterization can be seen, in some respects, as an improved version of that in Lakatos' introduction (1970/1978). The standard examples of pseudoscience, such as astrology, graphology, homeopathy, parapsychology, creationism, and ufology satisfy these conditions.[13] In all these cases, it is not only quite clear that central dogmas are the point of departure for unscientific research and application, but it is also rather easy to indicate how they could become the point of departure for serious research. Such research, however, is seldom started.

We do not, of course, claim that within the sphere of academic research pseudoscientific behavior does not occur, but it takes place less than outside that sphere. Marxist economics, psychoanalytic psychology, and evolutionary biology increasingly seem to follow the general rules and principles of scientific research. However, claims of this kind are highly controversial, as the works of Blaug (1980) and Grünbaum (1984) illustrate for marxist economics and psychoanalytic theory, respectively.

An interesting question concerns how theology and philosophy should be evaluated in this respect. So-called systematic theology certainly has scientific pretensions, but usually no empirical ones. Nevertheless, it has, directly or indirectly, pretensions regarding empirical reality. Hence, the question of whether theology has more resemblance to science than to pseudoscience is not easy to answer. In view of the unclear nature of the pretensions, this question cannot be separated from other issues involved in the scientific evaluation of other theological claims to knowledge, such as the following ones. To what extent are they in conflict with empirical scientific insights, and what would a straightforward empirical scientific approach to these claims, and their persistence, look like? So-called religious studies direct their attention primarily at the description of other than Western religious cultures, but it is also plausible to take into account in this connection systematic theology and to direct attention to the evaluation and explanation of religious belief claims.

The three general principles, PT, PSE and PI, including the latter's program version, do *mutatis mutandis* also apply to philosophy. Of course, in this case, testability and evaluation do not now refer to empirical matters, as far as there are no empirical pretensions. Successes of conceptual theories, for example, should be interpreted as satisfactorily treated cases and realized general conditions of adequacy. Counter-examples become problems arising from cases and aspects that are evidently wrongly dealt with. As a consequence, there is also in philosophy much room for making progress in the form of responsible dogmatism, and good philosophy can rather easily be distinguished from pseudo- or para-philosophy. In the case of pseudo-philosophy, philosophical pretensions are usually combined with unscientific dogmatism. Several forms of exegesis of great philosophers belong to it. Instead of trying to develop or replace research programs from earlier philosophers, an attempt is made to preserve all statements of 'the master'. The previous and the present chapter may be seen as the result of an attempt to improve some of the conceptual aspects in Popper's and Lakatos's work, instead of defending them at all costs.

FINAL REMARKS: THE CONTEXT OF EVALUATION

Although it may be conceded that *the* scientific method does not exist, such a statement does not yet imply that any method works, as Feyerabend (1975) suggested with his slogan 'anything goes'. It is more realistic to start with the distinction (to be specified pragmatically) between two aspects of scientific research, viz. invention and testing of theories. This distinction became known as the *Context of Discovery* versus the *Context of Justification*.

For the Context of Discovery it may well be the case that almost all conceivable methods, from inductive generalization to sleeping another night, work in certain cases: 'anything goes *sometimes*'.

Within the Context of Justification, a universal method may also be lacking. However, *pace* Glymour, the HD-method certainly is a dominant method. Unfortunately, the term 'Context of Justification', whether or not specified in a falsificationist way, suggests, like the terms 'confirmation' and 'corroboration', that the truth or falsity of a theory is the sole interest. Our analysis of the HD-method makes it clear that it would be much more adequate to speak of the *Context of Evaluation*. The term 'evaluation' would refer, in the first place, to the separate and comparative HD-evaluation of theories in terms of successes and problems. As we have indicated, it may even refer to the further evaluation of their relative merits in approaching the truth, or at least the observational truth.

As a consequence, the foregoing may not only be interpreted as a direct plea for restricting the test methodology to cases where our only interest is the truth question. As soon as we are interested in the success question, then the evaluation methodology is more adequate. That methodology can, moreover, be surprisingly enough justified, in terms of truth approximation.

Let us, finally, pay some more attention to the pro's and con's of our strict comparative approach. As has already been stressed, there are few theories that can be ordered in terms of 'more successfulness' according to our strict definition. The same holds for our basic and refined orderings in terms of 'more truthlikeness' developed in ICR. Hence, the limited applicability of our comparative notions might be seen as a serious shortcoming, supporting a more liberal comparative or even quantitative approach. To be sure, such liberalizations are very welcome as far as they are realistic concretizations. We already introduced the idea of 'almost more successful', which has a plausible analogue in terms of 'almost more truthlike', which we will, however, not pursue in any further detail. As indicated already, in ICR we also introduce (in Chapter 10) refined notions of 'more successfulness' and 'more truthlikeness', replacing the 'basic' ones, when relevant. The main idea behind the (further) refined notion of more successfulness is that one counter-example may nevertheless be better than another.

However, in ICR we question the general usefulness of quantitative liberalizations of 'successfulness' and 'truthlikeness', mainly because they need real-valued distances between models, a requirement which is very unrealistic in most scientific contexts. Hence, the applicability of liberal notions may well be laden with arbitrariness. For this reason, we want to focus on unproblematic cases, guaranteeing that we get the bottom line of progress and rationality. In ICR (Chapter 11) we deal with the succession of theories of the atom developed by Rutherford, Bohr and Sommerfeld, and argue that it is a sequence of increasing success and, potentially, even of increasing truthlikeness, both in the strict (refined) sense. Hence, although the strict approach may not have too many examples, it has impressive examples.

Finally, it is important to stress that the strict strategy does not lead to void or almost void methodological principles. If there is divided success between theories, the Principle of Improvement amounts, more specifically, to the recommendation that we should try to apply the Principle of Dialectics: 'Aim at a success preserving synthesis of the two RS-escaping theories', of course, with a plausible program-bound version. Similarly, for truth approximation aims: if there is reason to suppose that two theories cannot be ordered in terms of 'more truthlikeness' in the strict sense, the challenge is to construe a theory which is more truthlike than both. In sum, the restricted applicability of strict notions of comparative success and truthlikeness does not exclude the possibility of clear challenges being formulated in cases where they do not apply, on the contrary.

PART V

TRUTH, PRODUCT, AND CONCEPT APPROXIMATION

INTRODUCTION

HD-testing and evaluation are primarily concerned with descriptive and explanatory research. In the previous part, we have seen how this leads to a plausible definition of empirical progress. In this part, we will give a brief exposition of how empirical progress is related to truth approximation, at least in descriptive and explanatory research that aims at 'the nomic truth', and is therefore called nomological research. Moreover, we will show that there is an interesting partial analogy with progress in the 'product approximation' that occurs in design and explicative research.

In Chapter 9 we will see that, although nomological, design, and explicative research seem rather different at first sight, the suggested partial analogy between them derives from the fact that they can formally be represented in terms of either a target set of desired possibilities or a target set of desired features. This characteristic implies that their respective definitions of formal progress essentially coincide. The differences between the three types of research are due to the fact that 'determinable progress' requires specific definitions: empirical progress for nomological research and conceptual progress for explicative research. Only in the case of design research does determinable progress coincide with formal progress, as long as the target sets of desired and undesired features are determined beforehand. The analysis in this chapter is, as far as nomological research is concerned, essentially based on Chapter 7 and 9 of ICR, including the resulting explication of descriptive and nomological research programs. However, the present presentation of the main results in terms of (un)desired possibilities and features is, although concise, more transparent for purposes of application, extension, and comparison.

In Chapter 10 we will study in detail design research programs. The previous parts have been dominated by descriptive and explanatory research programs on the object level, and by some explicative programs analyzing certain aspects of such programs, such as truth approximation. In this chapter we will present the results of an explicative program with respect to research programs that aim at a certain product. The core idea is that design research programs attempt to bring together the properties of available materials and the demands derived from intended applications. Our point of departure is the diagnosis of a conceptual confusion in a first attempt to explicate this idea, the lattice model of Weeder et al. The structure and development of such programs, in other words, their logic of problem states and state transitions, including assessment criteria and heuristic principles, is described in set-theoretic terms, starting with a naive model comprising an intended profile and the operational profile of a prototype. Drug research will provide the main example. In a first fundamental concretization, the useful distinction between structural and functional properties is built into the model. Three further concretizations, dealing with potential applications, potential realizations, and potentially relevant properties are also presented. Next we will elaborate the partial analogy between 'product' and 'truth approximation' that I will already indicate in Chapter 9. The differences will turn out to be at least as important as the similarities. We conclude with some indications of the usefulness of our models for the ways in which products reach the market, in comparison to the so-called social construction of technology approach.

The Chapters 9 and 10 have been written in such a way that the second chapter can be read independently of the, somewhat condensed, first one.

CHAPTER 9

PROGRESS IN NOMOLOGICAL, EXPLICATIVE AND DESIGN RESEARCH

INTRODUCTION

Although philosophers of science almost always use to identify scientific research with description and explanation, this distinction is not only vague (see Part II), but, as we have stressed in Chapter 1, there are also other important types of research, notably design research and concept explication. Description and explanation may concern the actual or historical world, or they may concern the nomic world, that is, the world of what is nomically (physically, chemically, biologically, etc.) possible. In such case, they constitute nomological research. Whereas the previous parts contain much that is also useful for the research focussing on the actual world, the possibility of progress and 'actual truth' approximation in such research will be neglected in this chapter. The reader is referred to ICR (Chapters 7, 8 and 10) for brief discussions of these topics. This chapter deals first of all with nomological research, that is, by definition, research directed, by laws and theories, at the truth about (some part or aspect of) the nomic world, called the 'nomic truth'. The chapter deals, moreover, though only briefly, with design and explicative research. Recall that design research includes all kinds of constructive research intended to make or improve certain products or processes. Explicative research aims at the construction of a simple, precise and useful concept that resembles an intuitive concept as much as possible.

The claim of this chapter is that there is an illuminating partial analogy between nomological, design and explicative research, with complete correspondence between what we will call *formal progress* in these three areas. In all three cases there is a target set, and a particular theory, prototype or provisional explication may approach that target set better than another. Although the favorite kinds of terminology are *prima facie* different, hence hiding the analogy, they can be translated into each other, making the analogy explicit.

However, to conclude that formal progress has been made, one has to know the relevant target set explicitly, which is only the case in design research. Nomological and explicative research each have their own, though partly similar, way of determining *prima facie* formal progress, called *empirical progress* and *conceptual progress*, respectively.

We will start (Section 9.1) by characterizing formal progress in nomological research in terms of desired and undesired possibilities to be included in or excluded from a theory. These possibilities may be translated in terms of the desired and undesired features which the theory may or may not have, and also in terms of the necessary and sufficient conditions for being a desired possibility, which may or may not be necessary or sufficient conditions for being included in the theory. Of course, several combinations are possible. In Section 9.2, we will deal with empirical progress in nomological research and briefly indicate the main findings in ICR about its relation to formal progress. It will enable a further explication of the nature of descriptive and nomological research programs. In Section 9.3, we will first indicate how design research as well as (formal and determinable) progress in it can be described in terms of desired and undesired properties that a prototype may or may not have. Then we will address explicative research, which is usually characterized in terms of necessary conditions, the so-called conditions of adequacy that a provisional explication may or may not satisfy and in terms of the evident examples and non-examples of the intuitive concept, which the provisional explication may or may not respect. These criteria enable a plausible definition of conceptual progress. However, assuming that there is a unique solution to the explication task, it can then easily be formally characterized, along with the formal progress in it and using the same means as nomological research does: desired and undesired conceptual possibilities which belong or do not belong to the extension of a provisional explication.

In the sections about design and explicative research, these types of research will briefly be contrasted to nomological research and to each other, in order to indicate the aspects in which the analogies between these three types of research are not complete.

The crux of our analysis is formed by an equivalence theorem of a rather elementary nature, in particular as long as a structuralist rather than a logical (language oriented) formulation is given. For (dynamic) logicians it might be interesting to explore the linguistic sophistication and, for example, the articulation of the evident relation to Leibniz's theory of identities.

9.1 FORMAL PROGRESS IN NOMOLOGICAL RESEARCH

9.1.1 *Three Versions*

The point of departure for characterizing nomological research is a domain and a vocabulary enabling the formulation of a set of conceptual possibilities (also called potential models, i.e. structures of a certain similarity type), indicated by Mp. Moreover, it is assumed here that the representation of the domain in terms of that vocabulary amounts to a unique, (in some way or other) characterizable subset T of Mp consisting of the nomic possibilities, that is, the nomically possible conceptual possibilities. $Mp - T$ represents, of course, the nomic *im*possibilities, that is, conceptual possibilities which cannot be realized, whatever efforts we might do. T is the, by definition, not yet known target set of research, which may be identified with the (nomic) truth for reasons that will become clear soon. The aim of theory formation and revision is the actual characterization of T, where the elements of T are called the desired possibilities and the elements of its complement

$Mp - T = cT$ undesired possibilities. A theory X consists of a subset X of Mp with the strong claim '$X = T$'. If X includes T, X does not exclude any desired possibility. In that case, the weaker claim '$T \subseteq X$' is true. If $T \subseteq Y \subset X$, Y excludes more undesired possibilities than X and hence the corresponding weak claim '$T \subseteq Y$' is stronger and nevertheless true. In this sense theory T is the strongest true theory, and this is the plausible explication of 'the (nomic) truth' in the present context. The elements of X will be called (the) admitted possibilities (by X) and those of cX (the) excluded possibilities (by X).

The basic definition of formal progress in nomological research, also called (more) truthlikeness, or more precisely, of the claim that Y is at least as close to T as X, now amounts to the combination of a, relative to T, internal and external claim:

(DP) all desired possibilities admitted by X are admitted by Y: $X \cap T \subseteq Y \cap T$
(UP) all undesired possibilities admitted by Y are admitted by X: $Y - T \subseteq X - T$

Figure 9.1 represents this situation.

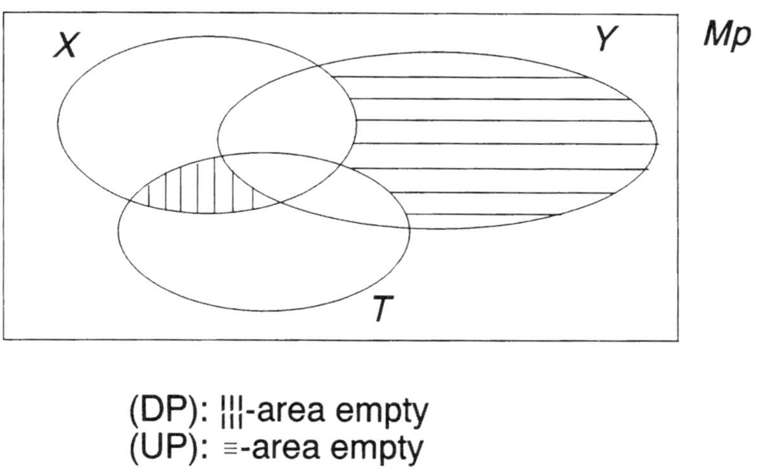

Figure 9.1 *Y is at least as close to T as X.*

Let us now turn to the translation of this definition in terms of desired and undesired features of theories. Our point of departure involves the features of possibilities. A feature of a theory is conceived of as a 'distributed' feature, that is, a property of all possibilities admitted by it. For example, a theory is called symmetric when all its possibilities have a certain symmetric character. A feature of a theory can be represented as the set of conceptual possibilities having the defining feature. It is important to note that this allows us to say that a feature of a theory (precisely) excludes the possibilities lacking the property.

The following definition of desired, undesired and other features is now plausible: desired features are features which include all desired possibilities (i.e. hence only exclude undesired possibilities), undesired features are features which include all undesired possibilities (i.e. only exclude desired possibilities). All other features exclude desired as well as undesired possibilities and will not play a further role. Figure 9.2 depicts the three kinds of features.

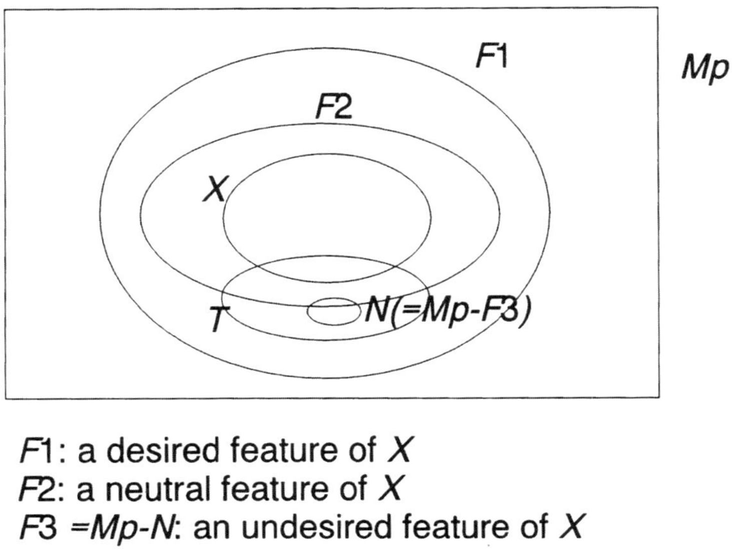

Figure 9.2 Three kinds of features.

The main equivalence theorem, which is now easy to prove, and hence left as an exercise for the reader, is that Y is at least as close to T as X if and only if:

(DF) all desired features of X are features of Y (equivalent to UP)
(UF) all undesired features of Y are features of X (equivalent to DP)

Note the pairwise similarity of the clauses and the crosswise character of the equivalences. The set-theoretical formalization of the new clauses can easily be given in terms of the 'co-powerset' $Q(A)$ of A, defined as the set of supersets of A (within Mp), that is, subset B of Mp belongs to $Q(A)$ iff A is a subset of B. In these terms we get:

$$\text{(DF) } Q(X) \cap Q(T) \subseteq Q(Y) \cap Q(T)$$
$$\text{(UF) } Q(Y) \cap Q(cT) \subseteq Q(X) \cap Q(cT)$$

Figure 9.3 represents these conditions.

From the equivalence theorem it follows that there can be given, besides *a possibility version* of the definition of formal progress in nomological research, *a feature version*, with a corresponding reformulation of the aim of theorizing: aiming at a theory fulfilling a target set of desired features, while lacking another set of undesired features. Figure 9.3 also represents this, where the figure could be restricted to the set of relevant features, i.e. $Q(T) \cup Q(cT)$, the union of the sets of

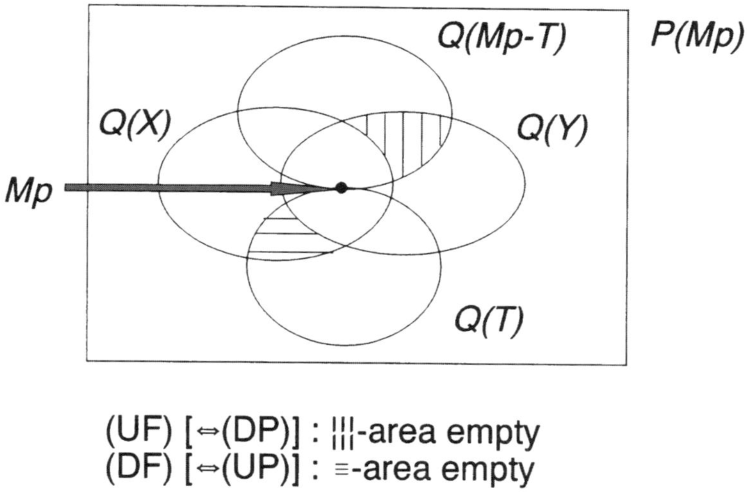

Figure 9.3 *Y is at least as close to T as X in terms of features.*

desired and undesired features, represented by $Q(T)$ and $Q(cT)$, respectively. Note that these sets have just one feature in common, viz. the tautology, represented by Mp itself. Note, moreover, that *a dual version* is possible in terms of desired possibilities and features, that is, the combination of (DP) and (DF), where the aim of theorizing now amounts to admitting the (target set of) desired possibilities, while fulfilling the (target set of) desired features.

So far for formal progress in nomological research. We will postpone the topic of empirical progress and truth approximation.

9.1.2 *Some Further Interpretations and Equivalences*

Let us first note that (DF) cannot only be interpreted in terms of desired features, but also in terms of true consequences, for a superset of *X* (within *Mp*) may well be seen as representing a consequence of *X* and a superset of *T* (within *Mp*) as representing a true claim. Hence, (DF) also amounts to the claim that all true consequences of *X* are (true) consequences of *Y*. Popper's bad luck with his definition, generally conceived as failing, roughly[1] in terms of more true and fewer false consequences, may now be reinterpreted in terms of a too liberal definition of undesired features, viz. all non-desired features, that is, undesired in our strict sense and otherwise. In terms of consequences, they may well be called false, since they are false for at least one desired possibility. Hence, we need a stronger interpretation of a false consequence, called 'strongly-false', defined by, false such that all its non-tautological consequences are false as well. It is easy to verify that (UF) can now be read as claiming that all strongly-false consequences of *Y* are (strongly-false) consequences of *X*.[2]

Since (DF) may be interpreted in terms of desired features or true consequences it is plausible that it can also be read in terms of necessary conditions: for all necessary conditions C to be desired (i.e. to be a desired possibility, $T \subseteq C$) which are also necessary to be a member of $X(X \subseteq C)$ are also necessary to be a member of $Y(Y \subseteq C)$.

This suggests the possibility of interpreting (DP)/(UF) in terms of sufficient conditions. Zwart (1998, p. 8, 52) has observed that these clauses are formally equivalent to a clause in terms of powersets:

$$P(X) \cap P(T) \subseteq P(Y) \cap P(T)$$

where he calls (the proposition corresponding to) a subset of X an *antecedence* of X. Unfortunately, whereas members of $Q(X) \cap Q(T)$ are plausibly called true consequences of X, there does not seem to be an intuitively appealing term for members of $P(X) \cap P(T)$ in terms of antecedences, that is, antecedences of X which are also antecedences of T. The term 'correct antecedence' might be introduced, leading to the following paraphrase of the relevant clause: all correct antecedences of X are (correct) antecedences of Y. However, an interpretation of the powerset version of the clause in terms of sufficient conditions is more easy to give: for all sufficient conditions C to be desired (i.e. to be a desired possibility, $C \subseteq T$) which are also sufficient to be a member of $X(C \subseteq X)$ are also sufficient to be a member of $Y(C \subseteq Y)$.

Finally, the powerset version of (DP)/(UF) suggests that there is also a powerset version of (UP)/(DF), which is indeed the case:

$$P(cX) \cap P(cT) \subseteq P(cY) \cap P(cT)$$

However, we do not know of a plausible reading of this clause, let alone an intuitively appealing one.

In summary, we have found that the second clause of the definition of formal progress in nomological research, formulated in terms of undesired possibilities, can be reformulated in terms of desired features or true consequences and necessary conditions, all based on the co-powerset version of the clause. The first clause, phrased in terms of desired possibilities, may be rephrased in terms of undesired features or strongly-false consequences, based on the co-powerset version, and correct antecedences or sufficient conditions, based on the powerset version.

9.2 EMPIRICAL PROGRESS AND NOMOLOGICAL RESEARCH PROGRAMS

9.2.1 *Empirical Progress and Nomic Truth Approximation*

Lack of space forces us to give only a sketch of empirical progress and its relation to nomic truth approximation, assuming that T is unknown and assuming a (relative) distinction between observational and theoretical terms. However, the basic (dual) definition in Chapter 8 of *potential empirical progress*, that is, of the claim 'theory Y is empirically more successful than theory X' is easy to rephrase in the present terms: (1) all established true observational consequences of X (i.e. general successes) are consequences of Y; (2) all observed counter-examples (i.e. individual problems) of Y (desired observational possibilities excluded by Y) are counter-examples of X; and (3) Y is better in at least one of these respects. Further testing of the comparative hypothesis that Y will remain empirically more successful than X may lead to the conclusion, using an unavoidable inductive jump, that Y is a case of *genuine empirical progress* relative to X.

It is not difficult to make plausible, on the basis of two theorems (the success theorem and the projection theorem), that the instrumentalist rule of success (RS), recall, the rule to choose for a genuinely more successful theory, is not only functional for approaching the observational nomic truth (formal progress on the observational level), but also for the theoretical and referential nomic truth, i.e. formal progress on the theoretical level and the referential 'level', respectively. Hence, instrumentalist criteria serve realist purposes, ironically enough, whether they are intended or not. For details, see ICR, in which it is also argued that the realist has some advantage over the instrumentalist in that the former can relativize counter-examples in a way the latter cannot.

Not only empirical criteria may play a role in theory choice, but also theoretical criteria, that is, criteria in terms of logical, conceptual or aesthetic features. In a recent paper (Kuipers, 1999a) a formal analysis is given of the relation between beauty, empirical success, and truth. It supports the findings of James McAllister in his inspiring *Beauty and revolution in science* (1996), by explaining and justifying them. First, scientists are essentially right regarding the usefulness of aesthetic criteria for truth approximation, provided they conceive them as less hard than empirical criteria. Second, the aesthetic criteria of the time, the 'aesthetic canon', may well be based on 'aesthetic induction' regarding (distributed) features of paradigms of successful theories which scientists have come to appreciate as beautiful. Third, they can play a crucial, dividing role in scientific revolutions. Since aesthetic criteria may well be wrong, they may retard empirical progress and hence truth approximation in the hands of aesthetic dogmatists but not in the hands of aesthetically flexible, 'revolutionary' scientists.

9.2.2 *Descriptive, Inductive, and Explanatory Research Programs Revisited*

The truth approximation analysis also gives the opportunity to reconsider the nature of descriptive and explanatory research programs. Recall that such programs presuppose, by definition, a domain, a problem, and a core idea, including a vocabulary, to solve that problem. Recall further that a *descriptive research program* uses an observational conceptual frame, and may either exclusively aim at one or more true descriptions (as for example in most historiography), or it may also aim at the true (observational) theory in the following specific way. In the latter nomological type of descriptive program, however, the goal of a true theory is supposed to be achieved exclusively by establishing observational laws. Given that this requires (observational) inductive jumps, it is plausible to call such a descriptive program an *inductive research program*.[3] It is easy to see that such programs 'approach the truth by induction'. For the establishment of observational laws, the micro-step of the HD-method may be applied, resulting in true descriptions which either falsify the relevant general observational hypothesis or are partially derivable from it. According to the basic definition of 'more truthlike', assuming that accepted observational laws are true, any newly accepted observational law guarantees a step in the direction of the true theory. For it is easy to verify (exercise) that if $S(t)$ and $S(t')$ indicate at time t and t' later than t, the strongest accepted law $S(t')$ is closer to T than $S(t)$. Hence, inductive research programs are relatively safe strategies of truth approximation: as far as the inductive jumps happen to lead to true accepted laws, the approach not only makes truth approximation plausible, it even guarantees it.

Let us now turn to the explication of the nature of explanatory or theoretical programs, which are by definition of a nomological nature. An *explanatory program* may or may not use a theoretical vocabulary. Even (nomic) empiricists can agree that it is directed at establishing the true observational theory. If there are theoretical terms involved, the referential realists will add that it is also directed at establishing the referential truth. The theory realist will add to this that it is even directed at establishing the theoretical truth. Scientists working within such a program will do so by proposing theories respecting the hard core as long as possible, but hopefully not at any price. They will HD-evaluate these theories separately and comparatively. RS directs theory choice and is trivially functional for empirical progress. However, although that rule is, moreover, demonstrably functional for all distinguished kinds of nomic truth approximation, it cannot guarantee, even assuming correct data, a step in the direction of the relevant truth. Though the basic notions of successfulness and truthlikeness are sufficient to give the above characterization of the typical features of explanatory research programs, they usually presuppose refined means of comparison, which are presented in Part IV of ICR.

9.3 PROGRESS IN DESIGN AND EXPLICATIVE RESEARCH

9.3.1 *Progress in Design Research*

As will be pointed out in the next chapter, design research may be set-theoretically characterized by the product that it targets and that realizes a set of desired properties and avoids a set of undesired properties. It may be tempting to think, in terms of the previous analysis, that this goal can be reconstructed as (materially) realizing a desired (conceptual) possibility. However, although a desired possibility realizes, by definition, all desired features, it satisfies also all undesired features, in the sense defined above, which do not exclude it. Hence, the parallel between product and truth approximation should not be sought on the level of conceptual possibilities. But an illuminating parallel arises by equating the intended product with the target set T in the case of nomic truth approximation. According to our previous definitions, *T* has all desired features and has no undesired features. Similarly, the intended product has, by definition, to satisfy all desired properties and to lack all undesired ones. Note that, in order to avoid confusion, we will remain speaking of features of theories and properties of (possible and realized) products.

In view of this parallel, formal progress in design research amounts to the claim that one prototype possesses more desired properties and fewer undesired ones than another. However, this claim is, of course, also the plausible definition of determinable progress in design research, assuming that the sets of relevant (i.e. desired and undesired) properties are fixed and finite. Finally, assuming a distinction between structural and functional properties, the possibility of functional equivalents arises in the same way as the possibility of (on logical grounds) observationally equivalent theories in the context of nomological research. However, this parallel immediately suggests a dissimilarity between the two types of research. In design research every reason to favor one functional equivalent to another can be rephrased in terms of an *additional* desired or undesired property. In nomological research, assuming a realist position, it may well be that one observationally equivalent theory is closer to the theoretical-cum-referential truth than another, without disposing of a similar extension possibility in some straightforward way.

Some other differences, which will be elaborated in Section 6 of the next chapter, are:

(1) a design target set is a free and changeable choice,[4] whereas given a domain and a vocabulary, the nomic target set is fixed;
(2) a design target set is known, whereas the nomic target set is not;
(3) the above enables a straightforward (comparative) evaluation of a (new) prototype, in contrast to the indirect evaluation of a (new) theory;
(4) to revise a prototype may be expensive for material reasons, to revise a theory is relatively inexpensive;
(5) for a fixed domain and vocabulary, nomic truth approximation is, ideally speaking, free from external influences, whereas design research is basically open to such influences.

However this may be, as has been indicated, there are strong formal analogies, simply due to the fact that design research, like nomological research, can be characterized that it aims at a product which satisfies a certain target set of desired properties, while avoiding another set of undesired properties.

9.3.2 *Progress in Concept Explication*

A brief exposition may also suffice for concept explication. Recall that concept explication starts from a vague, intuitive concept, with the intent of giving a simple, precise and useful definition of a concept that resembles the original. Like nomological research, this task may be represented in terms of conceptual possibilities. Let us further assume first that there is a unique solution and hence a unique set of desired possibilities. Let a provisional explication also be conceived as (determining) a unique extension of conceptual possibilities. Then it is plausible to define formal progress in explicative research formally in the same way as in the case of nomological research.

In real-life explicative research, however, the resemblance of a provisional explication has to be evaluated in other terms. In particular, such evaluation takes place in terms of evident examples, that is, evidently desired possibilities, evident 'non-examples', that is, evidently undesired possibilities, and, finally, so-called conditions of adequacy, that is, conditions to be fulfilled and which correspond to desired features. Hence, the definition of 'conceptual progress' in explicative research is straightforward. Provisional explication Y is better than provisional explication X, roughly speaking, if and only if Y treats more evident examples and non-examples properly and/or fulfills more conditions of adequacy.

The partial analogy of explicative research with design and nomological research is obvious, including the possibility of functionally equivalent explications. However, at least two differences with nomological research are very interesting. There is no analogue for evident non-examples, as this would require the realization of nomic impossibilities. Moreover, while nomological research is more or less bound to a unique solution, explicative research may well lead to the conclusion that two or more interesting explications can be given, which mutually exclude each other.

For design research, a similar non-uniqueness of the solution holds. However, like in nomological research, there does not seem to be an analogue for evident non-examples in the case of design research. Moreover, we have seen that formal progress in design research is relatively easy to determine, but not so in concept explication. Hence, although there is also a strong analogy between design and explicative research, i.e. both aim at a certain product, the analogy is not perfect.

CONCLUSION

The overarching analogy between the three types of research may even be characterized in terms of products: all three aim at a certain product, viz. a theory, an artifact, and a concept, all serving certain purposes. We have argued that these types of research, including formal progress in them, can in principle be described in terms of having desired and undesired features or properties. Hence, although scientists may feel strong differences in what they do, for example, due to strong differences in determining *prima facie* formal progress, they may also feel a strong kinship on a more abstract level of formal progress.

CHAPTER 10
DESIGN RESEARCH PROGRAMS

INTRODUCTION

In Chapter 1 we introduced four types of research programs: descriptive, explanatory, design and explicative research programs. So far the emphasis was on descriptive and explanatory programs. Moreover, several (groups of) chapters illustrated explicative programs in philosophy of science. Design research programs, however, only received a little attention in the previous chapter.

Let us briefly reconsider the main aspects of research programs and the reasons for introducing the concept of a design research program, as they were discussed in Chapter 1. The notion of a research program has proved to be useful for the description of scientific developments. Lakatos has characterized a research program in terms of a hard core of ideas and a positive heuristic, i.e. a set of heuristic ideas, e.g. generated by a model, to protect the core ideas against serious attacks. In our opinion there are many research programs that don't have a stable positive heuristic, but are nevertheless successful. Hence, we proposed to speak already of a research program when there is a hard core of ideas: programs based on an idea.

Research programs are supposed to deal with a certain domain of (empirical or conceptual) phenomena and with a certain type of goal with respect to this domain. In the scientific literature, it is clear that at least three types of goals occur: descriptive, explanatory, or explicative, and these three provide the bases for three types of research programs. It is certainly possible to describe large parts of scientific research in terms of such programs and their interaction, i.e. competition or co-operation. However, there are also many research programs which do not fit very well into one of these prototypes or a combination of them. For example, research programs directed at the design or construction of new materials, of medical drugs, of growing methods for agricultural products, of expert systems and other AI-systems, of new psycho- and behavioral therapies, etc. Hence, such research programs are directed at the design or construction of certain products or processes which have to satisfy previously determined demands and these demands are based on the intended applications. For brevity, we speak of products, also when processes are meant or the improvement of already existing products or processes. The hard core of such programs is frequently called the 'lead': the basic idea about how the product is to be composed or how it has to work.

In view of the fact that this type of research program occurs not only in the natural and technical sciences, but also in the social sciences and the humanities, we do not speak of technological research programs and prefer to speak of construction or *design* (research) programs: their internal goal is the construction of the intended product, which is based on one or more intended applications.

In this chapter we will analyze the structure and development of such programs in some detail and using set-theoretic terms. Roughly speaking, the main idea of a design research program is a systematic attempt to bring together the properties of available materials and the demands derived from intended applications. Our point of departure is the diagnosis of a conceptual confusion in a first attempt to explicate this idea, the lattice model constructed by Weeder et al. (Section 10.1). The logic of problem states and state transitions in such programs, including assessment criteria and heuristic principles, is described in set-theoretic terms, starting with a naive model comprising an intended profile and the operational profile of a prototype (Section 10.2). In a first concretization the useful distinction between structural and functional properties is built into the model (Section 10.3). In two further concretizations the inclusion of potential applications is motivated and described for the case of drug research as well as the inclusion of potential realizations for the case of complex products (Section 10.4). Next, another line of concretization of the naive model, the incorporation of potentially relevant properties, is sketched (Section 10.5). Then the partial analogy between 'product' and 'truth approximation' is indicated (Section 10.6). We conclude with a few remarks about the usefulness of our models for describing the ways in which products reach the market in comparison to the so-called social construction of technology approach.

10.1 THE LATTICE MODEL

The problem of how design programs develop must be studied in an explicative research program at the meta-level. In 1982 Pieter Weeder and Do Kester published a paper in Dutch (Weeder and Kester, 1982) which we would like to consider as the start of an important explicative 'program based on an idea' for the study of this issue. Some other members of the Science Studies Unit of the University of Groningen had been involved in the first development of the idea and the case study, viz. the Tenax project of the Akzo laboratory in Arnhem (Netherlands), in particular Philip Vergragt, but also Henk Bodewitz and Gerard de Vries. It was not until 1988 that the first English (improved and extended) version appeared, written by Henk Bodewitz, Gerard de Vries and Pieter Weeder (Bodewitz et al., 1988). In view of the fact that Weeder was the only common author we speak of the descriptive program (and the lattice model) of Weeder et al.

In characterizing the development of a descriptive program we have to make a distinction between the core idea on the one hand and the resulting description on the other. A similar distinction can be made for explanatory programs, roughly the explanatory idea and the resulting explanation. The present chapter is meant to contribute to the further development of the core idea of Weeder et al. A first formulation of this core idea is the following:

Core idea: The development of design programs can best be described as a more or less systematic attempt to bring together the properties of the available materials and the demands derived from the intended applications.

This core idea is specified in terms of a so-called lattice model, and it is convincingly illustrated by the Tenax case that the core idea is indeed very illuminating as the basic perspective for its development. Theoretically it can be seen to give rise to a plausible cyclic structure the R&D process in which a number of well-ordered strategic decisions play a crucial role. Finally, an interesting relation is suggested between the present type of analysis of the R&D process at the laboratory research level and the ideas on technology and technological innovation which are laid down in the evolutionary theory of economic change by Nelson and Winter.

For all this the reader is referred to the mentioned 1988 paper. The present chapter only intends to revise the specification of the core idea, claiming that the new model throws more light on 'the logic of problem states and transitions' in design programs. The informal lattice model of Weeder et al. does not enable the use of some standard formal description techniques, whereas our model enables the use of the techniques of elementary set theory. In order to motivate our set-theoretic model, we begin by quoting the concise presentation of the lattice model in the 1988 paper.

'Defined in broad terms, the aim of research and development in industrial laboratories is to contribute to the design of industrial products with technical characteristics which meet the purposes, desires and explicit requirements of customers as these are perceived within the firm. Consequently, we may view the R&D process as involving two sets of data, viz.

(1) the set of *materials* or classes of materials with certain relevant *properties*, especially of a physical-chemical nature;
(2) the set of *intended applications*, or tasks, incorporating technological and economic *demands*.

In an elementary form, the R&D process may be conceived of as an attempt to match these two sets, aiming at a product which realizes the intended applications.

Visually, the matching process may be represented in a "lattice", in which the research task consists in attempting to connect the set of demands $D(i)$ of an application A with the set of properties $P(i)$ of material M. Of course the properties are mutually dependent. The same holds for the demands $D(i)$ (see Figure 10.1).'

The figure shows immediately that it is difficult to visualize states and to assess state transitions such that 'the logic of the problem situation', to use Popper's favorite term, comes to the fore. The main cause of this lack of transparency stems, in our opinion, from the fact that two quite different intuitions are mixed in the property/demand-terminology, not only in the formulation of the core idea but also in the lattice model specification of it. On the one hand there is the distinction

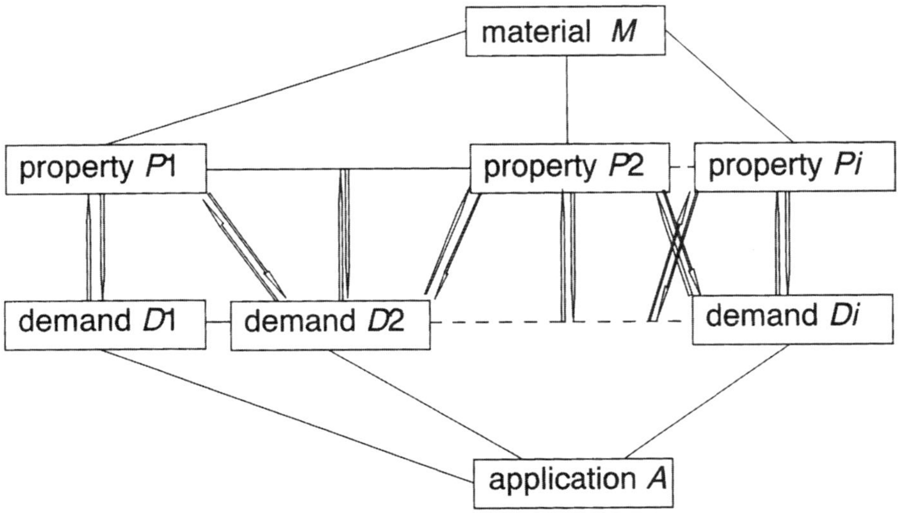

Figure 10.1 The lattice of properties and demands.

between the factual properties of the already available product and the desired properties of the intended product. On the other hand there is the distinction between technical or structural properties and service or functional properties, irrespective of whether they are factual or desired. The two distinctions are clearly of quite a different nature; they are perpendicular to each other. However, it is clear from the quoted passage that these distinctions are not differentiated in the lattice model. In this chapter we will show that the first distinction, between factual and desired properties, is the clue to the logic of the situation. Of course, the second distinction is also of fundamental importance, but it is to be introduced as a concretization.

10.2 THE NAIVE MODEL OF PROBLEM STATES AND TRANSITIONS

In this section we will introduce the naive model, by first characterizing problem states, and subsequently the evaluation criteria for state transitions. A survey of possible refinements for the naive model will be indicated.

Drawing upon a survey of the literature (Saren, 1984), it is interesting to see that existing models of design research, in particular within firms, where it is usually called 'innovation', tend to treat the actual design process as a kind of black box intermediate stage, within a more encompassing sequence of stages. In this chapter, we show how it is possible to analytically decompose this black box into a sequence of problem states and state transitions.

As a matter of fact, drug research (Vos, 1991, Section 2.7) provided the main type of example for our joint analysis of design research in (Kuipers, Vos and Sie, 1992), which forms the basis of the present chapter.

10.2.1 *Problem States*

Let *RP* indicate the set of *relevant properties* for the product to be developed.[1] For each element of *RP* it is assumed that its presence or absence is explicitly required in the specification of the intended product. Let the subset *W* of *RP* indicate the set of *(desired or) wished for properties* of the intended product; *W* will be called the *intended profile*. Of course, $RP - W$ is the set of undesired or unwanted properties.[2]

For each concrete candidate product x, henceforth called prototype x, it is important to determine which properties in *RP* it actually has. Let the subset $O(x)$ indicate the set of these factual or *operational properties* of x; $O(x)$ is called the *operational profile* of x.

In drug research (Vos, 1991, p. 62), the 'lead' can be identified with the intended profile and some idea about how to realize it. More specifically, the 'lead compound' comprises a chemical compound, with its operational profile, together with the intended profile. Assuming an interesting amount of overlap between the two profiles, but not yet a perfect correspondence, the challenge is to step-wise reduce the differences.

The problem situation in a certain state of development can clearly be depicted, in Figure 10.2, viz. the fact that the two profiles do not coincide: the problems consist of the two starred sets, $W - O(x)$ and $O(x) - W$.

$W - O(x)$ represents so to speak the unrealized 'positive' desires, and $O(x) - W$ the realized 'negative' desires.

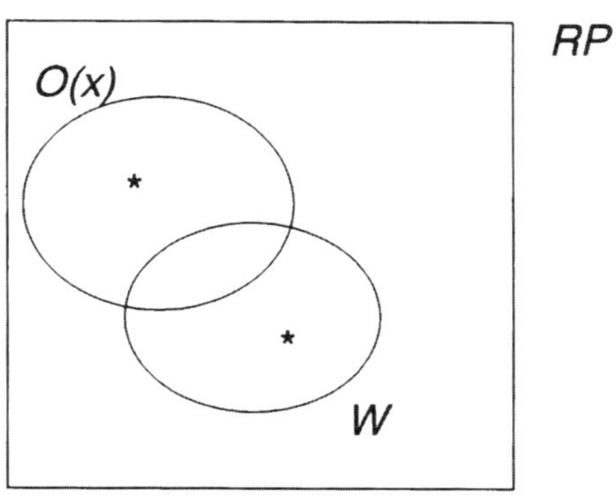

Figure 10.2 *A problem state.*

Let us introduce the plausible basic formal notion for comparing any pair of profiles P and P^*, i.e. any pair of subsets of *RP*. It is the so-called *symmetric difference* $P \triangle P^*$, defined as the union $(P - P^*) \cup (P^* - P)$. $P \triangle P^*$ indicates an elementary type of qualitative distance or dissimilarity between P and P^*. Hence, the smaller it is, the greater the similarity between P and P^*.

On certain occasions, it will also be attractive to have quantitative formal notions, in which case we have to assume that RP is finite, or at least all profiles to be considered. $|P|$ indicates the number of elements in P, hence $|P \triangle P^*|$ is a quantitative measure of the dissimilarity of P and P^*.

By consequence, $O(x) \triangle W$, the union of $W - O(x)$ and $O(x) - W$, the starred areas in Figure 10.2, represents the problem state in qualitative terms and $|O(x) \triangle W|$ in quantitative terms. More precisely, $O(x) \triangle W$ specifies the set of problems, i.e. the deviations from the claim that $O(x) = W$, whereas $|O(x) \triangle W|$ represents only the number of problems, i.e. the number of deviations.

10.2.2 *Transitions of Problem States*

$O(x) \triangle W$, indicating the set of problems in the problem state, forms the starting point for negotiation, the set of negotiation options, about what to do: trying to change x into some x' such that $O(x')$ becomes more similar to W, or changing W into some W' such that W' becomes more similar to $O(x)$, or both.

It is plausible to give a formal characterization of certain transitions of one problem state into another by the following definitions, which form the cornerstones of the present chapter ('$\subset$' is the sign for proper subset).

Definition 1
(a) Prototype x_2 is a *qualitative improvement* of x_1 in view of W iff $O(x_2) \triangle W \subset O(x_1) \triangle W$
(b) it is a *quantitative improvement* iff $|O(x_2) \triangle W| < |O(x_1) \triangle W|$

Definition 2
(a) Intended profile W_2 is a *qualitative concession* to prototype x compared with W_1 iff $O(x) \triangle W_2 \subset O(x) \triangle W_1$
(b) it is a *quantitative concession* iff $|O(x) \triangle W_2)| < |O(x) \triangle W_1|$

These definitions provide the basic assessment criteria for state transitions. The first definition enables us to evaluate potential improvements of the prototype, i.e. transitions from one prototype to the other in the face of a fixed intended profile. The second definition specifies how to evaluate potential concessions, i.e. transitions from one intended profile to the other in the face of a fixed prototype.

The first type of transition is an answer to one particular specification of the problem state, viz. how to bring the prototype closer to the intended profile? The second type of transition is an answer to the remarkable conversion of this problem specification, viz. how to bring the intended profile closer to the prototype? The fact that this problem specification is realistic is documented by Vos (1991) and hinted at in the title of that book: *Drugs looking for diseases*.

It is evident that in both cases the qualitative judgement implies the quantitative one, but not the reverse. The quantitative definitions are so rough that they even lead to linear orderings for prototype and intended profile transitions. However, the

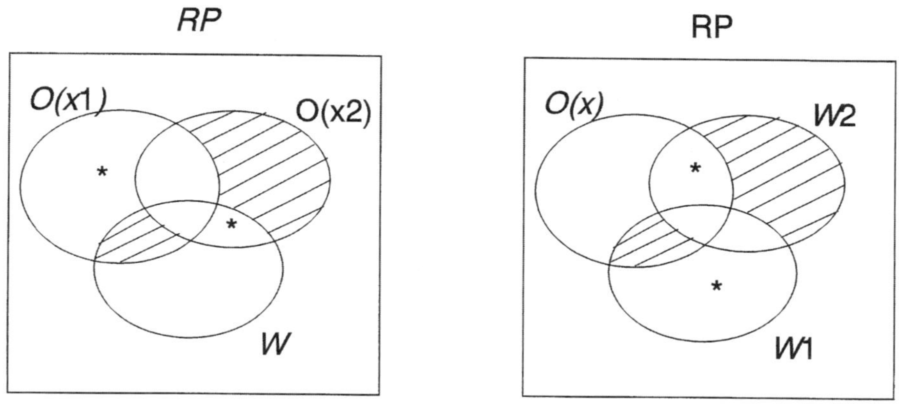

Figure 10.3 *A qualitative improvement of a prototype (left). A qualitative concession of the intended profile (right).*

qualitative definitions obviously lead only to partial orderings of both types of transition. But from the formal point of view, they represent the purest cases. They are depicted in Figure 10.3, where in both cases the two shaded areas are empty and at least one of the two starred sets is non-empty.

Some remarks about constraints for the domain *RP* of relevant properties have to be made. It is clear that there is no reason to have both a property in *RP* and its counterpart: if for example 'flexible' belongs to *RP*, and *W* or $O(x)$ is supposed to contain 'non-flexible' one simply excludes 'flexible' from either *W* or $O(x)$. This avoidance of duplication is even an advantage in that it restricts the formal constraints on profile pictures. An equally inconvenient formal constraint can be avoided by excluding the occurrence in *RP* of the combination of two properties in *RP* as a new property. If we would not exclude a combination of properties as a new property all profiles would have to be closed for combinations. There do not seem to be other plausible restrictions to RP. It is important to note that not every subset of *RP* needs to be nomically possible, i.e. not every conceptually conceivable profile needs to be realizable in reality. In other words, there will be all kinds of causal connections between (subsets of) the properties in *RP*.

From the presented point of view the development of a design research program is a succession of problem states, where problem transitions will, as a rule, be quantitative, and ideally qualitative, improvements or concessions. There may be different types of specific research involved. First direct experimental test research is involved. For any new prototype x and intended profile *W* the claim $O(x) = W$ has to be evaluated by experiments. Besides this direct empirical research descriptive, explanatory or explicative research that primarily belongs to other research programs may also be involved. In Zandvoort's terms (1986), the design program operates as *guide program*, and the others as *supply programs*.

10.2.3 *Possible Refinements*

The presented set-theoretic model is a naive model in many respects and its value depends largely on the degree in which it can be concretized in order to adapt to all kinds of realistic complications. In practice, one or more of the following refinements will be required.

(R1) Instead of having a simple yes/no character, properties usually have to be construed as functions with a range of more than two values, possibly even infinitely many.
(R2) Some properties may be more important than others, without the latter being negligible.
(R3) It is usually not immediately clear whether a property is relevant or not. In the course of product development their relevance may become clear, and the different actors in the process may start to negotiate about them.
(R4) In many cases there is a plausible distinction between structural and functional properties, such that the intended profile is primarily specified in terms of functional properties, whereas the available prototype is primarily known in terms of its structural properties.
(R5) In some cases, it is very helpful to include a set of potential (intended) applications explicitly in the model, as suggested in the lattice model of Weeder et al.
(R6) It may sometimes also be helpful to include a set of potential realizations, roughly corresponding to different materials in the lattice model.

Lack of space prevents us from dealing with all these refinements here. The first two refinements are essentially technical. Vos (1995) (see also Vos (1991)) describes in detail how the first refinement can be realized and suggests how to deal with the second. The refinements R3 and R4 are of fundamental conceptual importance. The distinction between structural and functional properties (R4) is introduced in Section 10.3. The importance of the further refinements R5 and R6 depends very much on the type of intended product. In Section 10.4 the set of potential applications (R5) will be seen to be particularly relevant in the case of drug research, viz. diseases. In this way, that section provides at the same time a concise application of our conceptual apparatus, based on Vos (1991), a book that can be seen as an illustration of the chapter in general. Section 10.4 also presents the introduction of a set of potential realizations (R6) for complex products, based on Sie (1989). The refinement of the naive model with 'potentially relevant properties' (R3) will be described in Section 10.5.

It will become easy to see that the R3-line of refinement is essentially compatible with the R4/5/6-line. Hence the question of how they can be integrated may be left as a technical exercise. The same holds for the combination of the technical refinements R1 and R2 and the others. In sum, we may conclude that we have a network of related models, with the naive model as the point of departure and the choice of a refined model to be determined by the context.

We would like to conclude this section by claiming that the presented naive model of problem states and transitions, conceived as an alternative specification of the core idea of the program outlined Weeder et al. is already an improvement on the lattice model. In Section 10.6, it will be shown that this naive model moreover enables a comparison of the differences and resemblances between design programs on the one hand and descriptive and explanatory programs on the other; to begin with, the former are product directed and the latter are truth directed. Finally, in concluding we will make some remarks about the usefulness of our models in describing the ways in which products reach the market in comparison to the so-called social construction of technology approach.

10.3 STRUCTURAL VERSUS FUNCTIONAL PROPERTIES

In many contexts, it is possible and customary to divide the set of relevant properties RP into a subset of technical or structural properties S on the one side and a subset of service or functional properties F on the other. S and F do not overlap each other and they exhaust RP.

We would like to mention an example taken from Saviotti (1996), where he uses the distinction, in his terminology, between technical and service characteristics, as the basis for a description of technological development, i.e. the rise and fall of products or technologies on the market. In the description of aircraft, e.g. helicopter technology, there are on the one hand technical properties like length, rotor diameter, engine power, engine type, number of engines, geometry and, on the other, service properties like maximum take off power, maximum speed, range. Note that these properties, in their present formulation, are not of the simple yes/no type, but that is irrelevant for our present purposes. It was an earlier paper by Saviotti (1988) that inspired us to use the same distinction in the description of the process of product development, but we prefer the more general terminology of structural versus functional properties.

It is plausible to call $O(x) \cap S$ the *operational structural profile* $OS(x)$ of prototype x and $O(x) \cap F$ the *operational functional profile* $OF(x)$. Of course, it follows that $O(x) = OS(x) \cup OF(x)$.

It is also possible to make the same division in the intended profile W: the *wished for structural profile* WS and the *wished for functional profile* WF. But it is not evident that the latter distinction is also realistic. It requires us to specify W beforehand, with the consequence that WS is simply equal to $W \cap S$ and WF equal to $W \cap F$. In practice it is usually the other way around. As a rule, WF is provisionally determined first, for we start by asking what the product is supposed to do in the intended applications. As soon as we have fixed WF the next question is how WF can be realized. It is of course not guaranteed that there is just one subset of S causally implicating precisely these functional properties. In other words, there need not be a unique WS; we have to leave room for 'functional equivalents'. We will call a set of structural properties *appropriate* or an *appropriate structural profile* for *WF* if it causally implies *WF*. Such a set will be indicated by *AS(WF)*, or simply *AS* when *WF* is clear from the context.

The foregoing discussion can be clarified by saying something more about the *S/F*-division. Of course, this distinction will have some arbitrariness. But it is plausible and helpful to assume that the *S/F*-division satisfies at least the following:

S/F-splitting principle of minimal causality
(a) for all x and x': $OS(x) = OS(x')$ causally implies $OF(x) = OF(x')$,
(b) all elements of S are necessary to make (a) true in general.

We call x and x' structurally equivalent (or imitations) if $OS(x) = OS(x')$ and functionally equivalent (or substitutable) if $OF(x) = OF(x')$. Now (a) says that structural equivalence must causally imply functional equivalence. In consequence, there is a function associating with each set of structural properties a unique set of functional properties. The reverse implication is not required, functional equivalence does not imply structural equivalence, and hence there is no similar function. The reason for clause (b) is the following. Without (b), the splitting principle would allow *S/F*-divisions such that '*S* causally over-determines *F*' and there is no reason for doing so. On the contrary, the larger *F* is, the more freedom exists in the choice of wanted and unwanted properties.

Note that we have not assumed that the splitting principle is sufficient for the determination of a unique division of *RP* in *S* and *F*. It leaves room for more than one possibility and the choice between them will be based on additional considerations. One further consideration may concern prices. It seems plausible to assume that the market price of a product is primarily determined by its functional properties, whereas the cost price is primarily determined by its structural properties. In other words, following the standard criticism of Marx's labor theory of value, it is assumed, and specified, that the cost price is determined by different aspects of the product than those which determine how much people want to pay for it. It may be useful to use such price considerations in the further determination of the *S/F*-division, for it is evident that cost and market price considerations play an important role in the R&D-process.

Let us now return to the intended and operational profiles. We have already noted that as a rule the desired functional properties of the intended profile are provisionally decided upon earlier than the desired structural properties. For the operational profile of a prototype it may be the other way around. To some extent its structural properties will be known beforehand, and the question is what additional structural properties it has, and what functional properties.

In this way we have differentiated the two intuitions contained in the property/demand-terminology of Weeder et al.: the distinction between factual and desired properties on the one hand and between structural and functional properties on the other. The resulting asymmetric model will be called the *S/F*-model and is depicted in Figure 10.4, in which the black arrow indicates causal determination.

It is plausible to assume that the comparison of prototypes in view of a fixed intended profile and of intended profiles in view of a fixed prototype is primarily a

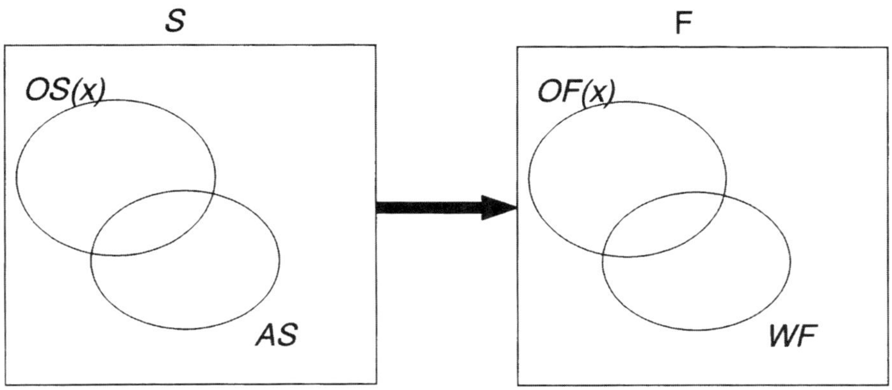

Figure 10.4 *A problem state in the S/F-model.*

matter of comparing functional profiles. In consequence, it is also plausible to conceive the restrictions of the original definitions of the assessment notions of improvement and concession to functional properties as defining the basic assessment notions of the *S*/*F*-model. Hence, from now on we assume for the *S*/*F*-model, and for its extended versions in the following section, that '*O*' and '*W*' in *Definitions 1* and *2* have been replaced by '*OF*' and '*WF*', respectively.

Moreover, it is now plausible to assume that preliminary estimations of judgements of functional improvements are suggested by ideas about possible relations between structural and functional similarity. It is easy to formulate some principles of which it is not only evident that they are formally *in*valid, but also that they play an important heuristic role. The basic idea is of course that structural similarity implies functional similarity, and vice versa. Restricting our attention to the qualitative cases we get the following heuristic principles, in which *AS* indicates an arbitrary appropriate structural profile for *WF*.

Heuristic principles

HP1: if x_2 is structurally more similar to *AS* than x_1 (i.e. $OS(x_2)\triangle AS \subset OS(x_1)\triangle AS$) then x_2 will (probably) be a functional improvement of x_1 in view of *WF* (i.e. $OF(x_2)\triangle WF \subset OF(x_1)\triangle WF$).

HP2: if x_2 is a functional improvement of x_1 in view of *WF* then x_2 will (probably) be structurally more similar to *AS* than x_1.

Intuitively speaking, *HP1* states that increasing similarity with an appropriate structural profile is likely to lead to increasing similarity with the intended functional profile, and *HP2* claims the converse. Because of the causal asymmetry *HP2* will have more exceptions than *HP1*.

These heuristic principles concern functional improvements and not concessions. In view of the fact that the question of whether a change of intended functional profile is a functional concession to a fixed prototype is entirely a matter of comparing sets of functional properties, there are no comparable heuristic principles for functional concessions.

10.4 POTENTIAL APPLICATIONS AND REALIZATIONS

First we will refine the notion of functional properties by introducing potential applications, with emphasis on those for drug research, viz. diseases. Then we will refine the notion of structural properties by introducing potential realizations in particular for the case that complex products are envisaged.

10.4.1 *The S/FA-Model for Drug Research*

It turns out to be worthwhile to introduce a set of potential applications. This is particularly so for design programs in which the intended profile is not fixed. This will be illustrated for the case of drug research.

The wished for functional profile *WF* will be normally formulated with respect to a particular disease or to a range of diseases. For example, a therapy suitable for the treatment of angina pectoris, a particular form of ischemic heart disease, should affect the mechanical and electrical processes in the heart in such a way that the pathological condition of the patient will be restored.

Therefore it is useful to include a set of *potential (intended) applications A*, in this case diseases, in the model. With the inclusion of this set results a model that deals with three types of profiles. It is called the *S/FA*-model. See Figure 10.5, in which $C(A)$ indicates the set of conceivable characteristics of potential applications, i.e. disease characteristics, while $C(y)$ denotes the profile of potential application y, i.e. the *disease profile* of disease y. Disease (and drug) profiles, in which knowledge about diseases (and drugs) is stored, are extensively discussed in Vos (1991). For a concise presentation, see also Vos (1995). Here, it suffices to state that disease profiles can indeed be considered as finite sets of disease characteristics.

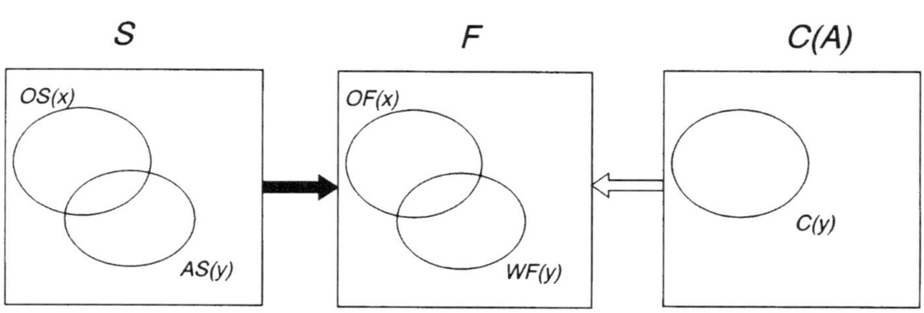

Figure 10.5 A problem state in the S/FA-model.

The distinction between functional characteristics of drugs and disease characteristics is crucial. Functional characteristics characterize the biological responses of drugs whereas disease characteristics form the essential features of the pathological condition. In analogy to the splitting principle from Section 10.3, we can state that an application profile $C(y)$ uniquely determines the wished for functional profile $WF(y)$, but the reverse need not be the case. For example, the reduction of the heart rate is desirable in the treatment of angina pectoris, but also in certain forms of cardiac dysrythm. The suggested *A*/*F*-splitting principle is, like the *S*/*F*-splitting principle, a causal principle in the sense that $WF(y)$ is causally necessary and sufficient to cure y, i.e. to free patients from the disease characteristics $C(y)$. Note that, whereas curing y is a causal effect of $WF(y)$, the initial determination of $WF(y)$ by $C(y)$, although based on this supposed curing effect, is not itself a matter of causal determination, for which reason we have used a 'white' arrow to indicate this initial determination in Figure 10.5.

The basic assessment notions of improvement and concession remain in the same way, as in the *S*/*F*-model, focussed on functional profiles. However, on the basis of the *A*/*F*-splitting principle we get in addition to *HP1* and *HP2* two new heuristic principles suggesting that similarity of disease profiles implies similarity of wished for functional profiles, and vice versa.

The advantage of the *S*/*FA*-model is that different developmental pathways in drug discovery (Vos (1991)) can be described. The curious history of verapamil illustrates this. When verapamil was introduced in the early 1960s as a new antianginal drug, this compound was the result of a common pathway in drug research. At that time the desired action of any antianginal drug was considered to be the dilation of the blood vessels in the heart, such that blood and oxygen supply to the heart will increase. This therapeutic principle of coronary vasodilation dominated pharmacology of the 1940s and 1950s, based on then available knowledge of the disease. From a series of analogues of papaverine, widely acknowledged as a powerful smooth muscle relaxant and coronary vasodilator, the substance D 365 (verapamil) was selected. In comparison with nitroglycerine, a very successful antianginal drug and the prototype of coronary vasodilating agents, verapamil seemed to be a promising drug.

During the 1960s, however, knowledge about angina pectoris changed, whence the wished for functional profile changed. Though this description oversimplifies the developments in the field of angina pectoris, an important phenomenon can be noticed. Due to transitions in the disease profile of angina pectoris, the wished for functional profile changed in such a way that verapamil, initially considered a promising drug, was judged negatively. In the late 1960s it was considered a 'dying drug'.

The reverse process, however, is much more interesting because it reveals a very important but neglected pathway of drug discovery. In this case the transformation of the disease profile, hence of the wished for functional profile, leads to a favorable reconsideration of a drug which had lost its therapeutic significance. Verapamil was reconsidered in such a way. During the 1970s, verapamil had become a prototype for an important class of anti-anginal drugs, and served as a tool in physiological and biochemical research on cardiovascular disease. One aspect of this complex history certainly was the distinction of two (sub)types of angina pectoris, hence of the wished for functional drug profiles. Verapamil was revaluated with respect to these profiles separately, and became a successful therapy for both forms of angina pectoris.

The history of verapamil shows that next to the common pathway of developing a new drug in view of a static disease profile, hence a fixed intended drug profile, a less-recognized alternative pathway exists, viz. given an existing drug profile, 'drugs are looking for diseases'.

10.4.2 *The RS/FA-Model for Complex Products*

As has been argued extensively by Sie (1989), it is sometimes also helpful to include a set of potential realizations, especially for design programs of complex products like complex machines and industrial production processes. Here the distinction between structural and functional properties and the inclusion of a set of potential applications, as outlined in the previous subsection, no longer suffice. A further refinement of the structural properties may then be helpful: the introduction of potential realizations R which are assumed to be directly related to the set of structural properties S.

For instance when we take a closer look at the development of aircraft technologies, we can identify technical properties like engine power, weight or loading capacity (volume). These may be distinguished as structural properties. But the engine power is a function of various variables such as the volume of the cylinders, the compression ratio and the type of motor. A certain weight may be realized using different kinds of materials, while some strictly required loading capacity leaves room for many different dimensions and geometries.

In fact, most technical or structural properties of products may be realized by different configurations, choices of material, dimensions, (chemical) structures, etc. Such properties will be called *material properties* and sets of them will be called material profiles or *potential realizations*. The *material profile* of a prototype gives a detailed specification of the product in the 'real world'. Structural properties are then properties or features that characterize the material profile in some more abstract way. Often they concern (physical or chemical) quantities expressed in their proper dimension. Another example is the operating pressure (or temperature) of some process line. These properties are characteristics which can be derived from the actual configuration (types and order of the machinery) and operation of the process.

Thus, once a potential realization is agreed upon, the corresponding structural properties are fixed. In consequence, we can state, in analogy to the splitting principle from Section 10.3, the R/S-splitting principle that material equivalence causally implies structural equivalence. The reverse implication, that structural equivalence implies material equivalence, is again not assumed. Of course, the R/S-

splitting principle suggests heuristic principles in both directions: similarity of material profiles implies similarity of structural profiles, and vice versa.

When the concept of potential realizations is used in combination with the inclusion of potential applications, as described in the previous subsection, the result is a model that consists of four profiles. This model is called the *RS*/*FA*-model. Its basic assessment notions remain, in the same way as in the *S*/*F*- and *S*/*FA*-model, focussed on functional profiles. The core of the model is depicted in Figure 10.6, in which, in addition to Figure 10.5, $M(R)$ indicates the set of material properties of potential realizations, $M(x)$ the material profile of prototype x, and $AM(y)$ a material profile that is adequate for $AS(y)$.

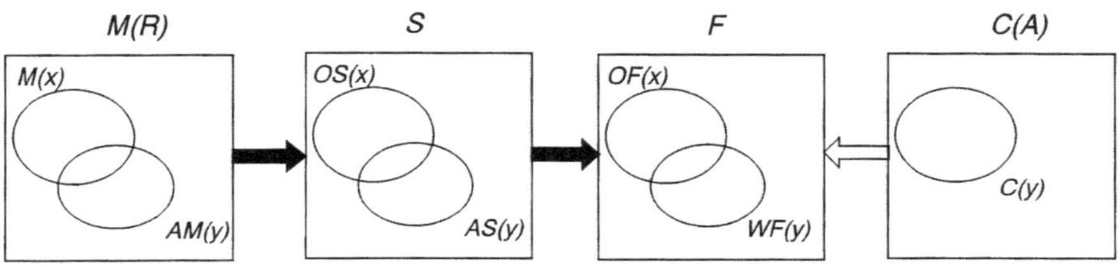

Figure 10.6 A problem state in the RS/FA-model.

The original sets of functional and structural properties are in this model refined respectively by the set of properties of potential applications $C(A)$, causally determining the corresponding intended functional profiles, and the set of material properties of potential realizations $M(R)$, causally determining the corresponding operational structural profiles. Hence, the *S*/*F*-splitting principle from Section 10.3 is enriched with similar splitting principles for the *R*/*S*- and *F*/*A*-interfaces and corresponding heuristic principles.

The *R*/*S*-division may even be such that the asymmetry between functional and structural profiles disappears, i.e. they may now causally imply one another. Functional properties like a 'light' or 'strong' aeroplane wing may be directly translatable into structural properties, such as the density of the material or the thickness of that wing. In other words, if the material has been chosen, there may remain no room for functional equivalent structural realizations of a certain functional property. Of course, the original asymmetry has been transposed from an asymmetry between the structural and the functional properties to one between the potential realizations and the structural properties.

The splitting up into four different profiles is especially useful when the product is very complex. For example, a full scale industrial production plant consists of many different parts. The design program that led to the development and construction of such a plant cannot be clearly expressed in terms of solely two or even three different profiles. When a design program is aimed at developing a plant with the functional properties 'safe' and 'efficient', difficulties may arise concerning the interpretation of these properties. What structural properties, such as operating pressure or emission rate, are required for these functional properties?

The same operating pressure or emission rate may be reached by using different types of pumps, so that different material properties are involved. Directly relating the type of pump to its safety or efficiency leaves out important questions in the design program of the plant. Important strategic decisions concern questions like 'At what operating pressure and emission rate is the plant safe and efficient?' and 'With which type of pump do we reach that operating pressure and emission rate?' Such strategic questions explicitly assume the distinction between material, structural and functional properties.

Looking back at the lattice model of Weeder et al., it is clear that the two ingredients of the present section were already essentially included in that model, though there it is hardly motivated by concerns about applications and materials. However, as we have already argued, the main conceptual problem of that model concerned the mixed intuitions involved in the property/demand-terminology, which we have differentiated in the S/F-model. In addition, we have shown why and how 'applications' and 'materials' have to be included.

10.5 POTENTIALLY RELEVANT PROPERTIES

The foregoing refinements could be presented as a natural sequence of successive refinements. We have to return to the naive model in order to deal with potentially relevant properties.

At a certain stage in a developmental process, it may be decided that a certain property which was, up to then, not considered relevant, is relevant after all. The reverse may also occasionally occur: a property may no longer be considered as relevant. Both transitions make it clear that we have to consider a broader class of *potentially relevant properties* (*PRP*) as the domain of properties. At each stage there is then a subset of *PRP* of relevant properties composed of the non-overlapping sets W and N, the set of wished for properties and the set of undesired properties. The rest of *PRP*, i.e. $PRP - (W \cup N)$, indicated by I, is, for the time being, the set of indifferent properties. The operational profile $O(x)$ is now of course the subset of properties in *PRP* which x actually possesses, and $PRP - O(x)$ contains the properties which x does not have.

I is the set of properties about which the different actors in the process may negotiate. There is not much to say about this set in formal terms. However, the definitions of improvement and concession have to be adapted in order to account for the occurrence of indifferent properties, of which we will restrict ourselves to mentioning only the qualitative cases.

To find the plausible adaptations let us start by noting that the problems in a given state of $O(x)$ and $W/I/N$-division essentially consist of two sets, $O(x) \cap N$ and $W - O(x)$. In Figure 10.7 the new problem situation is depicted.

Now it is plausible to call a transition from x_1 to x_2 in the face of a given $W/I/N$-division an *improvement* if the new sets of problems are subsets of the old sets, and at least once a proper subset; formally, if $O(x_2) \cap N \subseteq O(x_1) \cap N$ and $W - O(x_2) \subseteq W - O(x_1)$[3], and at least once proper. We have already used the easy-to-check fact

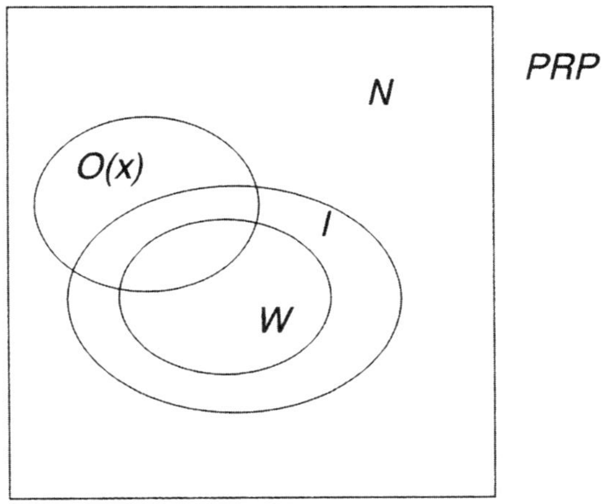

Figure 10.7 *A problem state with potentially relevant properties.*

that the two types of problems cannot interfere, assuming constant $W/I/N$, for N and W do not overlap. It may be instructive to realize that the second clause is equivalent to: $O(x_1) \cap W \subseteq O(x_2) \cap W$. Finally, it is easy to verify that the present definition reduces to the original *Definition 1(a)* by assuming I to be empty, with the consequence that N may be replaced by $(R)PR - W$. Note that the original definition is in fact introduced in terms of the union of the problem sets $O(x) \cap (PR - W) = O(x) - W$ and $W - O(x)$.

For the adaptation of the definition of concession we use the same problem decomposition. In this case it is plausible to define the transition from $W_1/I_1/N_1$ to $W_2/I_2/N_2$, with constant union RPR, in the face of a given x a *concession* if $O(x) \cap N_2 \subseteq O(x) \cap N_1$ and $W_2 - O(x) \subseteq W_1 - O(x)$ and at least once proper. Again the two types of problems cannot interfere by constant $O(x)$; there is an instructive equivalent of the second clause, viz. $O(x) \cap W_1 \subseteq O(x) \cap W_2$; and the definition reduces to *Definition 2(a)* when I is empty.

It is easy to see that the line of refinement presented in this section is essentially compatible with the sequence of refinements in the foregoing two sections. Hence the question of how they can be integrated will be left as a technical exercise. The same will be done with the combination of the two technical refinements mentioned at the end of Section 10.2 (R1 and R2) and the others. As a consequence, we may conclude that we have partially presented and partially indicated a network of related models, with the naive model as the point of departure and accounting for all six refinements mentioned at the end of Section 10.2. Of course the choice of model for a particular case or class of cases will depend on the context.

10.6 RESEMBLANCE AND DIFFERENCES WITH TRUTH APPROXIMATION

In the previous chapter we already addressed the interesting resemblance between product development and (nomic) truth approximation. This analogy already applies to the naive model, which provided the simplest formal specification of the core idea of Weeder et al. According to the naive model, product development can be described as a process of trying to increase the overlap between two sets, the set of operational properties $O(x)$ of a prototype x and the set of wished for properties W.

In ICR (Chapter 7), it has been argued extensively that the development of descriptive and theoretical (explanatory) research programs can also be described as a process of trying to increase the overlap between two sets. In particular, for the case of nomological (theoretical or theory-directed) programs in the natural sciences, it was argued that the ultimate aim is to characterize, for a given domain and within the set *Mp* of conceptual possibilities (or potential models) generated by the vocabulary of the program, the subset T of nomic possibilities. This set T represents the true theory or *the truth* (within *Mp*). An arbitrary theory X selects a particular subset of *Mp* and claims '$X = T$'. This claim will be false in all cases but one. The problems of theory X can be represented by the symmetric difference $X \triangle T$. The aim of theory development can now be seen as the attempt to decrease the symmetric difference of X and T, in other words, to increase the overlap between X and T.

The formal analogy with naive product development is obvious. In both cases the aim is to decrease a symmetric difference, that is, to increase the overlap between to sets. But the analogy is more than merely formal. The formal analogy follows directly from a common conceptual viewpoint, viz. the idea that product and theory development are both problem solving and hence problem reducing activities. In both cases the problem state can be represented by a symmetric difference. It was in fact the conjecture of the formal and conceptual analogy of product development with theory development which suggested the naive model for product development.

This last consideration exhausts the resemblances. In the previous chapter, we also hinted at some differences. Here we elaborate them, for they are at least as interesting and instructive.

(1) As long as the domain and *Mp* are fixed, T cannot be changed, whereas W can be changed, even if *RP* is fixed. The constancy of T follows directly from the assumption that nature does not fundamentally change in the sense that what is nomically possible and what is not remains constant. That W may be changed is obvious. The degree of changeability, however, is limited as long as we keep the intended applications constant, but it is of course considerable if we change the intended applications.

(2) T is unknown, W is known. In theory-directed research, with fixed domain and *Mp*, T can be considered as the great unknown: what one is looking for. A fundamental complication is that one can never know when one has reached this target (fallibilism). On the other hand, at every stage of product development, it is known, at least in the naive model, which properties are wanted (constituting the set W) and which unwanted. The fact that this division may change and that there may be indifferent properties for the time being, does not alter the fact that at each stage it is known on the basis of what division the operational profile has to be evaluated.

(3) A direct consequence of (2) is that it is impossible to evaluate in a direct way whether replacement of theory X by a new theory is a step in the direction of T. On the other hand, it can be judged directly whether a new prototype is an improvement 'in the direction of W'. Having the option of changing W, it is also possible to judge directly whether a change of W is a concession 'in the direction of $O(x)$'. Of course, we do not want to suggest that such evaluations are simple.

(4) In material respects it is relatively easy and inexpensive to change a theory. Similarly, for the intended profile. However, to change a prototype and hence its operational profile is usually a time and money consuming activity, similar to designing and performing test-experiments. In other words, changing theory X is essentially a paper-and-pencil-activity, which may of course nevertheless occasion much brain-racking. A change of prototype x (hence $O(x)$) essentially requires material transformation.

(5) For fixed domain and fixed Mp, the X/T approximation process is, ideally speaking, free from external influences, the $O(x)/W$ approximation process certainly is not. Given an explanatory research program, generating a fixed set of conceptual possibilities, theory development is something that can take place in the niche of that program. This is at least the case for the so-called internal phase of a program (see Chapter 1). In the application phase, theory development is directed to the solution of science-external problems or of problems raised by another research program. In the latter case, the first program functions as a supply program for the second, which plays the role of guide program. However, even in the application phase the cognitive factors influencing further theory development within the explanatory program are relatively easy to survey. On the other hand, from the network of models presented for design programs it is easy to see that there are many kinds of relatively external factors that influence the process of product development, making a survey in this case much more difficult.

For a number of these differences (1/2/5), the assumption that the domain and the vocabulary giving rise to Mp are fixed is crucial. We would like to stress that in the initial phases of descriptive and explanatory research programs the conceptual possibilities are not only not fixed, but that they are even negotiable to a considerable extent, of course interwoven with the negotiation about what one is trying to describe or explain, that is, the domain.

Despite the foregoing five differences, the resemblance is striking enough to show that theory and product development are both examples of problem solving activities which can be structured in such a way that the rationality of transitions from one problem state to the other can be defined in general and evaluated in particular cases.

CONCLUDING REMARKS

We will conclude by briefly indicating that the presented network of models cannot only play a role in the description (and evaluation) of the development of products, it can also be used in the description of the rise and fall of products that actually reach the market. Our primary claim is that the network of models is very useful for the first purpose. Vos (1991) gives an extensive illustration of this claim for drug research, in the present chapter we could only provide a glimpse of it.

However, Saviotti (1996), who introduced the distinction between structural and functional properties in the context of products reaching the market, and in particular Bijker (1995) strengthen our conjecture that the models are also useful for the second purpose. From Bijker's case studies on the bicycle, bakelite and fluorescent lighting, it becomes clear first of all that sharp distinctions cannot be made between products which reach the market and those which do not. Moreover, in Bijker's approach the distinction between different groups of potential users of a product in the making turns out to be of crucial importance. In our approach this leads immediately to the recognition that these groups have different intended profiles. (Note that their sets of relevant properties will hence also differ as a rule; this can be easily accounted for in terms of Section 10.5: indifferent, potentially relevant properties.) Hence, our approach gains new perspectives from this conceptual link with Bijker's.

The same holds for Bijker's description of cases and his so-called 'social-constructivist' theoretical analysis; both would have profited very much from this link, not in the least by demystifying some of Bijker's claims and notions. Unfortunately, we must restrict ourselves to a few remarks on this subject.

According to Bijker (pp. 73–77) inherent 'interpretative flexibility' of artifacts can be demonstrated by 'deconstruction' of an artifact into more than one artifact, each corresponding to a group. In our terms this pertains to nothing more or less than the recognition that different groups may have different intended profiles and hence that in view of a certain artifact, with a certain operational profile, these groups are confronted with different problem situations. Thus, the same idea can be expressed in a less magical way.

Also a number of Bijker's technical terms (pp. 84–88) can be explicated quite easily. His notion of 'closure' essentially comes down to the phenomenon that the changing intended profiles of different groups gradually become more similar, which may or may not be due to the fact that one of them becomes dominant. Moreover, 'stabilization' means that an operational and/or an intended profile becomes clearer and clearer. However, from our perspective it is surprising that Bijker does not introduce a third notion: the 'convergence' of an operational profile and an intended profile due to successive improvements and concessions. Sometimes it seems as if he includes something of this sort in his notions of closure and stabilization. But from our point of view it is clear that these three types of development are to be conceptually distinguished.

PART VI

CAPITA SELECTA

INTRODUCTION

In this final part of the book we will pay attention to a number of special subjects that merely illustrate the rich number of developments in philosophy of science that are taking place. The selection is very much determined by the personal research activities of the author. Subjects that are equally deserving of a chapter include: statistical hypotheses, causal analysis, default reasoning, measurement, systems theory, explicative research programs.

Two of three topics to be dealt with concern very important and productive research programs, viz. computational (design) programs in the philosophy of science, and the structuralist approach to scientific theories. We close with an exploration of default-norms for research ethics.

The first special topic deals with a particular type of design research program, known as computational philosophy of science. The products to be designed are computer programs that deal with discovery, evaluation and revision of hypotheses. This development may best be described as a recent research tradition, with several specific research programs. The binding ideas are, first, that discovery, contrary to traditional opinion in the philosophy of science, is accessible for methodological analysis and, second, that discovery, evaluation and revision should be seen as special cases of problem solving, computationally to be approached in ways developed in cognitive psychology and artificial intelligence. Deviations may arise from differences in the main objectives one is after, where one has at least to choose between historical, psychological and philosophical adequacy, and practical relevance. Moreover, the specific techniques to be used may diverge considerably. We will give impressions of some specific lines of research: the BACON-family of programs, in search of quantitative laws in physics, a family of programs in search of qualitative laws in chemistry, the program PI, of a rather different set-up, in which evaluation is at least as important as generation, and a variety of other programs, in which theory revision is dominant. We will also critically deal with the evaluative part of PI and the program ECHO. It will become clear that all the mentioned computer programs are partly competing and partly able to co-operate. Although it is too early for a general evaluation and integration of such programs, it is clear that this development is important as a stimulus for neo-classical philosophy of science and, possibly, for actual scientific research.

The second special chapter typically deals with an explicative research program in philosophy of science, viz. the study of the structure of scientific theories. There are two main approaches to the structure of empirical theories. The statement approach conceives theories primarily as sets of statements. This approach has long been considered as the only and obvious approach, e.g. by Carnap and Popper. However, it is also possible to conceive theories primarily as sets of models. One version of this so-called semantic approach is the set-theoretic or structuralist approach. It was already introduced by Suppes and refined by Sneed, Stegmüller, Balzer, and Moulines. Its basic idea is that theories amount to the specification of classes of set-theoretic structures satisfying certain conditions. After briefly discussing the attractive features of the structuralist approach in general it will be introduced stepwise, first without the distinction between theoretical and non-theoretical terms, then with that distinction in order to avoid circularity or infinite regress in measurement. The basic outline of the resulting representation of three examples will be given: classical particle mechanics, the periodic table of chemical elements, and psycho-analytic theory. Then some further refinements will follow, viz. absolute versus relative empirical content, various ways of determination of intended applications, relations between theories, theory-nets, and constraints. Finally, we will briefly consider the usefulness of the structuralist approach for non-empirical theories.

In the final chapter we try to contribute to the discussion about incorruptible professional behavior in scientific research. For this purpose we will take the Merton's famous norms as a point of departure, not in order to note that scientists do not always conform to them, but to conceive them as 'default-norms' by raising the question which deviations may be defensible and which ones are not. Accidentally, this line of inquiry also leads to an illustration of the fact that a choice between so-called deontological and utilitarian ethics is not plausible: a reasonable compromise between strict principles and (occasional or rule-guided) deviations based on possible consequences seems also in this case the only serious option. The emphasis will be on the norm of disinterestedness. We will conclude with a brief discussion of the possibility and usefulness of subsuming all this in a general professional code for scientific researchers.

CHAPTER 11
COMPUTATIONAL PHILOSOPHY OF SCIENCE[1]

INTRODUCTION

Computational philosophy of science is a collaboration between the philosophy of science and cognitive science. Cognitive science is itself a joint venture involving cognitive psychology, artificial intelligence research, neuroscience, linguistics, anthropology, and philosophy (Stillings et al., 1987), suggesting many interesting problems for philosophers of science (see Bechtel, 1988a). As illustrated in Chapter 6, Bechtel pays quite a lot of attention to reductionistic and non-reductionistic possibilities for co-operation between the disciplines involved.

In the computational philosophy of science, however, the roles have been interchanged. One tries to solve classical problems in the philosophy of science with means that have been particularly developed in cognitive psychology and artificial intelligence research. The kind of results aimed at are computer programs that enable certain cognitive tasks, or at least to simulate them, such as, discovering laws from data, designing hypotheses, evaluation and revision, concept formation, proposing experiments, etc.

The core ideas of the computational philosophy of science conceived as a collection of related, specific research programs are the following.

(1) The most revolutionary starting point, viewed from classical philosophy of science, is that not only the evaluation process, the *Context of Justification* (CoJ), but also the discovery process, the *Context of Discovery* (CoD), is considered to be methodologically examinable and perhaps partially programmable. One even holds the classical distinction of contexts as meaningless since in the CoD one anticipates the CoJ, whereas the CoJ bears all kinds of traces from the CoD. However, as has generally been acknowledged, it is also here necessary to weaken some of the traditional CoJ-aims of the logical empiricists in a Popperian direction: certainty about knowledge claims cannot be obtained, for all judgements are provisional, and the hypothesis which is preferred at a certain moment need not be the only one compatible with the experimental data.

(2) In the computational philosophy of science, scientific research is conceived as a form of problem solving. The co-operation between cognitive psychology and artificial intelligence has resulted in a general paradigm (developed by Newell and Simon) known as *heuristic search* (Newell and Simon, 1972). In a problem space of possible states, a problem is defined as the difference between an initial state and a goal state. State transitions are made by ''if... then *do* ---''-rules, called production rules. In these rules heuristic considerations are built in, for which reason they are

also called heuristic operations. A computer program for a certain type of problem hence essentially consists of a suitable specification of the problem space, in which the initial state (e.g. the data) and the goal state (e.g. a law describing the data) can be encoded, together with a set of heuristic operations.

(3) There are at least four claims one may associate with such programs: historical adequacy, the claim to be able to reproduce milestones in the history of science; psychological adequacy, the claim to be able to simulate in some relevant detail the processes in the minds/brains of scientists; philosophical adequacy, the claim to generate sound descriptive and explanatory theories that are increasingly successful, in empirical respects or even in approaching the truth; practical relevance, the claim to be of supportive value in the actual process of scientific discovery, evaluation, and revision. It is clear that these claims can be separately sustained, they do not need each other. However, as we will see, they may come into conflict with one another.

We will not pay attention to the implementation of programs, just a few remarks are to be made. Regarding programming languages, it should be noted that the choice will not only depend on the specific goal, since the use of languages that can serve different purposes (such as LISP) has become of increasing importance, viz. in order to be able to connect different programs. Regarding techniques of knowledge representation, many possibilities for so-called expert systems have been developed. Davis (1990) provides a broad introduction to them.[2]

In Section 11.1 we will give an impression of a number of computer programs, among others, BACON, which searches for quantitative laws in physics, GLAUBER, STAHL and DALTON, which search for qualitative laws in chemistry, and (extensively) Paul Thagard's program PI, in which evaluation is at least as important as generation. Section 11.2 will critically deal with the evaluation part of Thagard's PI, his later program ECHO and his evaluative claims about it. It will be argued that implementation of the simple evaluation matrix, developed in Section 8.1, would lead to at least as good results as far as historical and philosophical adequacy of the resulting theory selections are concerned.

11.1 IMPRESSIONS ABOUT PROGRAMS

Introduction

We will start with the BACON-family of 'quantitative programs' (Subsection 11.1.1.1), followed by three (mainly) qualitative programs (11.1.1.2). Both collections of programs are extensively treated in (Langley et al., 1987). Next (Subsection 11.1.2) we will present Thagard's PI (the abbreviation of Processes of Induction!) in some more detail (Thagard, 1988). In Subsection 11.1.3 we will give our impression about six examples of programs (presented in (Shrager and Langley (eds.), 1990)) in which theory revision is central. Finally, Subsection 11.1.4 gives an indication of the possible interaction between the computational philosophy of science and certain kinds of philosophy and sociology of science.

Table 11.1 gives a survey of the nature and goals of the programs to be dealt with. It is important to note that theory formation may be data-driven, but may also be, in addition, theory-driven, and that theory revision usually starts with an anomaly, that is, a conflict between new data and the theory, whether or not in combination with the body of background knowledge, in order to be guided thereupon by (parts of) these three elements. Finally, all programs have something like separate theory evaluation, but not all deal with comparative evaluation.

Table 11.1 Survey of nature and goal of programs.

	BACON	GLAUBER etc.	PI	Sect. 1.3
quantitative (mainly)	+	–	–	–
theory formation	+	+	+	–
theory revision	–	–	+	+
body of knowledge	–	–	+	+
comparative evaluation	–	–	+	+
further testing	–	–	–	+ / –

It is impossible to do justice in the following sketches to all the provisos that the authors (should) attach to their claims. For this reason, the exposition has a more optimistic tone than, strictly speaking, is justified. Nevertheless, we think that the programs illustrate a very interesting and important development.

11.1.1 *The Generation of Quantitative and Qualitative laws*

11.1.1.1 *In Search of Quantitative Laws using the BACON-Programs*

The aim of the oldest programs, the BACON-programs, is the discovery of quantitative laws summarizing the available data, at least approximately. The starting point (initial state) is a set of real values of a number of independent and dependent variables. The goal is approached by means of the explicit definition of new terms on the basis of heuristic criteria, and successively testing whether or not a constant function or a linear relation has been obtained.

The basic heuristic operations are the following: if the data of two terms are inversely related, that is, if one increases and the other decreases, then the product of the terms is formed; if on the other hand the two terms are directly related, that is, the data increase or decrease together, the quotient is formed. It is important to note that these operations have been invented while designing the program, which illustrates the creative results that the computational philosophy of science may deliver.

In the following tables (derived from (Langley et al., 1987), pp. 69 and 85) the relevant steps are represented for designing Kepler's third law on the basis of fictitious or ideal data that satisfy the law (Table 11.2.1) and on the basis of the original data (from Borelli) on the basis of which Newton checked whether the moons of Jupiter satisfied the law (Table 11.2.2). In both cases it is easy to ascertain

Table 11.2.1 Data obeying Kepler's third law of planetary motion.

Planet	Distance D	Period P	Term-1 $= D/P$	Term-2 $= D \times T1$	Term-3 $= T1 \times T2$
A	1.0	1.0	1.0	1.0	1.0
B	4.0	8.0	0.5	2.0	1.0
C	9.0	27.0	0.333	3.0	1.0

Table 11.2.2 Borreli's data for Jupiter's satellites.

Moon	Distance D	Period P	Term-1 $= D/P$	Term-2 $= D \times T1$	Term-3 $= T1 \times T2$
A	5.67	1.769	3.203	18.153	58.15
B	8.67	3.571	2.427	21.035	51.06
C	14.00	7.155	1.957	27.395	53.61
D	24.67	16.689	1.478	36.459	53.89

that the last term is equal to D^3/P^2 (D: the distance body-center, P: the period of the body's orbit), which Kepler's third law makes constant.

The second table makes it immediately clear that a program cannot work with real data without rounding off criteria, as genuine empirical data never satisfy laws in a perfect way for various reasons. However, if the error percentage is too low BACON does not find a law and when it is too high 'false' laws are found. (Yet Kepler was satisfied with 'D^2/P (Term-2) is constant' as a law for 10 years before he discovered 'D^3/P^2 (Term-3) is constant'). Such problems arising from the need to approximate data do not get as much attention in (Langley et al., 1987) as might seem necessary.[3]

The BACON-programs, to which ever more heuristic operations have been added, have 'rediscovered' a large number of laws, sometimes on the basis of real data (indicated below by *) but more often on the basis of ideal data:

- starting from the indicated basic program: the laws of Kepler (*), Boyle (*), Ohm (*), and Galilei;
- starting from more than two variables and the addition of a stratified method of searching in the data, viz. varying different pairs of variables keeping the rest constant: the ideal gas law, and the laws of Coulomb, Kepler (refined), and Ohm (refined);
- starting from quantitative and nominal variables and an introduction method of intrinsic properties coupled to the nominal variable(s),
  - by means of a 'have a try method': the laws of Ohm (still more refined), with resistance as intrinsic concept, Archimedes (volume), Snellius (indices of refraction), Black (specific heat), conservation of momentum (inertial mass), and the law of gravitation (gravitational mass);

  - or by means of searching a common divisor: the results of Cannizaro (atomic weights) and Millikan (charge of the electron);
- assuming a theoretical point of view, hence (data- and) theory-driven, such as the search of symmetries or laws of conservation: refined versions of the laws of Snellius and Black, and the laws of conservation of momentum and energy.

11.1.1.2 *In Search of Qualitative Laws and Models*

It is plausible that the quantitative BACON-programs deal primarily with examples from physics. It is equally unsurprising that programs directed at qualitative laws and models focus in the first place on examples from chemistry. The three most well-known programs are even called after main figures in their history.

GLAUBER starts from a number of chemical reaction equations and properties (taste, and the like) of the participating substances, and forms classes of substances and reaction equations in terms of these classes of substances. In this way the classes now known as 'acids', 'alkalis' and 'salts' were formed as well as the law stating that 'acids and alkalis produce salts', starting from the taste of eight substances and four reaction equations between them. The basic heuristic operations in the program amount to: form as large classes as possible, and quantify universally, if possible, and existentially otherwise.

STAHL also starts from a number of reaction equations, but is directed at the decomposition of compounds into their components. The heuristic operations are as simple as plausible: delete substances occurring at both sides of an equation; if at one side there is only one substance, then conclude that the other side specifies its components; and substitute a substance by its components in the remaining equations.

Finally, DALTON also starts from reaction equations (obeying Gay–Lussac's law of combining volumes, in spite of Dalton's personal rejection, see Chapter 1), and is directed at the determination of the (relative) number of molecules of each substance taking part in the reactions, and of the atomic formulas of the molecules of the participating substances. Here the heuristic operations are: the number of molecules have to be brought into agreement with Avogadro's hypothesis; the number of atoms at the left and the right side have to be equal; for the rest atomic formulas should be as simple as possible. It is clear that DALTON is strongly theory-driven and that new reaction equations may lead to a revision of atomic formulas. For example, from the single equation 'hydrogen and oxygen generate water(vapor)' and the heuristic operations follow the molecular equation (using some plausible abbreviations) '2Hm(H-molecule) + 1Om $\rightarrow$ 2Wm' and the atomic formulas: 1Hm = 1Ha(H-atom), 1Om = 2Oa, and 1Wm = 1Ha + 1Oa; after adding the equation for the production of ammonia, viz. 'hydrogen and nitrogen generate ammonia', DALTON revises its assumptions and finds the equations and formulas that we are used to.

The most plausible criticism of the programs presented so far is that the initial problem situation always presupposes a given conceptualization, a vocabulary, whereas the finding, or construction for that matter, of a suitable conceptualization usually is a very important part of the discovery process. However, the authors of the programs argue convincingly that the conceptualization that later turned out to be useful, was, as a matter of historical fact, in many cases available long before the discovery of the law was made. Hence, in these cases, the present type of (law-) discovery certainly has been of great historical importance.

It should finally be noted that the book by Langley et al. frequently makes nice excursions to so-called 'accidental discovery', and its programmability. Certainly with the techniques to be presented below, optimism in the latter respect is, on second thought, more realistic than at face value: notably, (parts of) programs for the detection of anomalies and the design of explanations by analogy can always be made sensitive to unsought findings.

11.1.2 *Hypothesis Formation and Evaluation by PI*

In this section we will first give an impression of the specifics of representation used in PI, followed by an indication of its various ways of induction. Next attention is given to the evaluation sub-program of PI. Finally, some ideas about Thagard's view of the philosophy of science from the computational point of view are given.

11.1.2.1 *Representation*

Whereas the type of programs dealt with in Subsection 11.1.1 belong to the mainstream in AI, Thagard's approach in his *Computational Philosophy of Science* (1988) is more related to cognitive psychology (CP). The point of departure is the distinctive representation of concepts, rules (for laws and hypotheses), theories and (explanatory) problems. A *concept* (e.g. sound) is characterized with the aid of super- and subordinate concepts (e.g. physical phenomenon, and music, respectively) and rules in which the concept occurs. In the following scheme (derived from Thagard, 1988, p. 17) the example of sound is represented:

In Scheme 11.1 the rules are mentioned that are represented in extenso elsewhere in the systems represented in extenso. A *rule* postulates an empirical relation between two or more concepts (sound is propagated/reflected), represents a prototypical expectation, and is provided with a report of individual (explanatory) successes and counter-examples, and a degree of confidence based thereupon. A *theory* is a system of observation concepts (sound, wave), theoretical concepts (sound-wave), corresponding rules, general (explanatory) successes and counter-examples. Finally, there are two types of *explanatory problems*, viz. of an individual and general nature: the initial state concerns a specific or unspecific object (a certain sound or an arbitrary sound) and the goal state is supposed to give an explanation of behavioral features (propagation and reflection).

Name:	sound
Data-type:	concept
Activation:	0
Superordinates:	physical phenomenon, sensation
Subordinates:	voice, music, whistle, animal sounds
Instances:	
Activated by:	

Rules:

Rule-0:	if x is heard then x is a sound
Rule-1:	if x is a sound then x is transmitted by air
Rule-2:	if x is a sound and x is obstructed, then x echoes
Rule-3:	if x is a sound and y is a person and x is near y, then y hears x
Rule-4:	if x is a sound then x spreads spherically
Rule-5:	if x is a sound then x is a sensation
Rule-6:	if x is a sound then x is physical phenomenon

Scheme 11.1 A representation of the concept of sound.

As all problems in 'CP-models' of heuristic search, explanatory problems are approached by PI in a quasi-connectionist way (see Subsection 11.2.1.2), by activating the concepts that occur in the problem, followed by activation of concepts coupled to the first ones and then of the rules ('rule-firing'), for which the appropriate concepts are activated. This procedure is more or less standard in CP, and a problem is solved when some kind of match arises between the initial and the goal state. The special aspect of PI, as opposed to non-learning expert systems, is that, if the match does not occur, and only then, different kinds of *induction*, i.e. kinds of hypothesis and concept formation, become active, viz. generalization, abduction and concept formation. Such inductions are of course followed by evaluation.

11.1.2.2 *Induction*

Induction amounts to taking non-deductive steps. When Ga is to be explained by the initial state Fa and if no match occurs, the rule 'if Fx then Gx' is formed by *inductive generalization* and subsequently evaluated on the basis of other information connected with F and G. When the rule is 'sufficiently and diversely supported', it is added to the 'knowledge base' with a corresponding degree of confidence (for more details, see Holland et al., 1986).

Abduction is a name covering four other forms of hypothesis formation, all followed by one and the same evaluation method, viz. a kind of so-called 'inference to the best explanation' to be discussed below. The four specific methods of abduction can be characterized as follows:

Individual abduction amounts to the proposal of the hypothesis Fa when Ga is to be explained and the rule 'if Fx then Gx' belongs to the knowledge base and is activated, since Ga can be explained by this hypothesis and the rule mentioned. If it has to be explained that a certain sound is propagated and if the law 'waves propagate' belongs to the base, the hypothesis is proposed that the particular sound is a wave.

Similarly, *rule abduction* amounts to the proposal of the hypothesis 'if Fx then Hx' when the law 'if Fx then Gx' has to be explained and 'if Hx then Gx' belongs to the active knowledge base. In order to explain 'sounds propagate' when 'waves propagate' belongs to the base, the hypothesis 'sounds are waves' is proposed.

It is clear that rule abduction is more or less a general form of individual abduction. Both would be examples of the fallacy known as 'affirming the consequent' if it were not the case that the hypotheses are merely proposed; they still have to be evaluated. The third form of abduction belongs to the same category of quasi-fallacies, the fourth form is of a quite different nature.

When the rule 'if $S(x,y)$ then Gx' belongs to the active knowledge base, and Ga is to be explained, *existential abduction* suggests the hypothesis 'there is a y such that $S(a,y)$'. Well-known examples are the postulation of the planet Neptune and of the substance phlogiston, and later oxygen. This form of abduction also occurs in metaphysics, think e.g. of postulating a human independent world.

The most surprising, and probably the least supported, hypotheses are formed by *analogical abduction* (AA): suppose that an individual or general fact F^* is to be explained and further that F^* is similar to F (according to certain activation patterns) and F is according to the knowledge base explained by G, then reconstruct the activation pattern around F and G is *-terms as much as possible, and postulate the corresponding analogue G^* of G as hypothesis.

Analogical abduction gives rise to a number of qualifying remarks. (1) If G^* passes the evaluation with success, PI subsequently forms an abstract scheme, with the pairs $\langle F, G\rangle$ and $\langle F^*, G^*\rangle$ as instances, which makes a generalized form of analogical abduction (GAA) possible in the future. (2) There is also a specific causal form of AA (CAA): if F and F^* are similar, and F and G are causally related, and G and G^* are also similar, then the hypothesis is formed that F^* and G^* are causally related. Of course, here too a generalized form is possible. (3) Besides this special form of causal heuristics, science in general, and hence also PI, is guided by the general causal heuristics that a statistical correlation suggests a causal (cor)relation. This kind of heuristics is rather different from the causal heuristics that, according to Thagard, dominates in many pseudosciences: a similarity suggests a causal relation between those things that are similar. Note that the role of similarity is here of a quite different nature from that occurring in causal analogical abduction.

This last point completes the sketch of the four forms of abduction, which was, after inductive generalization, the second form of induction distinguished. The third and last form of induction concerns the *formation of a theoretical concept and a corresponding theory*. Suppose that in the context of a general explanatory problem an initially activated concept and a secondary activated concept share some instances, such that the rules coupled to these concepts are partly incompatible. In that situation one can form the new 'combined concept' and corresponding theory about the initial concept with the following expectations: insofar as all rules of both are compatible they are included, and of two incompatible rules the one belonging to the initial concept is included.

Recall that in the case of rule abduction the example was mentioned that the concept of sound becomes related to that of wave (of fluids), which leads to incompatible expectations, viz. spreading spherically versus in a single plane. In this case the concept of sound-wave is formed, with spherical propagation attached to it, and the wave theory of sound is formed, i.e. 'sounds are sound-waves', with explanatory successes, among others, that sounds propagate and reflect.

11.1.2.3 *Evaluation*

PI will in general propose, if possible given the active knowledge base, several hypotheses and theories for the solution of an explanatory problem. These should be evaluated, where the knowledge base itself is also in discussion. More specifically, the evaluation by PI consists of a kind of 'inference to the best explanation' (IBE), in which not only the original problem takes part, but also the explanations that already belong to the base. Thagard deals with the theories of Darwin, Lavoisier and Fresnel in order to make clear that IBE has played an important role in the history of science.

Suppose that for the explanation of a new general fact at least one new hypothesis is formed. PI then determines all *competing* hypotheses (new or already in the base) and applies two criteria: *consilience* and *simplicity*. Thagard also argues that a third criterion, a kind of analogy score, should also be taken into account in an extended version of PI, but we have doubts about this, heuristics exceeding, role of analogy.

Two hypotheses are competing when they are logically independent and have overlapping success or when they are the ends of a chain of hypotheses with overlapping successes. All successes of all competing hypotheses are to be taken into account in the final evaluation.

A hypothesis is more consilient in proportion to the number (and the importance) of its successes. A hypothesis is the more simple the fewer auxiliary hypotheses need to be assumed to achieve its successes. If both criteria lead to the same ordering of the hypotheses, the most successful and the most simple hypothesis is chosen. If the orderings do not coincide, which is of course very well possible, the value of a hypothesis is defined as the product of its successfulness and its simplicity, and IBE now tells us that the hypothesis with the highest value is to be chosen.

Note that Thagard's notion of consilience is very similar to the notion of successfulness that we used in Chapters 7 and 8. Moreover, it is important to stress that Thagard uses 'simplicity' not in the usual sense as a matter of the structure of a hypothesis, but operationalizes it, very convincingly, in terms of the number of specific extra assumptions that have to be made for the successes. It is in this respect that Thagard also notes an important difference in his 'portraits' of science and pseudoscience.

IBE raises of course the question of its justification. Thagard spends two chapters on this problem. First he derives three criteria for the evaluation of systems of argumentation (consisting of normative principles, practices, background theories, and inferential goals) from methods of evaluation proposed for the evaluation of other systems of argumentation, among others, the method of 'reflective equilibrium' in Rawls' social philosophy. Thagard, (1988, p. 129) arrives at the following three criteria for the coherence of an inferential or inductive system:

(1) Robustness: to what extent do the normative principles account for inductive practice?
(2) Accommodation: to what extent do available background theories account for deviations of inductive practice from the normative principles?
(3) Efficacy: given background theories, to what extent does following the normative principles promote the satisfaction of the inferential goals?

According to Thagard, IBE is robust because it is used very frequently, as illustrated by the above mentioned examples. Deviations of IBE can be explained (accommodated) with the aid of psychological theories of motivation. And IBE is effective with respect to three goals: explanation, prediction and truth approximation. The first is evident, and the second is implied by the fact that prediction is a kind of potential explanation. For the third goal Thagard takes another detour by first arguing in favor of realism, at least for the natural sciences. He does so by applying IBE (in a non-circular way, or so he claims) on the relevant philosophical theories with respect to the explanation of three phenomena: technological applications, the accumulation of knowledge, and consensus about claims to knowledge. In view of our analysis of the relation between successfulness and truth approximation, the last detour, of which the non-circular character is not beyond doubt, is superfluous. More specifically, we have shown that IBE, provided it is revised in the form 'inference to the best theory as the closest to the truth', is demonstrably functional for truth approximation (ICR, Section 8.5).

11.1.2.4 *Varia*

Thagard's book is rich in ideas about the philosophy of science looked at from the computational point of view. Starting from a non-extreme form of theory ladenness of observation terms, Thagard arrives at a moderate kind of meaning holism and a corresponding methodological conservatism, leaving room for the practical commensurability that has been built into PI, as we have seen, for the evaluation of hypotheses.

So-called evolutionary epistemology is according to Thagard based on a bad analogy. Regarding the three main problems (the explanation of variation, selection, and transmission) the analogy gives the wrong answers. Hence, PI may be supposed to assign a low value to the 'evolutionary hypothesis' concerning knowledge development if that would even be formed by analogical abduction at all.

The two differences already mentioned between science and pseudoscience form elements of two Wittgensteinian profiles that Thagard distinguishes, leaving room for intermediate forms: correlation versus resemblance thinking, being directed at empirical evaluation or not, comparison with alternative theories or not, aiming at as simple theories as possible or not, aiming at progress, and to begin with, aiming at the improvement of theories or not. (See also Section 8.3 for a related diagnosis of pseudoscience).

Thagard also argues that, applied to cognitive science, the computational approach might help to make a real science of it, with real successes. However, for that purpose it is necessary that PI, or similar programs dealing with problem solving, not be restricted to induction and evaluation of hypotheses, but that the design of experiments should also be taken into account. However, most of the (directed) connections between the three relevant activities, viz. problem solving, induction, experimenting, still have to be elaborated.

Thagard emphasizes the importance of parallel computational systems, because it leads to better programs. Moreover, Thagard suggests that we consider and simulate a scientific community as a collection of interacting parallel computational systems. Here the idea of social adequacy comes into the picture. The suggested analogy raises the question of the possibility and the desirability of a further division of labor than the well-known one between theoreticians and experimenters. In this connection he thinks in particular of the division of labor between 'constructive dogmatists' and 'skeptical critics', to use our favorite terms for what is essentially the same idea.

Finally, in the course of the book, all three forms of adequacy (historical, psychological, philosophical), distinguished in the introduction, play a comparably important role. In his later book, *Conceptual Revolutions* (1992), to be critically discussed in Section 11.2, historical adequacy is the main goal (an example, revolutions in geology, is contained in the collection of Shrager and Langley (1990), the collection to be dealt with in Subsection 11.1.3, and the example in Subsection 11.2.2.2). Unfortunately, Thagard does not pay explicit attention to the fourth possible claim of computational philosophy of science, viz. practical relevance, although he regularly hints at the possibility of computer assisted discovery and evaluation.

11.1.3 *Forms of Theory Revision*

The first five chapters of the book edited by Shrager and Langley, *Computational Models of Scientific Discovery and Theory Formation* (1990), deal with variations and elaborations of what has been presented so far (two of the chapters have already been indicated). The last five chapters deal with topics central to cognitive psychology. Our focus will be on Chapters 6 to 11, which deal with important innovations in comparison to the foregoing, with emphasis on theory revision. The point of departure is always a knowledge base of data and theories, usually represented by the general method of representation of qualitative processes proposed by Forbus (cf. Davis 1990). It should be noted in advance that there is not much similarity with the so-called 'belief revision' research program of Gärdenfors (1988) and others. The main reason seems to be that the latter program does not attempt to deal with scientific knowledge. In particular, it is 'actual' rather than 'nomic world' oriented, and it is consistency driven.

Falkenhainer has developed a program (PINEAS) that realizes a far-reaching unification between four scenarios of explanation (deductive, using assumptions, generalizing, or using an analogy) and corresponding theory formation and revision by viewing the latter as aiming at the maximally feasible similarity between the explanandum and the explanans. The further selection of theories takes place on the basis of the degree of severity of the extra assumptions (in increasing order): no additions, addition of new properties of an acknowledged type to already acknowledged entities, addition of new entities of an acknowledged kind, addition of new properties of a new type to acknowledged entities, and finally the addition of new entities of a new type.

The 'Abduction Engine' (AbE) devised by *O'Rorke, Morris and Schulenburg* pretends to make revolutionary revision of a theory possible, starting from an observational anomaly between the theory, the knowledge base, and new data. After the recognition of such an anomaly, the theory is weakened to a basic theory, which is compatible with the new data. Next, this basic theory is strengthened, by abduction, roughly along similar lines as in Thagard's PI, to a theory that is able to explain the anomalous data. To be sure, the weakening and subsequent strengthening corresponds to 'contraction' and 'expansion' in the belief revision program of Gärdenfors et al. The AbE-program is developed on the basis of the transition of the phlogiston theory to the oxygen theory of combustion. Hence, it is no surprise that the program is able to reproduce that episode.

In the COAST program of *Rajamoney* theory revision is approached stepwise: (1) registration of an anomaly between new data and a theory in the knowledge base; (2) designing proposals for revision that transform the anomalous data into explanatory successes; (3) suggesting experiments; (4) first selection on the basis of the anomaly to be explained, the outcome of the experiments, and the degree in which previous successes are retained; (5) further selection on the basis of simplicity and predictive power. This program, and the following two, were guided in their development by examples from biochemistry.

The KEKADA program developed by *Kulkarni and Simon* registers surprising phenomena by comparing new experimental results obtained for 'normal' purposes with expectations on the basis of the knowledge base, and proceeds by dealing with such phenomena; that is, it tries to revise the relevant theory and proposes experiments to test such revisions. For this purpose five strategies are built into the program, each of which consists of a hypothesis generator, an experiment proposer and an evaluator. The core idea of, for example, the first strategy is to try to strengthen the surprising phenomenon by independently manipulating the variables of the system.

Karp's program (HYPGENE) considers hypothesis formation and revision as a matter of design, with constraints (a profile of desired properties) and operators that revise the provisional prototype (the profile of factual properties) in order to make it satisfy the constraints more and more. For a detailed localization of HYPGENE in the context of drug research, and its similarity to the models of design research presented in Chapter 10, see Bosch (2001, Section 9.5).

Finally, *Darden* considers the solution of an anomaly of a theory as a task for diagnostic reasoning, which has been developed for expert systems, guided by the tracing of a defect in a technical system. She shows, using Mendelian examples (see also Darden (1991)), that the decomposition of all presuppositions of a theory may provide the points of departure for solving the anomaly, and that its solution, as in Lakatos' famous analysis (1976), may or may not lead to fundamental theory revision, that is, a revision staying within a research program or breaking through the barriers of the relevant program.

11.1.4 *Some Connections with Neo-Classical Philosophy of Science*

As far as modern philosophy of science is directed at cognitive heuristic patterns in knowledge and method, such as this book is, there are many conceivable connections with the computational philosophy of science. On the one hand these points of contact concern the methodological topics of separate and comparative evaluation and truth approximation, on the other hand they involve cognitive patterns that could provide the heuristics for computer programs.

In the computational literature about theory evaluation and revision, there remains quite a lot of confusion as to what exactly are the units of successes and problems, and how they hang together. Our analysis of theory evaluation (Chapters 7 and 8) provides several suggestions for what precisely is to be conceived as an explanatory success, what can be conceived as a descriptive problem of theories and test implications, and how they can be represented. The plausible 'instrumentalist' goal of theory revision, increase of explanatory success, and decrease of descriptive problems, is also made more precise in this way. In the above sections, formulations that were sharpened from this perspective have already been used from time to time. Moreover, in the paper 'Abduction aiming at empirical progress or even truth approximation' (Kuipers, 1999b) we have elaborated the conceptual prospects for generalizing, on the basis of the analysis in ICR, the aim of abduction, showing how the standard goal of explaining surprising or anomalous observations can be integrated into a notion of empirical progress or even truth approximation. It turns out that the main abduction task then becomes the indicated instrumentalist task of theory revision with the intent of devising an empirically more successful theory, relative to the available data, but not necessarily compatible with them. The rest, that is, genuine empirical progress as well as observational, referential and theoretical truth approximation, is a matter of evaluation and selection, and possibly new generation tasks for further improvement. The paper concludes with a survey of possible points of departure, in AI (in particular, PI and the programs indicated in Subsection 11.1.3) and logic, for computational treatment of the instrumentalist task guided by the 'comparative evaluation matrix'. Below we will briefly indicate one of the logical approaches, viz. the semantic tableau approach.

We hinted already at the relevance of our correction of 'inference to the best explanation' and the possibility of directly justifying that corrected version as functional for truth approximation.

Finally, it is indeed illuminating, as Karp suggests, to conceive theory formation and revision as a kind of design research, with the ultimate purpose of constructing the strongest true theory within a given or adapted vocabulary. As we have indicated in Chapter 10, the differences between descriptive and explanatory research programs on the one hand and 'normal' design research programs on the other are at least as interesting as the similarities.

The heuristic patterns that neo-classical philosophy of science can provide concern in the first place standardized decompositions of different kinds of explanation and reduction (Chapters 3–5). These can easily be reinterpreted as generalized abduction schemes in the sense of Thagard. Moreover, they can be used for diagnostic reasoning in case of anomalies as proposed by Darden.

Other, more global, heuristic patterns are provided by the analysis of the structure and development of research programs and their possibilities for interaction (Chapter 1), the hierarchy of knowledge arising from the theory relative distinction between observational and theoretical terms (Chapter 2), and the internal structure of theories (Chapters 2 and 12). For instance, it will become quite clear in the next chapter that the structuralist analysis of theories can almost directly be used for the computational representation of theories.

Let us turn to the promised brief discussion of one possible logical approach[4] to instrumentalist abduction, viz. the semantic tableau approach. It was introduced by Mayer and Pirri (1993) (see also (Pirri, 1995)), and redirected in a more efficient way and one more relevant to the philosophy of science by Atocha Aliseda (1997), who focusses on novelty and anomaly guided abduction. In the case of a surprising observation, i.e. the situation that new evidence E is not entailed by the background theory B, the task is to expand the latter with some hypothesis H such that $B\&H$ entails E, but H alone does not. This is called *novelty guided abduction*. In the case of an anomalous observation, i.e. the situation that E contradicts B, the task is to revise B to B^* such that B^* entails E. This is called *anomaly guided abduction*. We will only indicate the basic idea behind Aliseda's algorithms for dealing with these types of abduction. These algorithms are as yet restricted to the semantic tableaux for propositional languages. A semantic tableau is normally constructed (see e.g. Smullyan, 1968) to find out whether a set of premises logically entails a conclusion by systematically trying to construct a counter-example, that is, a propositional model or structure in which the premises are true and the conclusion is false. If all branches generated by the (formula decomposing) tableau rules from the set of premises and the negated conclusion become 'closed', that is, contain a formula and its negation, the argument is valid. Open branches describe counter-examples to the argument. In the case of novelty guided abduction, there must be open branches in the tableau starting from the background theory *B* and the negation of the phenomenon description *E*. Systematically closing these branches produces equally many consistent proposals for abductive explanation of the novel phenomenon. In the case of anomaly guided abduction, the background theory apparently entails the negation of the anomaly; that is, the corresponding tableau starting from the theory and the phenomenon description is closed. Now the task is to systematically open the branches of this tableau by deleting parts of the theory (contraction), followed by adding new formulas (expansion) that lead to the closing of all branches when combined with the negation of the phenomenon description.

The above description neglects all nuances and versions of Aliseda's approach, but it may nevertheless already be plausible to conjecture that the approach can be generalized to instrumentalist abduction purposes. The main thing is to operationalize in tableau terms the possibility that one theory is 'more incompatible' with the data than another. Then the task is to generate theories that are not more incompatible with the available data than the original one, explain at least as much as the latter, and improve the latter in at least one of these respects. Some other adaptations will be required to make it suitable for scientific purposes. First, room should be left so that general facts can be used as data and, second, the original and resulting theories should be of a general nature. When restricted to propositional languages, simple implicative formulas will do the job in both cases, provided there is enough room for three kinds of truth: propositions may be actually true, 'physically true' and logically true. One plausible way to do this is using a modal propositional language (Zwart, 1998), but there may well be other ways.

From the above sketch of possible connections, concluding with one example in some detail, it will be clear that there is not much competition between computational and classical philosophy of science, as soon as one is willing to reject the dogma of the impossibility of fruitful philosophical-methodological analysis of the Context of Discovery. But there is interesting possible competition between the computational philosophy of science and modern sociology of science, witness for example the title 'Scientific discovery by computer as empirical refutation of the strong program' (Slezak, 1989). For, where is the dominant role of socio-cultural factors in computational discovery, evaluation and revision that many sociologists of science find so important? Besides such possibilities for competition, conciliatory positions are conceivable on both sides so that fruitful co-operation becomes possible. In particular, as mentioned in Subsection 11.1.2.4, Thagard proposes a few computer simulations of co-operation and competition between scientists that are intended to achieve sociological adequacy.

11.2 COMPUTATIONAL THEORY SELECTION AND THE EVALUATION MATRIX

Introduction

Four years after his pioneering book on the computer program PI (Processes of Induction) *Computational Philosophy of Science* (Thagard, 1988), Paul Thagard published a new book entitled *Conceptual Revolutions* (Thagard, 1992), on the ECHO program. In this last book he not only studies conceptual change in much more detail than before, he also replaces his 1988-approach to theory selection with a refined version.

Our claim in this section is that the architecture of theory selection in both cases seems to be on the wrong track. PI and ECHO impose a mixture of distinct considerations, such as explanatory value, simplicity, analogy, competition and the like. In our descriptive-cum-normative perspective, theory comparison is and should be essentially stratified, giving priority to explanatory evaluation. Hence, first the explanatory merits and problems should be assessed and compared. Only if that does not lead to the conclusion that one theory is explanatorily superior to the other, should additional considerations be brought into play in order to turn the scale, if one aims at selection at all in this case.

Thagard's primary aim is to design a set-up that is historically adequate, in the sense that simulation should lead to the same theory selection as scientists made in the past on the basis of the same elementary judgements. In his last book, Thagard presents in full detail five historically adequate selections made by the quasi-connectionist ECHO program: they concern the theories of Lavoisier, Darwin, along with three selections from geology. In this section we will show, by inserting the relevant input from Thagard in the simple *Evaluation Matrix (EM)* presented in Chapter 8, that our approach, i.e. explanatory merits first, also leads to the historically adequate selections, but now in a straightforward and transparent way. Hence, by way of meta-application of the stratified approach, we claim that our approach, being explanatorily as equally successful as Thagard's approach (as far as Thagard's explicit evidence is concerned), is to be preferred because of its simplicity.

In the last section we will discuss Thagard's reaction to the EM-approach, starting with Thagard's second aim of not only simulating the results of historical theory selection, but also the main processes, or at least the role of the main considerations involved, hence involving questions of psychological adequacy.

To be sure, this section is not a review of both of Thagard's books. For the 1988 book we refer to Subsection 11.1.2. Like that book, the 1992 book also presents an enormous number of fresh ideas and criticisms by looking at the philosophy of science from a computational point of view. However, our single aim in this section is to argue for a different approach to computational theory selection. In this respect, we want to illustrate the possibility of fruitful interaction between neo-classical and computational philosophy of science.

11.2.1 *Unstratified Approaches*

In this subsection we will show in some detail that both PI and ECHO deal with evaluation in an unstratified way, such that explanatory superiority may be sacrificed to simplicity.

11.2.1.1 *The Unstratified Architecture of PI*

As we have seen, Thagard's 1988-approach to theory selection is essentially a kind of inference to the best explanation (IBE), where the best explanation is determined by three criteria: *consilience, simplicity and analogy*. The implementation in PI is restricted to consilience and simplicity, and the relevant fragment of PI is straightforwardly of a classically computational nature. With respect to PI we will now, in addition to Subsection 11.1.2.3, (only) argue that it is essentially unstratified, despite some formulations made by Thagard suggesting the contrary. In particular, PI may well sacrifice explanatory superiority to simplicity. Our descriptive stratification claim, however, is that scientists have almost never made such a sacrifice. To be sure, Thagard does not explicitly claim to have historical cases of this kind. Moreover, the actual historical cases discussed by Thagard, (1988) are not fully worked out, a fact that makes them difficult to evaluate. Hence, our claim with respect to PI is only that as far as cases of explanatory superiority are concerned, PI is not only very complicated, but, assuming that our descriptive stratification claim is correct, it is also likely to lead in many cases to the historically wrong verdict. Note that we do not claim something for the case of (seriously) divided success. As far as divided explanatory success is concerned, PI may be on the right track.

Let us consider PI in some detail. The consilience of a hypothesis is defined as the sum of the relative importance, or weight, of its explanatory successes. The simplicity of a hypothesis is defined in such a way that it decreases when the number of auxiliary hypotheses required for the explanatory successes increases. PI selects the hypothesis with the highest value, defined as the product of simplicity and consilience. Of course, these arrangements are such that if both consilience and simplicity are favorable for the same hypothesis then that hypothesis is selected. However, consider the case that the consilience of one hypothesis is straightforwardly superior to that of another, in the sense that all explanatory successes of the latter are also explanatory successes of the former. Hence, in this case the sum of the weights of the successes of the latter does not exceed the sum of the weights of the successes of the former, whatever (positive) weights are assigned to specific successes. In other words, in this case the weights of the successes, of course a cause for debate, do not play a substantial role in the comparison of the consilience of the hypotheses. But the PI-arrangement is such that, if the hypothesis with higher consilience happens to be less simple than the other hypothesis, both the degree of simplicity and the total importance of the explananda come to play a substantial role. Hence, in that case it may occur that explanatory superiority is sacrificed to simplicity.

11.2.1.2 *The Theory of Explanatory Coherence (TEC) and the Architecture of ECHO*

In his 1992-approach to theory selection Thagard first develops a theory of explanatory coherence (TEC), governed by seven principles (pp. 65 and 66), 'that establish relations of explanatory coherence and make possible an assessment of the acceptability of propositions in explanatory systems'. The seven principles include: symmetry of coherence relations (P1), coherence obtained from explanatory relations, differentiated by simplicity considerations (P2), coherence obtained from analogical considerations (P3), 'incoherence' obtained from contradiction and competition (P5/6), acceptability of propositions due to data priority (P4) and due to coherence and explanatory impact (P7). According to TEC, a new theory will replace an old one if its hypotheses possess greater explanatory coherence. Though TEC, the underlying selection theory of ECHO, may be seen as a refinement of IBE, the underlying selection theory of PI, ECHO is essentially different from the relevant, classically computational, fragment of PI. Like many other fragments of PI and ECHO, ECHO implements TEC in a quasi-connectionist way by using the techniques of activation spreading and parallel constraint satisfaction. It is not straightforwardly connectionist, for each piece of data and knowledge is represented by a single node rather than distributed throughout the network. However this may be, as one may expect, 'ECHO-selection' is a non-transparent updating process, which may nevertheless lead, as a rule, to an unambiguous conclusion.

The input propositions are of the following kinds:
evidence-propositions, including *'negative' evidence-propositions*, and *hypotheses of a theory*, including *auxiliary hypotheses* and *additional facts*. Input claims are made with input propositions, of which there are four types: *explanation claims*, *analogy claims*, *contradiction claims*, and *data claims*. Data claims only concern evidence-propositions leading to evidence statements, as we will call them, and additional facts. The other claims may concern all kinds of combinations of propositions. For example, not only the explanation of an evidence statement by certain hypotheses is taken into account, but also the explanation of a hypothesis by some other hypotheses, or by some background additional facts.

Starting with activation of the evidence-propositions and zero activation of all other propositions, after a number of updating runs all propositions approach their asymptotic activation value. Those getting a positive value, including the propositions with a data claim, constitute a coherent set, and they are accepted. Those getting a negative value are rejected.

Our main point of criticism of ECHO, to be elaborated in what follows, again concerns the unstratified interplay of selection considerations.

11.2.2 *Priority of Explanatory Superiority (PES)*

In this subsection we will first restate the basic features of the evaluation matrix, then show, using five examples, that it leads to the same results as ECHO, and hence that it deserves further consideration to be used in evaluation programs.

11.2.2.1 *The Evaluation Matrix (EM)*

In our approach explanation is also crucial, even though we will restrict it to the explanation of evidence statements by a theory. A theory is supposed to consist of all relevant hypotheses, including auxiliary hypotheses and additional facts. Except when otherwise stated, we will follow Thagard in using a liberal, not strictly deductive, notion of explanation, leaving room for all kinds of special types.

Recall that we introduced in Section 8.1 the (comparative) *evaluation matrix* (EM) in which two theories were compared in terms of their ability to account for the (individual or general) facts. Although the definitions of positive, negative and neutral (evidential) facts for a theory were essentially just (conditionally) deductive, nothing prevents us from the suggested more liberal interpretation of the logical relation on the one hand, combined with an explanatory interpretation on the other. Only when our further conclusions depend on the deductive relation do we have to make a proviso.

The matrix for theories of *X* and *Y* and a certain fact was as follows:

		X		
		negative	neutral	positive
	negative	B4: 0	B2: –	B1: –
Y	neutral	B8: +	B5: 0	B3: –
	positive	B9: +	B7: +	B6: 0

The (comparative) evaluation matrix (EM)

From the perspective of *Y* the boxes B1/B2/B3 represent unfavorable facts, B4/B5/B6 indifferent facts, and B7/B8/B9 favorable facts. The numbering of the boxes was determined by three considerations: increasing number for increasingly favorable results for *Y*, symmetry with respect to the \-diagonal, and increasing number for increasingly positive indifferent facts.

The claim that *Y* is *more successful* than *X* in the light of the available facts was defined by: B1/2/3 empty (no unfavorable facts for *Y*) and at least one of B7/8/9 non-empty (some favorable facts for *Y*), with corresponding versions of the comparative success hypothesis and the rule of success, prescribing, for the time being, the choice of the theory that has so far proven to be more successful than others.

We defined the claim that *Y* is *almost more successful* than *X* in such a way that there is room for some 'Kuhn-loss': the new theory may no longer retain (or at least may no longer seem to retain) a success of the old one, without contradicting it (i.e.

a B3-fact). The condition is that some (unfavorable) B3-facts are allowed, provided there are (favorable) B8- or B9-facts or the number of B3-facts is (much) smaller than that of their antipodes, that is, B7-facts. For the detailed motivation, including calling B7-facts weakly favorable for *Y* and B9-facts strongly favorable, see Section 8.1.

From now on we will interpret EM and the two comparative definitions as means of dealing with explanatory successes in a liberal sense. Since EM only takes explanatory considerations into account, and not simplicity, it may be seen as the elaboration of the idea of *Priority of Explanatory Superiority* (PES), in the same way as ECHO was the implementation of TEC.

In the next subsection we will apply EM to the five cases extensively treated in (Thagard, 1992), where all formally indicated statements can easily be found.

11.2.2.2 *Five Applications of EM*

Lavoisier

We will start with the relevant input used by Thagard for the Lavoisier example. The phlogiston theory (PT) as well as the oxygen theory (OT) is summarized by Thagard in six hypotheses. Eight evidence (E-)statements are considered. Seven of them can immediately be classified on the basis of the input claims as follows: $2 \times$ B6 (neutral) and $5 \times$ B7 (weakly OT-favorable). According to Thagard, these seven E-statements were the major E-statements taken into consideration by Lavoisier. Hence, from Lavoisier's point of view OT is explanatorily superior to PT.

For some reason or other, Lavoisier did not investigate OT's explanatory relation to the eighth E-statement (named E2 by Thagard): 'Inflammability is transmittable from one body to another'. Following Thagard's input that E2 is explained by PT, it belongs to one of the boxes B1/3/6, but he makes no mention of his opinion about OT in this respect. However, it seems plausible to assume that it would have been easy for Lavoisier to argue that OT is at least compatible with E2, hence that E2 should be placed in B3 or B6. In other words, for Lavoisier E2 would have been at most a case of Kuhn-loss, to be compared with 5 opposite B7-cases. Moreover, it even seems that it would have not been very difficult to argue that E2 can be explained by OT, and hence that E2 belongs to B6. In sum, according to Lavoisier's account, OT is explanatorily superior to FT, and at least almost, if we also take E2 into consideration.

Darwin

Next we will consider the Darwin example. Thagard discerns 14 evidence statements (E11 is missing for some reason or other). Darwin's theory (DT) is summarized in 3 main hypotheses, 3 auxiliary hypotheses, and 7 additional facts. These additional facts will not play any role in our evaluation. The creationist theory (CT) is summarized in the statement: 'Species were separately created by God'. Of the 14 evidence statements, 12 can be immediately classified on the basis of Thagard's input claims as follows: 3 × B6 (neutral) and 9 × B7 (weakly DT-favorable).

Although the presentation suggests that E6, 'Once extinct, species do not reappear', should also be classified in the B7-box, one might claim that this statement is not only explained by Darwin's theory, but also by the creationist theory, at least by a deistic version. Hence it should be placed in box B6. But neither of the two alternatives changes the judgment that DT is explanatorily superior to CT, as long as we do not take E4 into consideration: 'Species when crossed become sterile'.

Thagard claims of course that E4 is explained by the creationist theory. However, Darwin is not explicitly reported to deal with E4 in the same way as with three other evidential statements explained by CT (E1,E2,E3) and also considered by Darwin as potential counter-evidence to DT. Darwin is able to explain these cases with his theory (using the auxiliary hypotheses). Hence, E4 seems to be of a similar nature as the exceptional case in Lavoisier's example. E4 is not really in conflict with DT, hence it should be placed in box B3 or B6. Further research might have led to the conclusion that it can be explained by DT (possibly with the aid of some extra auxiliary hypothesis), hence that it may be placed in B6, otherwise it is to be registered as a case of Kuhn-loss (B3), against 9 opposite B7-cases. In sum, the overall conclusion is essentially the same as for the Lavoisier example. According to Darwin's account, DT is explanatorily superior to CT, and at least almost, if we also take E4 into consideration.

Three geological cases

The third explicitly presented example by Thagard in fact concerns three different geological evaluations. Wegener's own evaluation of his theory of continental drift (WT) against the contractionist theory (CT), the same evaluation by his opponents, and finally Hess's evaluation of his later 'sea floor spreading' version of continental drift (ST) against CT. The evaluations produce both strongly favorable results (B9) and some negative evidence (NE-)statements as well.

Let us begin with Wegener. Thagard summarizes CT in 10 hypotheses, and WT in 12. It is important to note that Wegener borrows the so-called isostasy hypothesis (C8) from the contractionist, stating that 'Large sections of the crust are in gravitational equilibrium'. There are 20 evidence statements, plus two negative ones, NE6 and NE11, contradicting E6 and E11, respectively. All simply positive evidence statements, without corresponding negative ones, hence E1-E20, except E6 and E11, are easily classified: 9 × B6 (neutral) and 9 × B7 (weakly WT-favorable). Given that WT not only explains E6, but also that CT explains NE6, their combination E6/NE6 is a proper case of a strongly favorable result for WT (B9).

The situation is more complicated for E11/NE11. WT is simply reported to explain E11, but CT is not only said to explain NE11, but also to explain E11 (by some other hypotheses of CT, including the overlapping C8). Hence, the pair E11/NE11 may be placed in B6, B7 or even B9. However, none of these options influence the overall judgment that WT is explanatorily superior to CT.

Of the three options where to place E11/NE11, B7 seems the most reasonable for explaining E and NE, hence not-E, is comparable to explaining neither E nor not-E. In this case the final result is 9 × B6, 10 × B7 and 1 × B9. Another possibility, would be to split the placement of E11 and NE11, in which case E11 would come in B6 and NE11 in B9. But this procedure seems unjustified for unambiguous cases. e.g. when CT would just explain NE11, for then it essentially leads to a doubling of the results. Hence, in the ambiguous case of opposite explanatory claims of one theory with respect to E and NE we prefer to say that the theory is indifferent to that couple.

In the case of the opponents, the two opposing theories are reformulated in the light of points made by Wegener's critics and, as one may expect, of course some other evidence statements are taken into consideration, 5 simple positive ones and 13 couples of positive and negative versions. Placing pairs in the way described above, we obtain the following overall result from the perspective of the contractionist theory: 1 × B3 (weakly CT-unfavorable), 3 × B6 (neutral) and 1 × B7, 2 × B8, and 11 × B9 (weakly, moderately and strongly CT-favorable). Hence, apart from one case of Kuhn-loss (B3) to the effect that the contractionists do not claim to have an explanation for Wegener's primary observation: 'The outlines of continents on the opposite side of the Atlantic are similar' (E3), their theory certainly is explanatorily superior to Wegener's from their evidential point of view.

Note that there occur two cases of B8, which had up to now not been instantiated: negative evidence explained by one theory (WT) and indifferent for the other (CT). Just to give an idea, one of these two statement pairs says 'There are no mountains on the ocean floor' (E10), indifferent for CT, versus 'Rigid ocean floors would also be compressed and form mountains' (NE10), predicted by WT.

Let us finally consider Hess's evaluation of his seafloor spreading version of continental drift (ST) as opposed to the contractionist theory (CT), summarized in 17 and 19 hypotheses, respectively. There are 24 evidence statements, 5 coupled with a negative evidence statement. There is one ambiguous couple, which is, for similar reasons as in earlier cases, placed in B7. The overall outcome, showing the explanatorily superiority of Hess's theory, is: 12 × B6 (neutral), 8 × B7 (weakly ST-favorable), 4 × B9 (strongly ST-favorable).

11.2.2.3 *Further Elaboration and Motivation*

The general conclusion of the foregoing subsection is that the EM-simulation (by hand, hence by computer) of the five examples on the basis of PES is historically equally adequate as Thagard's ECHO-simulation based on TEC. Given that PES/EM is much simpler than TEC/ECHO, meta-application of PES/EM seems to prefer PES/EM, at least on the object-level. The point is not only that TEC/ECHO is much more fine-grained than necessary, its unstratified set-up is at least as problematic. Hence, there is sufficient reason to evaluate and elaborate (and program) the stratified PES/EM-approach further. To begin with the further evaluation, all four revolutions in physics informally dealt with by Thagard (Copernicus, Newton, relativity theory and quantum theory) seem cases of straightforward explanatory superiority. In the last section we will argue that this is actually the case for the examples of Copernicus and Newton.

With respect to further elaboration the natural question is, what to do in the case of equal or seriously divided explanatory success? One possibility, closest to TEC/ECHO, is to bring in other considerations, such as simplicity and analogy. Whereas we have doubts about any evaluative role of analogical considerations (of course, we are not against a heuristic role), we find it very plausible to bring in simplicity considerations, at least in the case of equal explanatory success. Suppose that on closer inspection in the explanatory evaluation of a theory one sub-hypothesis does not play any role at all. Then the theory obtained by omitting that sub-hypothesis, being equally successful as the original one, is to be preferred to the original one. This simplicity approach to equal success, may be extrapolated for divided success in one way or another, or it may be combined with the quantitative approach to divided success, as suggested below. (Of course, simplicity and analogy considerations may also be applied in cases of selection within a set of equivalent theories.)

Cases of divided success may be approached by some quantitative weighing of neutral and (un-)favorable results. For this purpose the reader is referred to Section 8.1, where a quantitative version of the evaluation matrix (QEM) was presented. QEM suggests, of course, that simplicity only comes into play in the case of quantitatively equal success, where the latter possibility is just a matter of sheer accident.

As far as programming is concerned, the implementation of PES/EM in the rest of ECHO (i.e. ECHO minus its specific TEC features) will be a relatively easy task. But PES/EM is not only a proposal for a local adaptation of ECHO, as already suggested in Subsection 11.1.4, we also conceive it as an explication of most of the explanatory and/or success evaluations built into other programs of theory selection and revision, in particular, most of those occurring in the collection edited by Shrager and Langley (1990) and dealt with in Subsection 11.1.3.

One reason to prefer the PES/EM-approach to the TEC/ECHO-approach was the way it deals with simplicity, the other reason is that PES/EM is directly suggested by our theory of truth approximation (ICR, Chapter 7). When restricting the attention to deductive explanatory relations, the core of PES/EM essentially corresponds to, a symmetric version of, the *Rule of Success* (*RS*), prescribing to choose (for the time being) the most successful theory. It has been demonstrated in ICR (Chapter 7) that RS, though not guaranteeing truth approximation, is nevertheless functional for truth approximation in a very transparent way, due to the non-trivial success theorem that being closer to the truth guarantees being at least as successful. To be sure, the theorem is almost trivial when no (relative) distinction between theoretical and observation terms is made. In that case it is a direct consequence of the definition of 'being closer (more similar) to the truth', of 'being more successful than' and the assumption that the data are correct. However, if there is a distinction between observational and theoretical terms the theorem is not at all trivial, and even needs some extra conditions. Recall that 'the truth' in the context of truth approximation is conceived as the most informative and, hence, strongest true theory that can be formulated with the available conceptual means. Here we confine ourselves to some specific remarks about PES/EM from this perspective.

First, it is important to note that PES/EM, like RS, can unproblematically deal with falsified theories, due to the room it allows for counter-evidence, i.e. boxes B1/2/4 for *Y*. To be precise, *Y* may well be preferable to *X*, even if *Y* has been falsified, provided these falsifications also apply to *X*, i.e. B4 is non-empty, but B1 and B2 are empty.

Second, this means that RS, as discussed in ICR (Section 7.5), and more generally PES/EM, conceived as a rule of inference, is a severely corrected version of Inference to the best explanation (IBE). IBE should not be interpreted as the assertion that the best explanation is *true*, but it is the *closest to the truth*. In this way it cannot only deal with falsified hypotheses, but the conclusion is equally relative to the available explanations as the premises are. The standard version of IBE is in both respects rather problematic.

11.2.3 *Process- or Only Product-Simulation?*

Paul Thagard, in correspondence on a previous version of this section, has noted that the EM-proposal is attractive, among other things, because it ties in with truthlikeness, but it 'strikes as only an approximation to what is involved in theory selection. It ignores factors that scientific practice convinces me are important to theory choice: simplicity, analogy, and explanation of hypotheses by higher order hypotheses. The merit of ECHO is that it takes these into account AND explanatory breadth...' Moreover, Thagard adds the interesting remark: 'The Newton and Copernicus simulations [by ECHO (Nowak and Thagard, 1992a/b)] are far larger than the ones in *Conceptual Revolutions*.'

The following reply will illuminate the points of discussion. We do not want to dispute at all that factors like simplicity, analogy and higher order explanation prima facie play an important role in historical theory selection. These factors are frequently explicitly discussed. Hence, historical adequacy may be restricted to product-simulation, but one may also aim at a *process-&*-product-simulation of historical theory choices. In the latter case, these cognitive factors should be included. In that respect, TEC/ECHO may very well be on the right track. Although Thagard primarily stresses in *Conceptual Revolutions* that TEC/ECHO can reproduce the resulting choices, there is from time to time also the suggestion that he wants to be realistic with respect to the choice processes as well, and both the above quoted reaction and (Nowak and Thagard, 1992a/b) make this very clear.

Our claims are restricted to product-simulation, hence without the aim of psychological adequacy. As far as the five examples dealt with in Subsection 11.2.2.2 are concerned, it is possible to reproduce the resulting choices only on the basis of the reported explanatory successes and problems. Hence, the other reported factors are redundant insofar as product-simulation is concerned. A minor point of criticism of Thagard's 1992 work becomes relevant here. He calls theory selection an essentially holistic affair, as opposed to piecemeal theory formation. This view is misleading, at least as far as ECHO-selection is concerned, for it is evident that ECHO can make distinctions between the hypotheses of a theory. But Thagard does not report historical or artificial cases of such differentiated evaluation of the hypotheses of one theory. However this may be, it is clear that (implementation of) EM can be applied to subsets of the hypotheses of a theory, and hence also be challenged for reproduction of differentiated evaluation. As suggested before, in that case, simplicity will come into consideration: if an extra hypothesis does not change the success, it will and should be left out.

The Newton and Copernicus cases present interesting challenges to the main product-simulation claim. In the first place, as Thagard stresses in his quoted reaction, their ECHO-simulations are much more complicated than the five cases mentioned in the book. On the other hand, determination of the EM-choice, although laborious, can still be done by hand. The second, and most important challenge is of course whether the resulting choice is, like the ECHO-choice, in agreement with the historical choice.

Only 45 minutes of calculation by hand showed the following, using the inputs of Nowak and Thagard. Copernicus against Ptolemy leads to the results: $51 \times$ B6 (neutral), $8 \times$ B7 (weakly favorable) and $2 \times$ B9 (strongly favorable). Newton against Descartes to: $2 \times$ B3 (weakly unfavorable), $17 \times$ B6 (neutral), $2 \times$ B7 (weakly favorable), $9 \times$ B8 (moderately favorable) and $8 \times$ B9 (strongly favorable). Hence, the choices of Copernicus and Newton are also easily reproduced by the EM-approach.

In general, the challenge of new cases is that they may lead to strong counter-examples of the claim that the EM-method reproduces the historical choices: the EM-method might prescribe the opposite choice. If there are such cases our stratified model is descriptively inadequate, i.e. even with respect to the simulation of products. For Thagard, (1992) such a counter-example to a general claim about historical choices would seem to be a too high price. However, following from Thagard's Chapter 7 of (1988), entitled 'From the Descriptive to the Normative', we would seriously consider the possibility of occasional deviations from the norm. Moreover, we would also like to accept the extra challenge of trying to show for the particular counter-example that the historically made choice was less adequate from the truth approximation perspective. Of course, given that we never know what the relevant strongest true theory is, the latter evaluation will only be possible in terms of our favorite theory at the moment. However this may be, the extra challenge is still waiting for a good counter-example, i.e. the challenge of new cases.

The latest potential counter-example involves the wave versus particle theory of light. In response to Achinstein's opposite claim (Achinstein, 1990), Eliasmith and Thagard, (1997) convincingly argue that in the nineteenth-century debate about the nature of light coherentist, and not probabilistic, arguments played a central role. They claim that their coherence account has two major advantages over

Achinstein's probabilistic account: computational tractability and psychological plausibility. Now ECHO and EM both provide specifications of a coherence account. Assuming that the EM-version also leads to the choice of the wave theory, the relative merits of the three accounts can be measured in terms of these two criteria. Eliasmith and Thagard make it very clear that the ECHO-account is computationally much more tractable than a probabilistic account. However, it is also evident that, whatever the outcome, the EM-account is computationally still much more tractable than the ECHO-account. Similarly, Eliasmith and Thagard make it also clear that the ECHO-account is psychologically more plausible, referring to psychological evidence in favor of connectionist models of mind on the one hand and low probabilistic skills on the other. We would, however, like to claim that the EM-account may well be still much more plausible, provided the evidence points more or less unambiguously in one direction. The latter is indeed the case, that is, the wave theory is indeed explanatorily superior to the particle theory on the basis of the presented reconstruction.

According to the Appendix to Eliasmith and Thagard's paper, there are 23 evidence statements. Apart from the main statements, the two theories (WT and PT) require one or more additional hypotheses for the explanation of one or more evidence statements. They both need one extra hypothesis, not the same one, for one and the same evidence statement. Moreover, PT needs three more extra hypotheses, each for a different evidence statement. Hence, WT is evidently simpler than PT, and if WT and PT would explain the same facts, this would play an important role. However, the supposition is quite counterfactual for the score is as follows: 1 E-statement apparently belongs to B5, 11 to B6, 10 to B7 and 1 to B8. Hence, WT is explanatorily superior to PT, without any reservation.

In the present case, the authors do not argue extensively for the claim that ECHO provides a process-reconstruction; there is just one quote from Young in which he stresses the importance of explanatory coherence much more convincingly than that of simplicity. However, even if they would argue this convincingly, there remains the possibility that simplicity arguments in case of explanatory superiority just serve as additional considerations that might convince people in doubt to agree with the choice. Moreover, if scientists, like ECHO, really intermingle explanatory superiority with simplicity considerations, we may raise the normative question: is the actual process to be criticized for being not as simple as possible? If so, computer assisted research should of course ultimately be guided by that criticism and not by actual processes.

CONCLUDING REMARKS

In the introduction to this chapter we mentioned four claims, viz. historical, psychological and philosophical adequacy, and practical relevance. In the last section it was shown that there may well be a tension between them. Even if ECHO is historically and psychologically adequate, it may well be that it is philosophically and practically questionable. On the other hand, it may well be that there are practically more useful alternatives that are philosophically adequate, historically adequate as far as product-simulation is concerned and that may even also be a first approximation of psychological processes.

To be sure, the possibility for the computational philosophy of science to be of considerable practical relevance is still far away. However, in principle the perspective of more or less standard computer assisted discovery, evaluation and revision need not remain science fiction. To begin with, Langley (1998) presents a systematic survey of seven specific examples of novel, computer-aided discoveries that have appeared in the scientific literature. They include a taxonomic discovery in astrophysics, a qualitative law in biochemical cancer research, a qualitative law in metallurgy, a quantitative relation in graph theory, a temporal law in ecology, a structural law in biochemistry, and, finally, a process law in catalytic chemistry.[5] Moreover, coupling uniformly represented knowledge bases from various fields would create an enormous general source for analogical reasoning and for producing surprising data on which we can proceed. In such specific and general ways the computational philosophy of science could join forces with the already flourishing practice of another type of computer assisted research, viz. computer simulation.

CHAPTER 12

THE STRUCTURALIST APROACH TO THEORIES[1]

INTRODUCTION

This chapter gives a systematic introduction to the structuralist reconstruction of empirical theories. In a number of chapters in this book and in ICR we have used the structuralist representation, without using its refined conceptual framework in all detail. Notably, in ICR (Chapter 11), where we deal with truth approximation in the old quantum theory and in capital structure theory, both (sequences of) theories were represented in the structuralist way.

In Section 12.2.4 we have already presented a first exploration of the nature and structure of empirical theories. Recall that a theory usually has a core of principles and a belt of auxiliary hypotheses. A theory is said to be ontologically stratified when there are two or more kinds of entities involved, where entities of one of these kinds are components of entities of another kind. It is then plausible to speak of a lower micro-level and a higher macro-level. In this case some principles will only concern the micro-entities, and their properties and relations, and are called internal principles, whereas some other principles will connect the different kinds of entities, and their properties and relations, and are called bridge principles. A similar distinction holds for auxiliary hypotheses.

Besides ontological stratification there is epistemological stratification: these frequently go together, but are essentially independent. A proper theory *X* was defined as an epistemologically stratified theory in the sense that it contains terms, and hence statements, that are laden with one or more of its principles, that are *X*-laden or *X*-theoretical, for short. The other terms of *X* are called *X*-unladen or *X*-non-theoretical. In contrast to proper theories, observational hypotheses were defined as improper theories, containing no theoretical terms of their own. A set of connected observational hypotheses was called an observational theory. It should be recalled that being *X*-non-theoretical is a theory-relative qualification of a term or a statement: such a term or statement may well be laden with underlying theories.

The main function of a proper theory *X* is the explanation and prediction of *X*-unladen, observational laws, i.e. true general hypotheses containing neither terms laden with these laws themselves nor terms laden with *X*. For this purpose the distinction between observational hypotheses and proper theories is of course crucial.

Recall, finally, that the principles of a theory, whether ontologically and/or epistemologically stratified or not, can frequently be distinguished in main or generic principles, claimed to be true for the whole domain concerned, and special principles, only claimed to be true for a certain subdomain.

In this chapter we will analyze the structure of theories in more detail. In Section 12.1 we will discuss the attractive features of the structuralist approach in general. Starting with the simple example of a slide balance we will first present, in Section 12.2, the structuralist representation without making the distinction between theoretical and non-theoretical terms. Reconsidering the slide balance in the light of attempts to measure quantities in a non-circular way, we will introduce, in Section 12.3, the structuralist representation with the distinction between theoretical and non-theoretical terms. We will also give the basic outline of this kind of representation for three examples: classical particle mechanics, the periodic table, and psycho-analytic theory. In Section 12.4 we will continue with some further refinements of the structuralist approach, viz. the distinction between absolute and relative empirical content, the possibilities of determining the intended applications, relations between theories, theory-nets, and cons traints. We will conclude by briefly considering which parts of the structuralist approach are useful for non-empirical theories, such as metaphysical, mathematical, conceptual and normative theories.

There might occasionally be some overlap with other chapters, which is unavoidable in order to enable the reader to read this chapter in isolation. Wherever possible we will use the same symbols if precisely the same concepts are involved, with one exception. In other chapters X, Y, and Z were variables for theories, whereas in the structuralist presentation M, M', M^* and the like are usual for this purpose, a practice which we also adopt in this chapter.

12.1 WHY THE STRUCTURALIST APPROACH?

There are two main ways of viewing the structure of empirical theories. The *statement* approach conceives theories primarily as sets of statements. In case of an axiomatized theory all these statements are logical consequences of a subset of so-called axioms. This approach has long been considered as the only and obvious approach, e.g. by Carnap and Popper. Kyburg (1968) illustrates the approach in great detail.

In the *semantic* approach theories are primarily conceived as sets of models. One version, the state-space version, goes back to Beth and is favored for instance by Suppe and Van Fraassen. Our favorite version, the set-theoretic or *structuralist* approach was introduced by Suppes and refined by Sneed, Stegmüller, Balzer and Moulines. Its basic idea is that theories frequently specify classes of set-theoretic structures satisfying certain conditions. Recall that a set-theoretic structure is an ordered set of one or more domain- or base-sets, and one or more properties, relations or functions defined on them, which satisfy certain conditions.

A biological family e.g. is a structure $\langle A, C\rangle$ with A as the set of members of the family and C as a ternary relation on A, i.e. C is a subset of $A \times A \times A$, such that $C(x, y, z)$ states that z is a child of x and y. A proper two-generation biological family exists if there are precisely two members x and y in A such that for all other z in A, $C(x, y, z)$ is true.

According to the structuralist view an axiomatized theory defines such a class of structures and the conditions imposed on the components of the structures are the axioms of the theory. The link with reality is made by the claim, associated with the theory, that the set of set-theoretic representations of the so-called intended applications forms a subset of the class of structures of the theory.

Unfortunately, there has been much debate about what the proper approach to theories is, whereas it is easy to see that the two approaches are not at all incompatible. At least for so-called first-order statement theories, i.e. theories formulated as a set of statements of a so-called first-order language, it is evident that the set of models of the theory, i.e. the structures for which the statements of the theory are true, is precisely a set of structures that might also have been introduced directly in the structuralist way. If we do not restrict ourselves to first-order languages, both approaches are still essentially intertranslatable. Hence, the choice is a pragmatic question.

The main advantage of the structuralist approach is that it is much more a bottom-up approach than the statement approach. It invites us, as it were, to represent and analyze a theory as close to the actual presentations in textbooks as is formally possible. As in scientific practice, all kinds of useful mathematics may be used for that purpose.[2] It does not mean that structuralist reconstruction of theories is an easy task. However, the statement approach is certainly more difficult for specific reconstructions. It is primarily useful in talking about theories in general and in studying logical, in particular model-theoretic, questions about the relation between sentences and their models. These questions and their answers become very complicated as soon as substantial mathematics is involved, e.g. real numbers. Happily enough, not all interesting theoretical questions need logical treatment. For example, as we have demonstrated in ICR (Chapter 10), theoretical questions concerning, for instance, idealization and concretization as a truth approximation strategy can be treated relatively easily in structuralist terms.

Another advantage of the structuralist approach is the 'systemic' perspective which takes the world to consist of many systems. Though the statement view could take up this perspective, it has not done so. However this may be, it is plausible that in this respect the structuralist representation is also much simpler than a statement version, if such a version would be made explicit. A third advantage is that syntactical features, like, for example, the type of a function, can be expressed in a realistic way, talking about functions, and not about building strings of symbols.

Given our intention to be as useful as possible for actual scientific research, we will restrict our attention to the structuralist approach. The best textbooks presenting the structuralist view on theories are Diederich (1981), Balzer (1982), Stegmüller (1973/1986) and Balzer, Moulines and Sneed (1987). We will present briefly a number of examples and the main general features, referring to extensive expositions when possible. We will not go into technical details which are not of primary importance for actual practice. Our main goal is to present by way of examples and general exposition the kind of entities that one may be looking for in theory formation and the ways in which standard questions about these entities can be explicated.

12.2 THE EPISTEMOLOGICALLY UNSTRATIFIED APPROACH TO THEORIES

Starting with the simple example of a slide balance we will first present the structuralist representation of theories and the corresponding terminology without making the epistemological distinction between theoretical and non-theoretical terms.

12.2.1 *Example: The slide balance*

Consider the slide balance as represented in Figure 12.1.

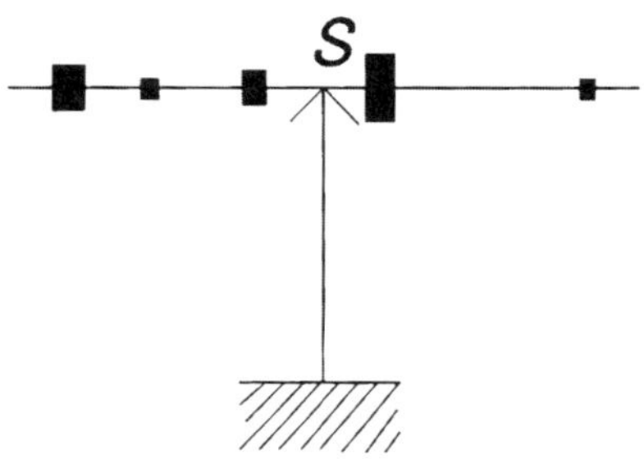

Figure 12.1 Slide balance.

On either side there can be placed a finite number of objects of various weights at all possible distances from the turning-point S. The balance is assumed to be completely symmetric, the equal arms are as long as necessary, and the objects are point masses, i.e. dimensionless particles. Our domain of interest consists of the equilibrium states, i.e. all possible distributions of objects resulting in equilibrium.

A plausible way to represent the equilibrium states, the intended applications, is as follows. We start by characterizing a possible or potential equilibrium state by a structure of the form $\langle P, Pl, d, w\rangle$, where P is the finite set of particles involved and Pl the subset of P such that Pl and $P - Pl$ represent the particles to the left and to the right of S, respectively. For every particle p in P, $d(p)$ indicates the distance of p from S and $w(p)$ the weight of p. Technically speaking, d and w are positive real-valued functions on P. Let us call the set of structures $\langle P, Pl, d, w\rangle$ satisfying all formal conditions the set of potential equilibrium models of our theory about the equilibrium states of the slide balance, indicated by SBp.

Accordingly, our conceptual claim is that the equilibrium states can be represented as members of SBp, i.e. there is a subset E of SBp representing the nomic, that is, the nomically possible, equilibrium states: the SBp-set of intended applications.

The ultimate purpose of theory formation now is to try to characterize E explicitly by one or more additional conditions. As is well known, the adequate condition in the present case is specified by the so-called law of the balance: the sum of distance times weight of the objects on the left should be equal to that sum involving the objects on the right. Let us call the subset of SBp of members satisfying this condition the set of equilibrium models of our theory, indicated by *SB*. The proper empirical claim of the law of the balance can now be formulated as '$E = SB$', and this claim is (generally supposed to be) true.[3]

As we will see in other cases the relevant claim need not be as strong as in the present case. The claim might just have been that E is a subset of SB. It will be helpful for later examples to add a more formal presentation of the naive theory of the slide balance (IR^+ indicates the set of positive real numbers).

The naive theory of the slide balance $\langle SBp, SB, D, E\rangle$

	contains $\langle P, Pl, d, w\rangle$	iff
	(1) P is a finite set	(particles)
	and Pl a subset of P	(the particles on the left of S)
	(2) $d: P \rightarrow IR^+$	($d(p)$: the distance of p from S)
SBp	(3) $w: P \rightarrow IR^+$	($w(p)$: the weight of p)
	(4) *the law of the balance*	
SB	$\Sigma p \in Pl\ d(p) \cdot w(p) = \Sigma p \in P - Pl\ d(p) \cdot w(p)$	

$SBp - SB$	*empirical content* (to be explained)
$E \subseteq SBp$	*conceptual claim*: the intended domain of applications D can be represented as potential models, the intended applications
$E \subseteq SB$	*(naive weak) empirical claim* the intended applications are equilibrium models (naive strong claim: $E = SB$)

Later we will see that the empirical claims are naive, because it appears to be impossible to test them in a non-circular way. But first we will present the general structuralistic set-up for unstratified theories.

12.2.2 *Unstratified Theories*

Let there be a given domain D of natural phenomena (states, situations, systems) to be investigated. D is supposed to be circumscribed by some informal description and may be called the *intended domain of applications*. Although D is a set, its elements are not yet mutually well distinguished. For this reason we do not yet speak of the domain of intended applications.

In order to characterize the phenomena of D, a set Mp of *conceptual possibilities or potential models* is construed. Technically speaking, Mp is a set of structures of a certain type, a so-called similarity type. In practice Mp will be the conceptual frame of a research program for D.

The confrontation of D with Mp, i.e. D seen through Mp, is assumed to generate a unique, time-independent subset $Mp(D) =_{\text{def}} I$ of all Mp-representations of the members of D, to be called the Mp-set of *intended applications*. Apart from time-independence, this assumption is a conceptual claim. Of course, since nomic impossibilities can, by definition, not be realized (see Subsection 9.1.1), I will be a subset of the (Mp-)set of *nomic possibilities* (indicated by T in Chapter 9), but it may be a proper subset, i.e. a more specific set of intended applications satisfying certain additional (more or less precise, but relatively observational) conditions. Assuming that the set of nomic possibilities is a proper subset of Mp, i.e. not everything which is conceivable is nomically possible, I is also a proper subset of Mp. In certain cases I may be a one-element set, in particular when we want to describe 'the actual world' in a certain context, that is, a realized (hence nomic) possibility, e.g. the description of conditions and results of a particular experiment. In ICR about truth approximation, the attention is focussed on the special case that I is the set of nomic possibilities.

A specific theory about D is concentrated around an explicitly defined subset M of Mp, the *models* of the theory. More specifically, a specific unstratified theory is any combination of the form $\mathrm{UT} = \langle Mp, M, D, I\rangle$ with, beside the conceptual claims that M and I are both subsets of Mp, the (weak) empirical claim that I is a subset of M. Sometimes the strong empirical claim is made that I is equal to M, but in this chapter we take the weak claim as standard. It is plausible to call UT true when its claim is true, and false otherwise.

The general set-up of the structure of epistemologically unstratified theories will now be presented in a scheme. Such a theory is a meta-structure of the following form:

	$\langle Mp, M, D, I\rangle$ is an *epistemologically unstratified theory* iff
Mp:	potential models: a set of structures of a certain type
$M \subseteq Mp$	models: the potential models that satisfy all axioms
$Mp - M$:	empirical content (to be explained)
D:	the intended domain of applications
$I \subseteq Mp$:	intended applications: conceptual claim: Mp-representation of D leads to the subset I of Mp
$I \subseteq M$:	empirical claim (strong claim: $I = M$)

$\langle Mp, M\rangle$ is sometimes called the theoretical core of the theory, and $\langle D, I\rangle$ may be called the application target of the theory.

The unstratified set-up of theories seems to be rather adequate for observational theories, a combination of one or more observational hypotheses, which contain by definition only terms that are understood independently of the theory concerned.

12.2.3 *Basic Terminology*

Before we go over to stratified theories, we would like to present some useful basic terminology, which can largely be seen as a structuralist explication of Popperian 'statement terminology' (Popper, 1934/1959). We will neglect all necessary provisos, in particular in regard to the complications arising from underlying theories. To use Lakatos's term (Lakatos, 1978), we explicate naive falsificationism, first unstratified, later stratified.

When the claim of theory $UT = \langle Mp, M, D, I \rangle$ is false $I - M$ is by definition non-empty, in which case it is plausible to call its members instantial mistakes or (empirical) *counter-examples* of *UT*. Note that being a counter-example in this sense does not imply that it has been realized already and registered as such. The set of counter-examples $I - M$ is by definition a subset of $Mp - M$. Hence, $I - M$ can, whatever *I* is, only be non-empty when $Mp - M$ is non-empty. In other words, the members of $Mp - M$ may be called the *potential* counter-examples of the theory and, as has already been stated, the set $Mp - M$ itself the *empirical content* of *UT*. From the present point of view, Popper had similar things in mind with his notions of 'potential falsifier' and 'empirical content'.

Other plausible explications of Popperian terminology (which will however not be used in the sequel) are for instance: *UT* is *falsifiable* (or empirical) if and only if $Mp - M$ is non-empty, and UT^* is *better falsifiable* than *UT* when $Mp - M$ is a proper subset of $Mp - M^*$. The latter condition is equivalent to: M^* is a proper subset of *M*. In its turn, this is equivalent to stating that the claim of UT^* implies that of *UT*, and not conversely, that is, UT^* is stronger than *UT*.

The verification/falsification asymmetry also arises naturally in the present set-up. To *verify* theory *UT* it would be necessary to show that all members of *I*, that is, all *Mp*-representations of *D*, belong to *M*. In interesting cases, this demonstration will always be an infinite task, even in the case that *I* is finite, for the task is only finite when *D* is finite. To *falsify UT*, it is 'only' necessary to show that there is *at least one* member of *I* not belonging to *M*. Hence, if a theory is true, verification will nevertheless not be obtainable if *D* is infinite. On the other hand, when a theory is false, falsification is attainable in principle, viz. by realizing one counter-example. If an attempt to falsify fails in such a way that the experiment provides an (*empirical*) *example* of *UT*, i.e. a member of *M*, this is called *confirmation* (or corroboration) of *UT*.

In the present set-up Popper's distinction between universal and existential statements is thus interpreted as the distinction between the general claim of the theory ($I \subseteq M$) that all intended applications are models of the theory, and the negation of this claim, the existential claim that at least one intended application is not a model ($I - M$ is non-empty).

A *basic statement* (see Chapter 2 for Popper's specific idea of a basic statement) becomes a claim to the effect that a certain intended application *x* in *I* belongs to a certain subset *F* of *Mp*, defined by a certain condition being imposed on potential models, i.e. $x \in I \cap F$. An *accepted* basic statement presupposes of course that the relevant intended application has been realized.

The basic statement $x \in I \cap F$ is in conflict with theory *UT* if it can be demonstrated on conceptual grounds that $F \cap M$ is empty. Such basic statements may be seen as a more direct explication of Popper's idea of 'potential falsifiers', compared to 'potential counter-examples'. However, it is easy to show that the suggested statement concept of potential falsifier has become essentially redundant. It is easy to verify that the discovery of a true potential falsifier of *UT*, i.e. the discovery of an *x* in *Mp* for which '$x \in I \cap F$' is true, implies that $I - M$ is non-empty and hence that *x* is a counter-example of *UT*. Conversely, the existence of counter-examples of *UT* is easily seen to imply that there must be true potential falsifiers. As a consequence, empirically demonstrating the existence of a counter-example, i.e. realizing a potential counter-example, goes hand in hand with demonstrating that there is a true potential falsifier. Hence, the statement concept of potential falsifier is not needed in the face of the concept of a (potential) counter-example.

12.3 THE STRATIFIED APPROACH TO THEORIES

Starting by reconsidering the slide balance in the light of attempts to measure the relevant quantities in a non-circular way, we will introduce the structuralist representation of theories and the corresponding terminology with the epistemological distinction between theoretical and non-theoretical terms. We will give the basic outline of this kind of representation for three examples: classical particle mechanics, the periodic table, and psycho-analytic theory.

12.3.1 *The Slide Balance Reconsidered*

The problem with the slide balance is that it might be impossible to test the claim in a non-circular way without leading to an infinite regression. For, to test the claim, it is necessary to measure the distances and the weights. Whereas distance measuring does not require something like a slide balance, weight measuring may not only actually be done by using a slide balance, there might even be no other possibility. If the weight of a particle is measured by a slide balance the law of the balance is obviously presupposed. Hence, assuming that the weight of a particle can only be measured by a slide balance, the concept of weight is *SB*-theoretical and leads to the so-called *problem of theoretical terms*. A non-circular test of the claim of the theory would presuppose that the weights of the particles have been measured before with the same or another slide balance. Hence, we get either an infinite regress or circular testing, if we stick to '$I \subseteq M$' as the empirical claim of the theory. There is, however, a way-out of this dilemma, by restricting the empirical claim to *SB*-non-theoretical terms. Later we will see that for two reasons the situation in the present example is not as dramatic as suggested, but this did not exclude the fact that the example could, by way of a thought experiment, be transformed into an instructive example of genuine theoretical terms.

In order to formulate a new empirical claim we introduce the set of potential partial equilibrium models *SBpp*, being the structures of *SBp* without the *SB*-theoretical weight component and the corresponding 'status-condition', viz. Clause (3). By consequence, there is a restriction or *projection function* π from *SBp* onto *SBpp* projecting every potential model on the potential partial model arising from deleting w and Clause (3). Hence, for $x = \langle P, Pl, d, w \rangle \in SBp$, the projection of x, $\pi(x)$, is equal to $\langle P, Pl, d \rangle \in SBpp$. For an arbitrary subset X of *SBp*, πX, the projection of X, is defined as the subset of *SBpp* containing precisely the projections of the members of X.

For stratified theories we assume that the set of intended applications E no longer represents the equilibrium states seen through *SBp*, but seen through *SBpp*. Hence, the corresponding conceptual claim that E is a subset of *SBpp* is not laden with our theory about *SB*.

Now it is easy to see that the claim that E is a subset of πSB is not laden with the weight term, in the sense that it does not presuppose that the weights of the particles have been empirically determined. Hence, this revised empirical claim can be tested in a non-circular way.

It is plausible to call the members of $E - \pi SB$, if any, counter-examples of the theory. It is clear that they have to come from $SBpp - \pi SB$. Hence, it is now plausible to call this set the *empirical content* and its members *potential counter-examples*. Note that the empirical content reduces to the empirical content of the unstratified theory $(SBp - SB)$ when *SBpp* and *SBp* are identical and π is, by consequence, the identity-function.

Unfortunately, the new claim is not only non-circular, it is also vacuous, for the empirical content is empty. The claim says in fact that all intended applications can be extended to models of the theory. To be precise, the claim is that every $\langle P, Pl, d\rangle$ in E can be supplied with a positive real-valued function w on P such that $\langle P, Pl, d, w\rangle$ is in *SB*. But it is easy to check that this is possible for every member of *SBpp*. In other words the empirical content $SBpp - \pi SB$ is empty.

However, the situation changes when we take so-called constraints into consideration: in the present case we have to require also that the weights assigned to the same particle, occurring in different applications should be the same. In contrast to the distance of the objects from the turning-point, our concept of weight is such that the weight of particles is constant in different applications. The formal treatment of constraints, however, will be postponed to Subsection 12.4.4.

Before we return to the general exposition we will summarize the formal features of the theory of the slide balance, leaving out the plausible specification of the projection function π:

The refined theory of the slide balance $\langle SBp, SBpp, SB, \pi, D, E\rangle$

$\rightarrow$	contains $\langle P, Pl, d, w\rangle$	iff
$\rightarrow$	contains $\langle P, Pl, d\rangle = \pi\langle P, Pl, d, w\rangle$	iff
	(1) P is a finite set and Pl a subset of P	(particles) (the particles on the left of S)
SBpp	(2) $d\colon P \rightarrow IR^+$	($d(p)$: the distance of p from S)
SBp	(3) $w\colon P \rightarrow IR^+$	($w(p)$: the weight of p)
SB	(4) *the law of the balance* $\Sigma p \in Pl\ d(p) \cdot w(p) = \Sigma p \in P - Pl\ d(p) \cdot w(p)$	

$SBpp - \pi SB$	*empirical content* (without w-constraint empty, with w-constraint non-empty)
$E \subseteq SBpp$	*conceptual claim*: the domain of intended applications D can be represented as potential partial models
$E \subseteq \pi SB$	*empirical claim* the intended applications can be extended to models (strong claim: $E = \pi SB$)

By way of digression, it is interesting to note that, assuming the weight-constraint, the SB-theory explains the following observational, i.e. SB-unladen, *factor slide law*: if, starting from an equilibrium, the distances of all objects are multiplied by the same factor, there is again equilibrium. For it follows trivially from the law of the balance that this law remains satisfied.

As a matter of fact, in the present case it is not difficult to formulate an observational law such that the notion of weight can be explicitly defined, apart from a proportionality constant, on its basis. The law referred to states the following: given a unit object at a unit distance at one side of S, every other object p has a 'unique equilibrium distance' $d_u(p)$ at the other side. The weight $w(p)$ is then defined as $1/d_u(p)$, hence, such that in the relevant cases the law of the balance is satisfied by definition. Hence, for these cases the law cannot be tested in a non-circular way. But there is no regress, let alone infinite regress. For, given the definition, the rest of the law of the balance is a straightforward empirical claim which can be directly tested.

As a consequence, the theory of the slide balance does not, on closer inspection, give rise to the problem of theoretical terms, when certain observational laws are taken into consideration. Of course, this does not affect the instructiveness of the SB-theory as an almost proper theory. Moreover, it illustrates an interesting way in which a seemingly proper theory may on closer inspection be a sophisticatedly formulated observational theory, in the present case: the conjunction of the 'unique equilibrium distance law', the weight-definition on its basis, and the law of the balance.

There is still one other reason why the problem of theoretical terms is not so dramatic in the case of the slide balance: there are other ways of measuring the weight of objects than by using a slide balance. But let us now turn to the general set up of stratified theories, designed for proper theories.

12.3.2 *Stratified Theories*

The general set-up of the structure of epistemologically stratified theories can now directly be presented in a scheme. Such a theory is a meta-structure of the following form:

	$\langle Mp, Mpp, M, p, D, I\rangle$ is an *epistemologically stratified theory* iff
Mp:	potential models: a set of structures of a certain type
Mpp:	potential partial models: the substructures of Mp restricted to non-theoretical components
$M \subseteq Mp$	models: the potential models that satisfy all axioms
$\pi : Mp \rightarrow Mpp$:	the projection function (from Mp onto Mpp)
	$\pi X = \{\pi(x)/x \in \mathrm{X}\}$, for $X \subseteq Mp$, implying $\pi X \subseteq Mpp$
πM:	projected models
$Mpp - \pi M$:	empirical content
D:	intended domain of applications
$I \subseteq Mpp$:	intended applications (non-theoretical):
	conceptual claim: non-theoretical representation of D leads to the subset I of Mpp
$I \subseteq \pi M$:	empirical claim (strong claim: $I = \pi M$)

Now it is plausible to call $\langle Mp, Mpp, M, \pi\rangle$ the theoretical core of the theory and $\langle D, I\rangle$ remains the application target. Figure 12.2 illustrates the refined empirical claim: the shaded area, representing $I - \pi M$, should be empty. To be precise, $I - \pi M$ should be empty on conceptual grounds, that is, the conceptual characterization of I and πM should not leave room for conceptual possibilities in $I - \pi M$ (let alone for actual intended applications).

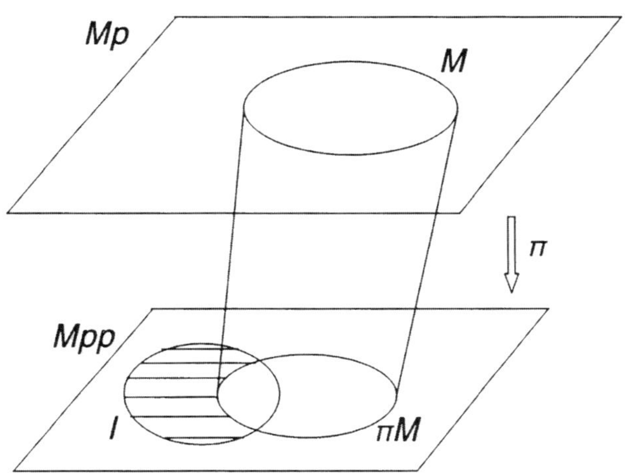

Figure 12.2 Refined empirical claim: shaded area empty.

12.3.3 *Examples*

In this subsection we will give the theoretical core of the structuralist reconstruction of three well-known theories, viz. Newton's classical (gravitational) particle mechanics, Mendeleev's and the refined theory of the periodic table of chemical elements, and Freud's psycho-analytic theory. The presentation will always start with the survey followed by a brief elucidation. For details of the theory and the reconstruction, the reader is referred to the original or other publications of the reconstructions. The theories (more precisely, the theory cores) will be named by their basic class of models.

From the fact that Freud's theory can be reconstructed in the structuralistic way it follows that this way of reconstruction is, like the statement approach, applicable to qualitative, non-mathematical theories. From the other examples, it is evident that the present approach is also well suited for quantitative theories, a kind of theory for which the statement approach leads to all kinds of complications.

In a sense it is a trivial claim that every empirical theory can be reconstructed in structuralist fashion. Hence, there should be additional reasons to do so in particular cases. A general reason frequently is the desire to get a better insight into the theory; besides that, one may be interested in particular questions, such as whether the theory has empirical content, whether it is an observational or a proper theory, what its precise relation is to another theory, etc. The examples to be presented are supplied with some comments to illustrate both reasons of reconstruction. But the main function of getting acquainted with the structuralist approach in general and by way of examples is of course the heuristic role it may play in the construction of new theories.

After the presentation of the three examples we will continue in the next section with general matters, such as the distinction between absolute and relative empirical content, the possibilities of determining of the intended applications, relations between theories, theory-nets, and constraints.

Classical particle mechanics for one dimension (with gravitation as specialization)
$CPM = \langle CPMp, CPMpp, \pi, CPM, GCPM \rangle$.

	—> contains $\langle P, T, s, m, f \rangle$	iff
	—> *contains* $\langle P, T, s \rangle = \pi \langle P, T, s, m, f \rangle$	iff
	(1) P is a finite set	(particles)
	(2) T is real interval	(time-interval)
	(3) $s: P \times T \rightarrow IR$	(position)
	giving rise to the first and second time derivaties:	
	$v: P \times T \rightarrow IR$	(velocity)
CPMpp	$a: P \times T \rightarrow IR$	(acceleration)
	(4) $m: P \times T \rightarrow IR^+$	(mass)
	(5) $f: P \times T \times P$	(force)
CPMp	$f(p, t, q)$	(force from q on p at t)
	(6) *second law*, implying the *first law* in this formulation, for all p in P and t in T $\Sigma_{qeP} f(p, t, q) = m(p) \cdot a(p, t)$	
	(7) *third law* (action = – reaction) for all p and q in P and all τ in T	
CPM	$f(p, t, q) = -f(q, t, p)$	

	(8) *the law of gravitation* there is a universal real constant t such that for all p and q in P and t in T
GCPM	$f(p,t,q) = +/- \gamma\,[m(p)\cdot m(q)]/[s(p,t) - s(q,t)]^2$

As is well known, Newton's theory of gravitation is based on the generic theory of particle motion, i.e. classical particle mechanics (*CPM*), of which the core is formed by three interrelated so-called laws of motion: the first law: the law of inertia, the second law: $F = m.a$, and third law: action is minus reaction. This general or generic theory can be specialized by adding the special law of gravitation(al force), but there are other well-known specializations, e.g. Hooke's law of spring-force and Coulomb's law of electrostatic force.

The following remarks may elucidate the survey to some extent. For detailed expositions the reader is referred to Sneed (1971), Zandvoort (1982) and Balzer et al. (1987).

It is clear that mass and force are treated as *CPM*-theoretical components; although this treatment may not be strictly necessary, it is always a safe option in case the situation is unclear; but a vacuous claim may be the result.

CPM concerns the generic theory, (6) and (7) are the proper generic laws/principles. *GCPM* is, due to the addition of the law of gravitation, a subset of *CPM*, that is a specialization of *CPM*.

πCPM and $\pi GCPM$ provide the projected models of *CPM* and *GCPM*, respectively. $CPMpp - \pi CPM$ and $CPMpp - \pi GCPM$ constitute the empirical content of *CPM* and *GCPM*, respectively. Note that the former is a subset of the latter, just as it should be, for *GCPM* is stronger than *CPM*.

As long as the identity constraint for mass is not taken into consideration *GCPM* has no empirical content, let alone *CPM*. With this constraint *CPM* still lacks empirical content, but *GCPM* gets it.

The intended domain of applications of *CPM* concerns in the first place that of *GCPM*, for instance, planetary orbits, falling stones, paths of projectiles, etc. but also movement of objects by spring or electric forces. Moreover, it contains compound applications, i.e. applications in which two or more force types operate, e.g. three in the case of an electrically charged ball on an isolated vertical spring on a charged table.

The periodic table of chemical elements (naive and refined) $NPT/SPT = \langle PTpp, PTp, p, NPT/SPT\rangle$

	—> contains $\langle E, m, \approx, z\rangle$	iff
	—> contains $\langle E, m, \approx\rangle = \pi\langle E, m, \approx, z\rangle$	iff
	(1) E: a finite set	(chemical elements)
	(2) m: $E \to IR^+$	(atomic mass)
PTpp	(3) $\approx$: equivalence relation on E	(chemical similarity)
PTp	(4) z: $E \to IN^+$	(atomic number)

(5)a. range $(z) = 1, 2, \ldots, \max(z)$ — z is onto $\{1, \ldots, \max(z)\}$
b. $m(e) < m(e')$ iff $z(e) < z(e')$ — z increases with m
c. $z(e) = z(e')$ implies $e = e'$ — z is a one-one function

NPT (6N) *naive periodic law*
$e \approx e'$ iff $|z(e) - z(e')|$ is a multiple of 8

resp.

RPT (6R) *refined periodic law*, elegant, but complicated; core: if $e \approx e'$ and if there is no element with z-number between $z(e)$ and $z(e')$ then $|z(e) - z(e')|$ can be written as $2n^2$, i.e. 2 or 8 or 18 or 32 etc.

For a detailed exposition the reader is referred to Hettema and Kuipers (1988, 2000).[4] The following remarks highlight some crucial points.

As is well known, Mendeleev developed the periodic table on the basis of the observations that the chemical elements can be classified in groups of elements with chemically similar behavior and that ordering of the elements by increasing atomic mass roughly leads to a matrix in which the groups appear as columns. To explain the system in this matrix he introduced the concept of atomic number and formulated the (naive) periodic law (*NPT*), which was later refined by others (*RPT*).

In the present example the intended domain of application concerns the chemical elements taken together, such that the conceptual claim states that this domain can be represented by just one potential projected model, say $\langle E^*, m^*, \approx^* \rangle$. Mendeleev's empirical claim was that this pp-model belongs to πNPT and the modern empirical claim localizes it in πRPT. Or, equivalently, there is z^* such that $\langle E^*, m^*, \approx^*, z^* \rangle$ belongs to *NPT* and *RPT*, respectively.

It is not difficult to verify that both theories have empirical content. In fact both claims are false. To fulfil the claims as much as possible, we must allow counter-examples to the three technical conditions imposed by Clause (5), with plausible names: (5a) missing elements, which may be discovered later, and some have been; (5b) order disturbers, having greater mass and lower atomic number than others or vice versa; and (5c) isotopes, i.e. different elements with the same atomic number.

Note that the notion of a counter-example is used here on a lower level than in the general set-up. This is possible because there is only one overall intended application, viz. $\langle E^*, m^*, \approx^* \rangle$. If that does not fit into πNPT or πRPT, this failure must be due to lower level counter-examples, i.e. specific elements. There may be systematic or just local counter-examples. In this sense *NPT* has both types of counter-examples, whereas *RPT* has only local counter-examples.

The history of *PT* provides marvelous examples of all four combinations of theory (un)laden and theory (un)directed observation, as described in Chapter 2. A successful search for missing links, for instance, means theory directed but theory unladen observation.

The quantum mechanical theory of the atom provides a reductive explanation (see Chapter 3) for *RPT*, by means of identification of z with the number of electrons of the atom concerned. In view of the fact that this number can be measured in *RPT*-independent ways, *RPT* is in fact an observational theory. Of course, Mendeleev's *NPT* was a proper theory, with z as a theoretical term. *RPT* was developed hand in hand with atomic theory, in which process it transformed from a proper theory into an observational theory.

Psycho-analytic theory $PA = \langle PApp, PAp, \pi, PApart, PA\rangle$

	—> contains $\langle T, E, L, \leq, ASS, B, N, A, U, REAL\rangle$	iff
		iff
	—> contains $\langle T, E, L, \leq, ASS, B, N\rangle = \pi\langle \ldots A, U, REAL\rangle$	
	(1) T is an interval of real numbers; variable t, t^*, etc.	(life span of a person)
	(2) E is a non-empty set	(experiences)
	(3) L is a proper subset of E	(painful experiences)
	(4) $\leq$ is a weak linear ordering on T	(not later than)
	• $<$ by definition $\leq$ and $\neq$	(earlier than)
	(5) $B(t)$ is a non-empty subset of E	(consciousness at time t)
	(6) $N(t)$ is a subset of $B(t)$ and L	(negative exp's at t)
	(7) ASS is a relation on E	(associated experiences)
PApp	• $ASS(e, e)$, i.e. ASS is reflexive	(self association)
	(8) A is a non-empty set; $A \cap E = \emptyset$	(unconscious impulses)
	(9) $U(t)$ is a non-empty subset of A	(unconsciousness at t)
	(10) $REAL$ is ternary relation on $E \times A \times T$: $REAL(e, a, t)$	(e is a realization of a at t)
	• if $REAL(e, a, t)$ then e in $B(t)$ and a in $U(t)$	
	• not for all t, e in $B(t)$ and a in $U(t)$: $REAL(e, a, t)$	
PAp	• if $REAL(e, a, t)$ and $REAL(e', a, t')$ then $ASS(e, e')$	
PApp	+ (11) *repression axiom* repression of negative experiences, incl. associated ones:	
PApart	if e in $N(t)$ and $ASS(e, e')$ and $t < t^*$ then e' not in $B(t^*)$	
PAp + (11)	+(12) *main axiom* every unconscious impulse is realized sooner or later: for all t and for all a in $U(t)$ there are e in E and t^*	
PA	such that $t \leq t^*$ and $REAL(e, a, t^*)$	
PA	+ (13) *neurosis axiom* having a neurosis generating experience for an impulse: there are t_0, e_0 in E, a_0 in A such that	
PAN	$REAL(e_0, a_0, t_0)$ and e_0 in $N(t_0)$	

For a detailed exposition the reader is referred to Balzer (1982), Stegmüller (1986) or, for the most refined one, to Balzer and Marcou (1989). The general psycho-analytic theory *PA* is intended for all human beings, they are all supposed to repress negative experiences (*PA*-11) and to satisfy the main axiom *PA*-10 that all unconscious impulses are sooner or later realized. Note that *PA*-11 does not use theoretical terms, so it makes sense, as indicated, to introduce (non-theoretical or observational or) *partial models* (*PApart*) as potential partial models satisfying this non-theoretical but substantial axiom. In the next section we will generalize this idea and investigate its consequences.

The psycho-analytic theory of neurosis *PAN* is a specialization to people with a neurosis generating experience, as implicitly defined by *PA*-13. It will be illuminating to define some additional notions:

'e_0 is *repressed* after t_0' iff
e_0 is in $B(t_0)$ and for all $t > t_0$ e_0 is not in $B(t)$
'being *neurotic* with respect to a_0 after t_0' iff
for all $t > t_0$ there is no e in E such that $REAL(a_0, e, t)$

Now it is easy to prove the following

Theorem: if one has had at t_0 a neurosis generating experience e_0 with respect to impulse a_0, e_0 is repressed after t_0 and one is neurotic with respect to a_0 after t_0.

Note that 'being neurotic with respect to a_0 after t_0' is formally almost in conflict with the main axiom, but the point is that the realization of a_0 required by the main axiom has already taken place at t_0, in particular as a result of a negative experience.

A serious problem in the present formulation is what happens when a_0 recurs in the unconsciousness, because the main axiom requires repeated realization. But here we will not deal with the necessary refinements for this and other reasons. We will just mention one other example of a further refinement of the theory of neurosis, which can be obtained by integrating it with another specialization of the general theory, viz. the theory of sublimation.

We conclude this section with references to some further examples: special relativity theory (Balzer, 1982); Jeffrey's theory of decisions (Stegmüller, 1986); Jakobson's theory of literature (Stegmüller, 1986) Classical and relativistic collision mechanics (Balzer et al., 1987) Lagrangian mechanics (Balzer et al., 1987) Daltonian stoichiometry (Balzer et al., 1987) simple equilibrium thermo-dynamics (Balzer et al., 1987) the Arrow-Debreu theory of individual and collective demand (Janssen and Kuipers, 1989), folk psychology and connection-ism (Bickle, 1993, 1998). Several psychological theories are reconstructed in (Westmeyer, 1989, 1992). The variety and unity of modern genetics is structuralistically reconstrued in (Balzer and Dawe, 1986, 1997). Finally, (Balzer et al., 2000) contains a representative sample of those above, as well as others.

As already mentioned, we have used the structuralist representation a number of times already. Not only in ICR (Chapter 11) about the old quantum theory and capital structure theory, but also we used it formally, for example, when dealing with the ideal gas law, in several places (Chapter 2, Appendix 1, and ICR, Section 10.4). Moreover, it is easy to ascertain that other major examples could have been represented in the structuralist way. For example, the atomic theory (Chapter 1), the kinetic theory, the genetic theory, and Olson's theory of collective goods (Chapter 3). Of course, such a representation may not be necessary for the purposes at hand. However, for detailed questions, such as 'Does a certain theory have empirical content?' such a representation is almost unavoidable. In the next section we will introduce a number refinements of the structuralist approach that enable us to answer such refined questions.

12.4 REFINEMENTS

Now we will continue with some further refinements of the structuralist approach, viz. the distinction between absolute and relative empirical content, the possibilities of determination of the intended applications, relations between theories, theory-nets, and constraints.

12.4.1 *Absolute and Relative Empirical Content*

It is always possible to divide the axioms into, on the one hand analytic (A) and synthetic or substantial (S) axioms and, on the other, non-theoretical (N) and theoretical (T) ones. As a result, there are four types of axioms: NA, TA, NS, TS. We do not mean to suggest that the two distinctions are unproblematic. We have discussed extensively in Chapter 2 and the previous section how the N/T-distinction can be made. The A/S-distinction is at least as notorious, and it is undoubtedly partly a matter of conventional decision where the boundary is drawn. See Niiniluoto (1999, Ch.5) and Sober (2000b) for lucid accounts, against Quine's well-known challenges, of the tenability of this distinction. However, here, and with respect to the N/T-distinction, it is advisable in case of doubt to choose the cautious classifications, i.e. S and T, respectively.

The following survey of sets and names of their elements will now speak for itself.

Mpp	(potential partial models):	NA
Mp	(potential models):	NA + TA
Mpart	(partial models):	NA + NS
M	(models):	NA + TA + NS + TS

It is clear that $\langle Mpp, Mpart \rangle$ is the (theoretical) core of a partial theory, i.e. an unstratified, hence observational theory, constituting a substantial part of the full theory. The empirical content of the full theory was defined as $Mpp - \pi M$, let us call it more specifically the (total or) *absolute empirical content* (AEC). The empirical content of the partial theory, the *partial empirical content* (PEC), is of course *Mpp* – *Mpart*. Given the trivial fact that πM is a subset of *Mpart*, the partial empirical content is automatically a subset of the absolute empirical content. The interesting question is whether the full theory has something to add to the partial theory, i.e. whether the (extra or) *relative empirical content* (REC), defined as *Mpart* – πM, is non-empty. Figure 12.3 depicts the three kinds of content.

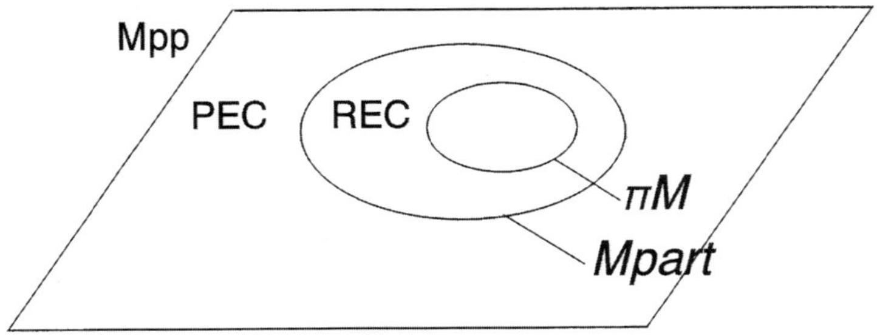

Figure 12.3 Three contents of a theory: AEC = PEC + REC.

It is easy to check that the absolute empirical content (*Mpp* – πM) is the union of the partial empirical content (*Mpp* – *Mpart*) and the relative empirical content (*Mpart* – πM). In consequence, if a theory has relative (and/or partial) empirical content it has absolute empirical content.

Conversely, however, a theory may have absolute empirical content without having relative empirical content, in which case the absolute empirical content coincides with the partial empirical content.

Balzer (1982) claimed that the general psycho-analytic theory has partial, and hence absolute, but no relative empirical content. Stegmüller (1986), however, is able to prove that it also has relative empirical content. Stegmüller then continues with the interesting observations that for this proof it is not necessary to take constraints and/or special laws into consideration and that, as we already noted, classical particle mechanics CPM has only (relative) empirical content when constraints and special laws are taken into consideration. Hence, according to the relative content criterion for empirical impact Freud's theory is in a sense even superior to that of Newton.

But we would like to add that the *CPM* example makes it clear that a generic theory (including constraints) need not have relative empirical content in order to be useful. The important research question is whether a generic theory can be supplemented with special laws (leading to specializations, see below) which have relative empirical content (cf. Bunge, 1977b).

Recall that we remarked in Subsection 12.3.1 that the status of utility theory is to some extent comparable to that of classical particle mechanics. Now we can say that the generic versions of both theories have no empirical content. However, although it is quite clear that the former can be specialized so as to have empirical content, this is not beyond dispute for the latter.

12.4.2 *Intended Applications Reconsidered*

In this subsection we will begin by makinf a few general remarks about the set of intended applications *I*, then we will formulate three different ways of determining *I*, and conclude by elaborating, to a certain extent, the problem of theoretical terms.

I was introduced as '*D* seen through *Mp*' and later revised as '*D* seen through *Mpp*', i.e. *I* represents the intended domain of applications with the conceptual means of *Mpp*. We will restrict our formulations to the refined case, when not otherwise stated, for it includes the extreme case that $Mp = Mpp$.

It is evident that *I* is *Mpp*-dependent, and *Mpp* is manmade. Hence we subscribe to a fundamental form of conceptual relativity. But this need not imply an extreme form of relativism: empirical claims are objectively true or false, for their truth or falsehood depends on nature, assuming that they have empirical content.

In its turn, the objective character of empirical claims does not imply that *D*, *Mpp* (and hence *I*) and *Mp* are fixed beforehand, and that the task remains to formulate a subset *M* of *Mp* leading to a true empirical claim. As a matter of fact, in practice, the determination of *D*, *Mpp*, *Mp* and *M* is a complicated dialectical interaction process, guided by the desire to formulate informative and true empirical claims. Unfortunately, it seems difficult to discern general patterns in this interaction process, without making some important idealizations.

However, if we assume, by idealization, that *Mpp* is fixed, the determination of (*D* and hence) *I* can be governed by at least three different principles.

If we are interested in all relevant nomic possibilities, *I* coincides with the set of nomic possibilities at the *Mpp*-level. Let *To* indicate this subset of *Mpp*. Although we may not have an explicit characterization of *To*, there is a clear empirical criterion for membership: *x* in *Mpp* belongs to $I = To$ iff *x* can be realized. If we are interested in a well-defined subset of the set of *To*, i.e. nomic possibilities satisfying some explicit condition, membership determination is not fundamentally different. In both cases we will speak of the *empirical determination* of *I*.

In this case, the obvious target of theory development is an explicit characterization of *I*, i.e. a set of models *M* is sought for which the strong claim holds: $I = \pi M$, such that *M* may be called *the true (Mp-)theory about I*, or simply, *the (conceptually relative) truth*. In ICR, Chapters 7 and 9, respectively, the formal structure of truth approximation by unstratified and stratified theories is studied in detail.

In ICR (Chapter 7) the notion of a nomic possibility is presented as an absolute qualification, although there may well be cases where it makes good sense to distinguish levels of nomic possibility, e.g. as suggested by the following sequence of 'lower' to 'higher' levels: the physical, chemical, biological, psychological, cultural-socio-economical level. Being nomically possible at a higher level then implies being nomically possible at a lower level, but not the converse. Another example concerns the idea of nomically possible states of an artifact, assuming that it remains intact, which means a severe restriction to its physically possible states, including broken ones. Such refinements can easily be built into the empirical determination of I, as long as the boundaries between the different levels of nomic possibility may be assumed to be sharp.

In many cases, however, the interest is directed to a proper subset of To, of which the membership is not sharply defined. One important way in which I can then have been circumscribed is by so-called *paradigmatic determination*.

Definition: I is *paradigmatically determined* if there are PAR and SIM such that

(1)	I is a subset of To	the intended applications
(2)	PAR is a finite subset of I	the paradigmatic examples
(3)	SIM is a binary relation on Mpp	a similarity relation
(4)	for all x in $I - PAR$ there is y in PAR such that $SIM(x, y)$	

Figure 12.4 depicts the relations between the various sets in the case of paradigmatic determination of I. The elements of PAR may, for instance, be determined by the founding father of the theory and corresponds to one of the meanings Kuhn (1962) had in mind with the term 'paradigms' and which he later called 'exemplars'. Of course, the main source of vagueness is the notion of similarity, for it will not as a rule be possible to define this notion sharply, at least not at the beginning of the research process.

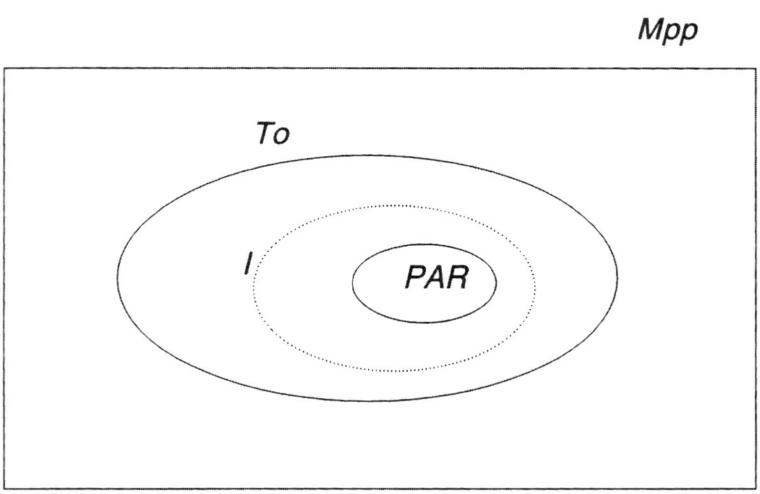

Figure 12.4 Paradigmatic determination of I.

In both cases of determination, assuming that the theory has (at least absolute) empirical content, the empirical claim of a stratified theory will not be trivial. As is easy to verify, the empirical claim becomes trivial in the third way of determination of I, so-called *auto-determination*: for x in To, x belongs to I iff x belongs to πM, i.e. is the projection of a model of a stratified theory. In the case of auto-determination, the theory in question is typically not something to be tested, but it will have been designed for other purposes.

In the case of an unstratified theory, the set of intended applications is of course a subset of the set of nomic possibilities on the *Mp*-level, which is then in its turn a subset of *Mp*. For the further determination of *I* there are again the same three possibilities of empirical, paradigmatic and auto-determination. Where there are proper theoretical terms involved, all three forms of determination result in problems.

Let us briefly restate and elaborate the background of (epistemological) stratification of a theory in *T*-theoretical and *T*-non-theoretical terms. Let there be an unstratified theory $UT = \langle Mp, M, D, I \rangle$ and assume that *UT* has non-empty empirical content $Mp - M$. The term *t* occurring as a component in *Mp* is said to be *T*-theoretical iff every known method of measuring *t* in a specific intended application results in a model of *UT*. It is otherwise *T*-non-theoretical. Let *UT* contain at least one *T*-theoretical term and let us first assume that *I* is supposed to be empirically or paradigmatically determined, in which case the empirical claim '*I* is a subset of *M*' is non-trivial. However, testing this claim is impossible, for it leads demonstrably either to circularity or to an infinite regress. In the case of auto-determination, the problem is that determination of the membership of *M* leads to circularity or infinite regress.

The remedy for these problems is the epistemological stratification of the theory in terms of a partial theory containing only and all *T*-non-theoretical terms. Assuming that the stratified theory has non-empty (absolute) empirical content $Mpp\text{-}\pi M$, the empirical claim or auto-determination is non-trivial, depending on whether *I* has or has not been fixed in advance.

The indicated definition of *T*-theoriticity is a pragmatic one, due to the 'every known method of measuring'-clause and goes back to Sneed (1971). Given the fact that the class of known methods can only increase, the definition is perfectly compatible with the advice to classify a term as *T*-theoretical in case of doubt. However, it is tempting to look for an intrinsic definition of theoriticity. Gähde (1983) has put forward an intrinsic definition. This proposal is not only highly technical and restricted to quantitative terms, it has also been criticized for other reasons (cf. Schurz, 1990). However, Balzer (1996) has rebutted this criticism and has, moreover, proposed an essentially simpler formal criterion than Gähde's is.

12.4.3 *Links Between Theories, and Theory-Nets*

Theories which are roughly about the same domain are frequently related. At each moment they may constitute a network of theories, i.e. an ordered set of theories that are directly or indirectly related. Such a network depicts the synchronic situation, the succession of networks indicates the diachronic development.

Let us first define the main relations between theories, also called (inter-theoretical) links. We will presuppose that all theories considered are stratified, hence the theories are of the form $ST = \langle Mp, Mpp, M, \pi, D, I \rangle$. It is easy to derive from the definitions what links result if the stratification is assumed to disappear ($Mp = Mpp$ and π is the identity function).

We have already indicated some specializations of theories, in the case of the theories of Newton and Freud.

ST^* is *specialization* of ST iff
(1) $Mp^* = Mp$ and $Mpp^* = Mpp$ and $\pi^* = \pi$
(2) M^* is subset of M and
D^* is a subset of D (and hence I^* is a subset of I)
and at least one of the subsets is proper.

Specialization is one of the main research activities within a research program starting from some basic generic theory.

A new theory may add new (non-)theoretical components. It may be a genuine superposition, such that the old theory remains completely intact. Balzer (1982) describes the example of the classical kinematical theory built on the classical space-time theory. When at least one new theoretical component is introduced this type of link is called (conservative) theoretization and can be defined as follows:

ST^* is a *theoretization* of ST iff
(1) all (non-)theoretical components of Mp remain (non-)theoretical components of Mp^*
(2) Mp^* adds to Mp one or more new (non-)theoretical components, at least one theoretical one, all of which are stripped off by the function f from Mp^* onto Mp
(3) for all x^* in $M^* f(x^*)$ belongs to M

Of course, it is possible to define non-conservative links between theories when new components are added.

The third important link between theories is that of reduction.

ST is *reducible* to ST^* iff there is a relation r on $Mp \times Mp^*$ such that
(1) for all x in M there is x^* in M^* such that $r(x, x^*)$
(2) if $r(x, x^*)$ and x^* in M^* then x in M
(3) for all y in I there is y^* in I^* such that $r_\pi(y, y^*)$,
where r_p indicates the restriction of r to $Mpp \times Mpp^*$

The relation of this definition to the definition of reduction by a theory in Chapter 3 is not straightforward, but plausible. Roughly speaking, the present definition captures the explanation of one theory by another when the models and intended applications can be formally related in a way which is typically possible when one or more of the three basic types of reduction distinguished in Chapter 3 apply.

The last type of link between theories to be mentioned is that of (idealization, or conversely) *concretization*, of which a precise definition has been given in ICR (Section 10.4). Recall that concretization played a crucial role in the truth approximation analysis of the transition of the theory of ideal gases to that of Van der Waals, and various transitions in the old quantum theory and capital structure theory.

It is not difficult to check that all defined links generate partial orderings and we assume that all links to be considered are of this type. Let two theories be called directly related when one such link applies and let them be called related when there is a chain of directly related theories between them. Of course, 'being related' is again a partial ordering. A *theory-net* is defined as a set of theories related in this way such that there is a *basic theory* Tb, in the sense that all other theories are directly or indirectly related to this theory. Figure 12.5 depicts such a theory-net.

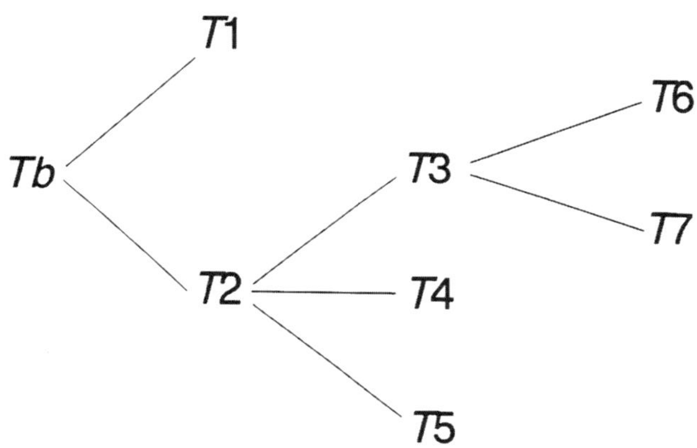

Figure 12.5 A theory-net.

For the succession of theory-nets it is plausible to distinguish two basic types, leaving room for mixed cases. A transition to a new net may be *conservative* in the sense that the new net retains all theories of the old one, but one or more theories to the old net are added (inserted). A transition is called *corrective* if one or more theories in the old net are replaced by new theories which are considered to be improvements.

12.4.4 *Constraints*

We have already referred several times to so-called constraints. Whereas laws and axioms in the normal sense lay down restrictions on individual potential models, constraints impose restrictions on sets of potential models. A particular type of constraint is a so-called identity-constraint, guaranteeing that a function assigns in different potential models, with some common base-sets, the same value to the same individuals. The weight-function in the case of the slide balance as well as the mass-function in the case of classical particle mechanics are cases in point.

A constraint can be formally defined in a very general way.

Definition: C is a *constraint* on the set S iff
(1) C is a set of subsets of S
(2) the union of the sets in C exhausts S ($\cup C = S$)
(3) if X is in C and Y is a subset of X then Y is in C subset-preservation)

It is easy to prove that all singleton sets $\{x\}$, for x in S, belong to C, hence a constraint does not exclude any individual potential model.

Let us now first concentrate on the typical role of a constraint C on Mp in a stratified theory $ST = \langle Mp, Mpp, M, C, \pi, D, I \rangle$. The standard empirical claim was 'I is subset of πM', which could be paraphrased by saying that all members of I can be extended with theoretical components to genuine models, i.e. there is a subset X of M such that $pX = I$. Taking the constraint into consideration this claim is strengthened: there is a subset X of M *belonging to* C such that $\pi X = I$. Hence, now both M and C restrict the degrees of freedom for the supplementation of theoretical components. In the corresponding versions of the strong claim the clause 'X is a subset of M' is simply replaced by '$X = M$'. It is clear that a stratified theory may even be a pure *constraint-theory*, in the sense that C is non-trivial and M is trivial, i.e. $M = Mp$.

In ICR we deal with truth approximation by theories by assuming that theories are sets of structures, with or without the distinction between theoretical and non-theoretical terms. It is not too difficult to check that, when the truth is a constraint-theory, similar (basic and refined) definitions may be given of the claim that one constraint-theory may be closer to the truth than another. Of course, it is then also possible to deal with truth approximation by theories consisting of a 'normal' and a constraint part. However, we will leave the elaboration of the suggested possibilities to the reader.

Taking a constraint into account, the following reformulation of the standard claim is instructive. Let $A(ST)$, the *application space* of ST, be defined as the set of projections of all subsets of M satisfying C (formally: $A(ST) = \pi(P(M) \cap C)$). The standard claim now comes down to: I is in $A(ST)$.

It is also now plausible to define the *(absolute) empirical content* $AEC(ST)$ as $P(Mpp) - A(ST)$, i.e. the subsets of Mpp which are excluded by M and C. Note that $AEC(ST)$ reduces to $p(Mpp) - \pi(P(M))$ when C is trivial, i.e. when $C = P(M)$, and to $P(Mp) - P(M)$ when π is trivial, i.e. when π is the identity function. It is easy to verify that $AEC(ST)$ is empty in these respective cases iff the originally defined empirical contents $Mpp - \pi M$ and $Mp - M$ are empty. Hence, the suggested new definitions of empirical content reproduce the original ones on the level of sets of sets of potential (partial) models.

Similar relations hold for the plausible definition of the *relative empirical content* $REC(ST)$: $P(Mpart) - A(ST)$. And again it follows almost trivially that non-empty $REC(ST)$ implies non-empty $AEC(ST)$, but not the converse.

Constraints also make sense in other cases, e.g. in partial theories and in unstratified theories. For the first case, let $Cpart$ be a constraint on Mpp. Like $Mpart$, $Cpart$ represents empirical restrictions. We may say that $Mpart$ captures the standard observational laws, whereas *Cpart* captures *constraint observational laws*. Of course, the associated claim states that *I* is a subset of *Mpart* belonging to *Cpart*. It is important to note that many empirical laws are constraint laws, or mixtures of standard and constraint laws.

If the indicated partial theory with constraint is isolated from the full theory, it is clear that we have an unstratified theory with constraint. In general, an unstratified theory with constraint is of course of the form $\langle Mp, M, C, D, I\rangle$, where C is a constraint on *Mp*.

CONCLUDING REMARKS

In this chapter we have dealt with empirical theories, but let us finally consider the question of which structuralist concepts are also useful for non-empirical theories. Following Popper, non-empirical theories are by definition theories which are not (intended to be) falsifiable. One may distinguish at least four types of non-empirical theories:

metaphysical theories are supposed to make claims about reality without assuming any particular conceptualization or, equivalently, they make claims generalizing over conceivable conceptualizations of reality,
mathematical and logical theories deal with defined abstract objects, i.e. mental constructs, e.g. the theory of groups,
conceptual theories concern ways of looking (perspectives) at a certain domain,
normative theories deal with what is (supposed to be) ethically, legally, aesthetically (in)admissible.

It is evident that almost all technical ingredients presented for empirical theories are also useful for non-empirical theories. In fact, Suppes (1957) invented the structuralist representation of empirical theories by transferring, as far as possible, the standard way of presenting mathematical theories, such as the theory of groups, to empirical theories. The crucial difference is that non-empirical theories do not make general empirical claims. The claims which are associated with them typically are either conceptual (logical, mathematical etc.) or restricted to individual intended applications. A typical claim around a mathematical theory is a mathematical theorem to the effect that the models of the theory can be proven to have a certain explicitly defined property. A typical claim of a specific conceptual theory is that a certain intended application is (or is not) a model of that special theory. Of course, generic theories, i.e. theories with vacuous empirical claims, are conceptual theories.

The ‘structuralist theory of the structures of empirical theories’ is a perfect example of a theory which is primarily intended as a conceptual theory (although one may strengthen it to a genuine empirical theory by adding a substantial claim). As a consequence, the foregoing exposition not only provides an elaborate example of a conceptual theory, it can also convince the reader of the usefulness of conceptual theories.

CHAPTER 13

'DEFAULT-NORMS' IN RESEARCH ETHICS

INTRODUCTION

Science and ethics have a complex relation. They cannot easily be separated. Although pleas for completely autonomous science as well as for extreme ethical dirigism still occur, an interaction model for science and ethics is more and more widely propagated. This model is not only presented as characteristic of the factual relation, but also as desirable, that is, in the interest of science and society, of course, under rationally defensible conditions. It concerns in particular moral dilemmas with respect to:

- Incorruptible professional behavior in scientific research.
- The present and future well-being of humans and (laboratory, domestic, and wild) animals and the state of their environment.
- The communication between science and society.

In the last twenty years all kinds of initiatives have been taken in order to shape the interaction between science and ethics on three levels. On the individual level, universities organize courses for students and post-graduates in an effort to activate their moral sensibility. On the institutional level codes of the professional behavior of scientists and codes governing the behavior of the scientific and non-scientific staff in private firms have been designed, suggesting or prescribing rules of behavior. On the societal level, public debates are organized between laity and experts about possible and desirable scientific and social developments.

In this final chapter we like to contribute to the discussion about the integrity of scientists as researchers, that is, about the possibility and limitations of incorruptible professional behavior in scientific research. For this purpose we will take the famous Merton norms as a point of departure, not in order to note that scientists do not always conform to them,[1] but to raise questions about which deviations may be defensible and which ones are not. Accidentally, this inquiry will also illustrate the fact that a choice between so-called deontological and utilitarian ethics is not plausible: a reasonable compromise between principles-whatever-the-consequences and deviations based on possible consequences seems also in this case the only serious option. After presenting the four norms in Section 13.1, we will briefly consider three of them as 'default-norms'. In Section 13.2 we will discuss the disinterestedness norm in some more detail. We will conclude with a brief discussion of the possibility and usefulness of subsuming all these concerns in a general professional code for scientific researchers.

13.1 MERTON'S NORMS CONCEIVED AS 'DEFAULT-NORMS'

As is well-known, Robert K. Merton (1942) formulated in the forties four norms which scientists, ideally speaking, should live up to and which have become known as the CUDOS-norms, after the initials of their names: Communism, Universalism, Disinterestedness, and Organized Skepticism. We will start by briefly characterizing them:

'*Communism*': scientific knowledge is the product of common effort, for which reason newly acquired knowledge should be made public and should be considered as a collective good, with explicit recognition, of course, of the discoverer or inventor.
Universalism: one should judge the work of others irrespective of persons ('blind refereeing') and in the selection among applicants for a position sex, race, nationality and the like should not play a role.
Disinterestedness: personal and group interests should be subordinated to the interests of research, which excludes evident kinds of fraud, such as plagiarism, forgery, and feigning.
Organized skepticism: new research results and methods should be presented to colleagues, with an open mind for criticism.

It will be clear that these norms need not be exhaustive and that they will overlap now and then. Moreover, although Merton was of the opinion that the norms were in the interest of science, he was also well aware that they did not have rigorous *des*criptive validity. We will argue that they even cannot have rigorous *pre*scriptive validity, because there are, besides evidently non-defensible deviations, various kinds of conceivably well-argued deviations. Moreover, and at least as important, there appears to be a large 'gray area' where every researcher has to find his own way in every particular case. Precisely due to the existence of this gray area it is difficult to conceive of a detailed prescriptive professional code.

Merton's norms are clearly principles with a deontological flavor in the sense of deontological ethics: whatever the consequences in specific cases, they prescribe what to do or not to do. Defensible deviations are, of course, based on utilitarian considerations pertaining to the consequences of a dogmatic application of the norms in concrete cases. In other words, these norms are 'default-norms'. Depending on whether or not the deviations are supposed to be catchable in subrules, default-norms have, besides their deontological reminiscence, also the flavor of so-called rule-utilitarianism or act-utilitarianism. In the subsequent sections we will try, for each norm, to explore the possibilities for defensible and non-defensible deviations, though without presenting them in each case explicitly in terms of their consequences and without investigating whether or not defensible deviations can be caught in subrules. For 'disinterestedness' we start a new section.

13.1.1 *'Communism'*

> *Scientific knowledge is the product of common effort, for which reason newly acquired knowledge should be made public and should be considered as a collective good, with explicit recognition, of course, of the discoverer or inventor.*

Merton intended with his 'communism' norm primarily that new scientific knowledge, as soon as it is sufficiently trustworthy, should be made public, not only for colleague researchers, but also for society.

Secrecy and Publicity

This claim raises, first of all, the question of whether there can be good reasons for secrecy. Kaiser (1996) distinguishes three kinds of interest that might be considered as justifying secrecy: military, commercial, and public interest. Reservedness about new findings that might be used on a large scale for military purposes can at least count on good arguments of a utilitarian nature. The case for secrecy is less convincing when only commercial interests are involved, and commercial and contract research is a rapidly growing area of concern in this respect. Temporary secrecy, for example, waiting for the approval of a patent request, may be defensible. Permanent secrecy of a scientifically interesting finding, however, does not seem to be ever justifiable. We may hope that recipes which are cautiously kept secret, such as the one for Coca-Cola, are not scientifically interesting.

That public interest does not easily justify secrecy may be illustrated by what happened in the early seventies in the Norwegian industry of salmon fish farming, an example extensively discussed by Kaiser and here restricted to the crucial facts. A team of scientific researchers advising the salmon farming industry kept one of their discoveries as a secret of the team at a great risk to public health. They had discovered that a generally used food supplement for the salmon was carcinogenic for the human consumers of the salmon, despite initial good reasons to assume that this could not be the case. The team faced the dilemma between a general health interest and a general employment interest; the latter was especially important since so many people were directly or indirectly involved in the salmon industry. The team leader decided to let the employment interest overrule the health interest. More precisely, he decided that first a functionally equivalent replacement should be found or developed for the food supplement before use of the health endangering one would be discontinued. As Kaiser (1996, p. 220) reports, the leader's message to his team was: 'I urge you to keep this strictly secret, until we have developed for substance X a replacement that is more desirable. Just imagine how this finding would stir up the public if it becomes known.' This consideration was apparently convincing to the other members of the team; that is, at least no 'whistle blowers' appeared. The new supplement was found and the old one replaced without public uproar. Despite our possible inclination to want to believe that such things could not happen anymore in the twenty-first century, the example suggests that more or less comparable cases could still occur.

Timely Publication

It is clear that in most cases there are no good reasons for secrecy. Nevertheless, publication of new findings at the proper time and in the proper way is a matter of great concern. Regarding the proper manner of publication, Solomon (1996) stresses that, in case of experimental results, enough should be specified in order to enable the repetition of the experiment by others and that, in relevant cases, data, samples and, we add, computer programs should be made available upon request.

Regarding timely publication, it is important to distinguish between the informal communication network of friendly colleagues, locally and otherwise, the official scientific publication circuit and, finally, the media for the general public. The problematic cases of untimely publication concern primarily the general media. Notorious examples in the 20th Century history of science are N-rays (around 1900), polywater (late sixties) and, more recently, cold nuclear fusion (Ponns and Fleischmann), memory molecules (Benveniste) and, quite a row at least in The Netherlands, an Aids-inhibiting substance (Buck and Goudsmit). It is important to note that untimely publication does not cause debates when it happens to become a success story. The first messages about superconductivity, for example, came also in the newspapers, with speculations about possible applications, before the phenomenon was considered to be well-established by the community of physicists.

Untimely publication is stimulated, among other factors, by interest in being the first, and to be recognized as such, by the aim of obtaining as soon as possible new sources of money or other means for further research, by the sincere belief that the discovery renders society a good turn, by vanity even to the extent of having Nobel prize expectations and, last but not least, by dubious science journalism. The most important factor preventing untimely publication is likely to be the potential loss of face among colleagues when things go wrong. For the rest, standard checkpoints like the referee system, repetition of experiments and proof checks by oneself and others, prevent new results from being officially published in a too early stage. For the general media, adequate science journalism is a prerequisite. Good science news reporting prevents, first of all, the expectations of many vulnerable groups from being unjustifiedly raised. Not only Buck and Goudsmit gave false hope to Aids-patients but the Dutch public television news was also responsible for that. Moreover, in general, it would be a good thing if not only scientists but also science reporters and the media that publish their stories were to lose credibility whenever a lame-duck discovery is reported with great fanfare.

13.1.2 *Universalism*

One should judge the work of others irrespective of persons ('blind refereeing') and in the selection among applicants for a position sex, race, nationality and the like should not play a role.

The use of authority arguments is strictly speaking a deviation from the universalism norm because somebody's prestige is brought into play. But in everyday and scientific practice we cannot live without such arguments. Consider, for example, the mutual trust one needs to have for doing efficient interdisciplinary research. However, we are inclined to make a difference between defensible and non-defensible authority arguments, for example, depending on whether or not the scientist in question makes a statement within the domain of his recognized competence. Well-known examples of problematic transgressions concern genuine and near Nobel prize winners making or supporting claims in a field in which they are incompetent, and, frequently, cannot be contradicted by real experts, as there are none in the area concerned. In the latter kinds of areas, e.g. regarding certain long term future developments about which everyone is of course free to speculate, utterances by laureates are taken more seriously than those made by other, equally incompetent, people.

Another type of problematic deviation from the universalism norm concerns all kinds of (supposed) discrimination, in particular, of women and ethnic minorities. The policies of 'positive discrimination' or 'positive action' are considered by its proponents as defensible in the face of the consequences of the normal 'laissez faire' policy. One may compare it with forbidding political parties that are against democracy, in order to protect democracy against attacks from inside. In the US, however, positive action in favor of minorities is in several states again being retracted. The same development seems to be taking place in The Netherlands regarding extreme forms of positive discrimination in favor of women, such as appointing a sufficiently qualified female candidate on an academic position in the presence of (much) better qualified male candidates. Nevertheless, it has to be noted that, at least in The Netherlands, it turns out to be very difficult to get a reasonable number of women on higher academic positions without positive action. Besides the admission of hidden differences in treatment of female candidates and male-biased limitations to part-time work, flexible-time work and telework, the notion that respective differences in ambition also play a role seems to be gaining ground. For women with appropriate ambitions, however, besides fighting against all kinds of unequal treatment and male-biased limitations, new kinds of networks and mentorship may need to be found.

13.1.3 *Organized Skepticism*

New research results and methods should be presented to colleagues, with an open mind for criticism.

Probably the best known relativization of the idea of (organized) rigorous skepticism comes from the historical findings of Kuhn and Lakatos, who concluded that some dogmatism has been rather productive in the history of science. However different their elaborations, they showed that persistent tenacity in supporting a paradigm or research program should not always be considered as dubious dogmatic behavior, but may also be considered as evidence for fruitful perseverance, seemingly against better knowledge. 'The function of dogma in science' is the telling title of an article by Kuhn. Of course, not all dogmatic research becomes respectable in this way. Within the boundaries, that is, the dogmas, of a research program, one may or may not systematically aim at empirical progress and even truth approximation. The presence of this striving is a characteristic difference between examples of fruitful 'dogmatic science' and static pseudoscience. For example, the attempts and debates aiming at improving the specifications of the evolutionary theory have no serious analogue in 'creation science' (see e.g. Sober, 2000a). For a general elaboration of the claimed contrast, see Chapter 8.

This point is associated with a latent, very productive, division of labor among scientists. Besides the manifest division of labor between theoreticians and experimenters, without, of course, neglecting the double talented exceptions, there seems to be another division operative that is less well known. In particular, among theoreticians there are not only many constructive researchers, who build on their own research program or, more frequently, on the work of others, but also researchers whose strength it is to play a critical role. One may even question whether the current ideal type of a researcher, who is alternately constructive and destructive, is the proper ideal type for modern science. Precisely because science is a co-production of many people, it is more plausible that a functional division of labor is more fruitful than a homogeneous group of researchers, alternately playing the roles that have to be fulfilled. Instead of a deviation of Merton's fourth norm, we come here across a possibility for realizing it, of which Merton himself did not seem to think.

13.2 DISINTERESTEDNESS, AND ITS CHALLENGES

Personal and group interests should be subordinated to the interests of research, which excludes evident kinds of fraud, such as plagiarism, forgery, and feigning.

Regarding offenses of the disinterestedness norm, one thinks in the first place of evident cases of fraud, such as plagiarism, forging of data, feigning of experiments and their results. One frequently reads that, happily enough, such cases of fraud are rather exceptional. The classic *The betrayers of truth* (Broad and Wade, 1982) and the Dutch variant *Valse vooruitgang* (*False progress*, Van Kolfschoten, 1993) confirm this belief. Nevertheless, it is worthwhile to pay attention to the possibility of such occurrences in academic education, because sooner or later one may be confronted, directly or indirectly, with such a case, and one will be forced to have an opinion about it. Moreover, there are all kinds of semi-fraud, in particular semi-plagiarism. Van Kolfschoten distinguishes three categories: conscious plagiarists, unconscious plagiarists, and synthesizers who give poor acknowledgements. Unconscious, and hence in good faith, plagiarism not only occurs, but perhaps even more frequently than one is inclined to think. One way to minimize the risk is by answering the question to oneself, after finishing a manuscript, whether some of the ideas in it may have been taken from others, learnt by reading or conference visiting, in particular in the period before actually working on the manuscript. Finally, when conceiving and writing a synthetic work, it is extremely difficult to indicate all the sources that may have played a role, let alone, to specify that role. Apologizing in advance for this possibility, and thus relativizing ones own apparent originality, may give some comfort.

In assessing manuscripts, e.g. when writing referee reports, in editorial work, in advising or working for a publisher, all kinds of variants of fraud and semi-fraud occur (see Lafollette, 1992). Referees may be tempted to lie in a referee report or, at least, to manipulate information, they may postpone the completion of a report unnecessarily long in view of some personal interest, and they can even steal ideas or even fragments from the manuscript. Members of editorial boards and publishers may feign or forge referee reports and they can lie about the referee process. All this occurs, probably not on a large scale, but there are variants, which are worthy of some further attention to.

We will restrict ourselves to the gray area of strategic behavior aiming at the allocation of personnel and material means: about influencing the number of publications and citations, about writing research proposals and their assessment and, finally, about the formation and extension of networks. In contrast to evident cases of fraud, one frequently encounters behavior that is difficult to classify as objectionable or not. In these cases everybody will use his own standards, with the risk of being judged in a specific case by others, using other standards.

In popular empirical science research (Latour (1987), Woolgar (1988), and others) it is not only suggested, rightly, that strategic behavior plays a large role in circles of scientific research, but it is frequently presented in a way which unconditionally sanctions and stimulates it, although at ones own risk. Such convictions even grant scientists a license to luxuriate in their own power. To be sure, scientists not only aim at cognitive goals like empirical success or even the truth of their theories, but they also have social aims like recognition and power, and hence means to reach such aims. And although these goals frequently strengthen each other, such convergences by no means implies that the conscious pursuit of these social goals is good for science. With regard to this question, we will consider five examples of strategic deviations from the norm of disinterestedness.

Number of Publications

The average number of published articles and books is an important measure for assessing individual scientists and groups of scientists. 'Publish or perish' became a well-known saying. Originally, only the pure number of publications was counted, but gradually one started to differentiate between scientific or scholarly publications and 'professional publications', that is, publications for non-experts. Whereas, at least in The Netherlands, professional publications are almost neglected, a practice which is certainly debatable in some areas of the social sciences and the humanities, scientific publications are divided into categories with different weights. Usually, but not always, a book is higher rated than an article, a local, e.g. Dutch, publication lower than an international one. Moreover, one distinguishes between a conference abstract, a paper in conference proceedings, a chapter in an edited book and an article in a blindly refereed international journal.

However this may be, it is frequently easy to increase the number of publications, a practice which need not be solely for the purpose of padding one's curriculum vitae. Of course, this is almost excluded in top journals. However, in other contexts it is plausible to modify, combine and extend earlier work. For example, invitations are frequently extended to established scientists to deliver specific contributions based on earlier work. In addition, translations frequently originate in this way. Another possibility for increasing the number of publications, in particular in the natural and medical sciences, is by being a co-author, where one and the same article is frequently counted as one publication for each co-author.

Most of the suggested types of publications are as such unproblematic and even useful for spreading ones ideas and findings. The only problem is that others can interpret them as attempts to multiply the number of publications, leading to a feeling of guilt. However, nobody accuses scientists for giving two different versions of essentially the same talk for totally different audiences. The same should be the case for versions of the same publication in different media. One way to diminish the feeling of guilt of multiplication behavior in the latter case is by supplementing the usual, at least in The Netherlands, minimum norm of about two scientific publications as the principal author per year with a kind of maximum of, say, three. That is, above three scientific publications, all the other kinds of publications should also be taken into account. Such a practice would leave the primary publication task of scientists untouched, but provide substantial room for other kinds of publications. Incidentally, one may doubt whether such norms should be used as requirements for individual scientists. In the first place, it is not so clear that one should prefer three scientific publications per year above one very influential publication in three years. This uncertainty makes the norm of at least two scientific publications on average per year and per group member much more appropriate than requiring such production from each researcher separately. Regarding other publications, a functional division of labor within a group may be very useful, depending on everybody's strong and weak points. Recall that we have already met other kinds of functional divisions of labor in the context of organized skepticism.

Number of Citations

Citations may of course be problematic, in the first place because they can be misleading or even outright fraudulent, but this is not our topic here. The increased importance of citations is frequently summarized by a variant of the 'publish or perish' slogan: 'Be cited or perish'. Many publications are seldom read, let alone quoted. Publications frequently have another function. In the first place one may want to reach primarily the group of kindred colleagues by way of the publication itself or by its reprints, in order to get comments that may be useful for a version intended for a renown journal. This is of course also possible with manuscripts, preprints and, recently, in an electronic way. However, in the case of published materials the effective potential readership for a publication is usually somewhat larger, in particular due to a greater possibility of unintended readers. A publication, in effect, is placed into one of many reading circuits, which can be classified to the extent that they are centered on authoritative journals, authoritative and popular authors, journals paying in particular attention to certain new developments and, finally, a frequent consequence of this, research program bound journals. Strategic citation behavior is partly related to this variety of circuits.

More important for our present purposes is it to note that it is more effective to quote an authoritative author or journal than a further unknown researcher who published the idea earlier in a rather unknown journal. The fact of the matter is that the average reader is, consciously or unconsciously, sensitive to authority arguments. The result is the Matthew-effect: who is frequently cited, will be cited still more. Such citation popularity may of course be legitimate or illegitimate, that is, authors may become popular for good and for bad reasons. One is also inclined, when one has a choice, to quote from a more recent publication than from an older one; otherwise it may appear that one does not read recent publications. For both strategies, there may be some justification, but two other strategies are not justifiable: 'one good turn deserves another', and 'don't quote countrymen or (other) scientific opponents'. The strong focus on American authors and journals, not only by many American researchers, but also by many European ones, is in general already a problem, but particularly insofar as citation behavior is concerned.

It is clear that much of the indicated citation behavior forces one to relativize the use of numbers of citations as a measure of quality or impact, the more so because there are considerable differences in the degree of strategic citation behavior among scientists. Only self-citations and 'in-group' citations are easily removed, all other problematic citations are difficult to exclude by some clear criterion. A better alternative for measuring impact seems to us to be the following: let researchers present from time to time a verifiable survey of the substantial use of their own publications by others. To be sure, citation indexes remain very important for this purpose, and should preferably be as complete as possible.

Writing Research Proposals

The writing of research proposals has become an important part of the work of scientific researchers. It may concern application for special university funds, for special national funds, and for contract research paid by governments, profit and non-profit organizations. Henk Withaar (1979) argued that, besides the so-called Context of Discovery and Context of Justification, there should be distinguished a Context of Persuasion in which research plans are set up and defended. However this may be, an accepted research proposal is seldom realized precisely in the way in which it was proposed. Three factors are important in this respect: the difficulty to plan research, exaggerated optimism of researchers, and misleading formulations in the proposal. It is generally acknowledged that it is difficult to plan research and that it is wise to change plans when other routes seem better. Moreover, as we noted already in Chapter 6, researchers usually are rather optimistic about what they can achieve in a given amount of time. This shortage of realism is not problematic by definition. On the contrary, if researchers would know in the advance the actual amount of time that it will take to complete some project, they would frequently lose their motivation. Finally, misleading presentation is the third and here most important reason why research plans are not realized in detail.

In research proposals one can frequently read about the likely social relevance and the relevance for other disciplines. Moreover, frequently a strong relation is claimed with an area that is announced as a focus for the relevant research fund or with some recent development that received much media attention. Also claimed connections with the work of likely reviewers of the proposal is sometimes impressive. Experienced reviewers of research proposals are usually alert for such kinds of inflationary statements. However, since they are mostly but not completely nonsensical, it is difficult to evaluate such expositions. Moreover, the future results of the research may well be worthwhile, despite the fact that many of the suggested expectations are not fulfilled. As far as misrepresentation is really problematic, it usually has, in contrast to fraud, no further negative consequences for science. Nevertheless, claims to social relevance, may only be pertinent when they are one of the main motives for the research. Misleading claims that come close to being fraudulent seem to occur most frequently in requests for continued support of research. In the sensational book written by Köbben and Tromp (1999), some revealing Dutch examples of this nature are presented. The temptations are considerable for the research manager of an independent research institute who is also an entrepreneur and who is trying to serve the continuity of the institute. Moreover, the employees of such an institute use to have much to lose and not much to gain by whistle blowing. However, at least in The Netherlands, the conditions for whistle blowers have recently been improved.

There is a totally different way in which research proposals are sometimes not executed after their acceptance; simply because the research has already been done before, and with success. The background of such behavior is the following. Research funds frequently ask for clear research plans with clear methods and phases and expected results. Scientists, however, may want to focus on new topics for which it is impossible to provide the required clarity. Hence, one way to beat the system in such a case, is to propose to do research that has already been successfully performed, but not yet published. The proposal can then be written with hindsight on the way it could in fact be realized and, if it satisfies the relevant referees and committees, the research grant is in fact used for the really new, but vague research plans, with the safe perspective that the promised results can be reported. Of course, we leave it to the reader to judge to what extent this type of behavior is defensible.

Assessment of Research Proposals

The assessment of research proposals varies strongly with the type of funding. In the case of university and governmental funds (e.g. NWO in The Netherlands, DFG in Germany, NSF in the USA) usually first a few referees are appointed to write a report of varying length. Then a committee or board prepares and/or takes the main decisions: whether the proposal is subsidizable and, if so, which degree of priority it is assigned. Partly on the basis of our own experience with NWO, it is our impression that the Matthew-effect also occurs here, perhaps to a more limited extent: who has great means will get still more, who has only little means will get less. To be sure, this effect is to some extent justified, but there is also likely a point where it becomes problematic. Implicitly or explicitly a number of factors play a role here, apart from the merits of the proposal itself. The reputation, according to the assessors and the committee, of the researcher or research leader or research team definitely plays a role. When a proposal is, as such, not of the highest quality, reputation may increase the appreciation. Moreover, the value assigned to the number of publications and citations, the so-called 'past performance', tells something, but not very much about the value of future research. Last but not least, sometimes one gets the impression that administratively active researchers, despite systematic exclusion from discussion about their own proposals, are somewhat more successful with their proposals than researchers who do not play an active role in the assessment circuits. Be it, because they are supposed not to be able to perform such a role, or because they feel themselves above spending time on such things.

Networks

Participation in administrative and assessment circuits is one of the ways in which one can build a network that is functional for one's striving at truth, recognition and power. However, building other types of networks, with researchers in the country and abroad and with people potentially interested in applications of research findings, is of course at least as effective. In all cases, the personal and didactic qualities play a role that is at least as important as the purely scientific qualities. For instance, the early 20th-Century philosopher Bolland from the University of Leiden could present, in a very charming way, views which are now not only considered to be nonsensical, but nobody can understand how they could be taken seriously by so many scholars and laymen at the time. The fact that he was taken seriously suggests that, in philosophy and other disciplines, there may always be new Bollands around, which have not yet been identified as producers of humbug. The problem is, of course, that there are also several exemplary researchers who are able to present their views in an excellent and charming way and so build a useful network. Hence, it is only important to always be alert to the fact that somebody may sell ideas which are not as yet recognized as humbug, and build further on a network to promote these ideas.

CONCLUDING REMARKS

It will be clear that Merton's norms are not principles from which one should deviate under no circumstances. However, they may well function as 'default-norms'. That is, norms of which deviations have to be specifically motivated in terms of the expected differences between the consequences when they are followed and when they are violated. Deviations, which are difficult to motivate in this way, may be considered as problematic. Our claim is not to have given a complete and thorough survey of defensible and indefensible deviations. However, in particular, we aimed to show the existence of gray areas.

Despite these areas, it makes sense to address the question of how Merton's norms may optimally function as default-norms. Sensibility for gray areas needs, of course, to be stimulated, and an informed judgements should be made in such cases. The desirability of this type of norm consolidation will be the subject of the final paragraphs.

In general, consolidation of norms is primarily a matter of socialization in the relevant community, supplemented with explicit attention to the awakening of the individual's awareness of these norms. However, rules for behavior will only function properly if the relevant reward system does not itself promote problematic deviations. As far as the assessment system of researchers can be characterized truthfully by slogans like 'publish or perish' and 'be cited or perish', the relevant scientific bureaucracy is to our opinion co-responsible for the increase in dubious publication and citation behavior. When discussing, in Subsection 13.1.1, deviations from the subnorm of timely publication, a number of factors preventing untimely publication were mentioned: the threat of loss of face, the need of repeating experiments and checking proofs, the referee system and responsible science journalism. These and similar factors function of course also to maintain the other norms.

The ideal way to internalize the norms, however, requires that the aims of the knowledge be legitimate. The more legitimate the research goals, the easier is the self-imposing of the norms, at least as far as one subscribes to that legitimization and believes that the norms are functional for these goals. In terms of knowledge goals, one may think in the first place of the search for truth, or at least for empirically successful approaches. From Part 4 it is immediately clear that, for example, fabrications and fraud will retard genuine confirmation and empirical progress. Moreover, from ICR it then follows that truth approximation is retarder as well. We also like to quote Ilkka Niiniluoto (1999, p. 300) on this subject:

> 'But it is important to realize that many norms in science, in both ethics (e.g. rules against fabrication of data and fraud) and methodology (e.g. rules for testing statistical hypotheses), are *truth-centred*: they are designed to guarantee the reliability and self-corrective nature of scientific inquiry. In other words, these social norms are epistemologically, not only sociologically, highly relevant and interesting.'

However, scientific, technological and societal relevance may also function as self-chosen supplementary signposts, including sensitivity to societal problems originating from the application of scientific findings, e.g. several environmental problems. As already stated, the more one subscribes to such goals in the interest of science and society and the more one believes that the norms are functional for these goals, the more easily will one conform to these norms.

The question remains whether this 'ideal approach' is sufficient. Let us have no illusions about the possibility of totally excluding evident kinds of fraud. Like in any other social sector, there will always be a number of more of less pathological individuals who insist on going wrong. A more serious problem is the demarcation of the gray areas between clearly defensible and indefensible deviations. In particular, it may be worthwhile to consider a general code of conduct for incorruptible scientific research, with a clear statement of its status, and covering such 'gray-area' practices. To be sure, the subject of incorruptible research is usually discussed in the various codes of conduct for specific groups of researchers, such as chemists and psychologists. However, as far as we know, there are no initiatives for formulating a general scientific code. Such a general code should be distinguished from general brochures that try to make scientists sensitive to ethical issues. The first example of this nature probably was *On Being a Scientist: Responsible Conduct in Research* of the National Academy of Sciences in the USA, published in 1989 (NAS, 1989). Other countries followed sooner or later. For instance, the Royal Dutch Academy of Sciences (KNAW) published in 2000 the booklet (KNAW, 2000) *Wetenschappelijk onderzoek: dilemma's en verleidingen* (*Scientific research: dilemmas and temptations*). In contrast to such valuable brochures, the still incomplete draft resulting from the unsuccessful attempts of the ICSU (the International Committee of Scientific Unions) to install a 'Hippocratic oath for scientists', suggests the kind of general code indicated above, but it focusses on social consequences of science and almost entirely neglects the matter of incorruptible research.[2] According to the most recent Web-information (aaas.org/news/oath, 25–01–01), the same applies to the debate within the AAAS (American Association for the Advancement of Science) about a possible oath. However, we hope to have made it clear that it makes much sense to consider the possibility of a general code of incorruptible conduct in research and to include it perhaps in an encompassing oath. In particular, the existence of gray areas asks for general recognition. They do not seem to differ very much among various disciplines, but the attitude to them differs very much among scientists. Hence, these gray areas will have to be addressed with much prudence, which almost excludes a strictly prescriptive code. It is telling that the 'Office of Scientific

Integrity', founded in the USA in 1989, was transformed already in 1992 into the 'Office of Research Integrity', with weakened terms: investigations of accusations of undesired behavior were no longer to be followed by ethical judgements (KNAW, 2000, p. 14). Certainly without disposing of a general code, taking gray areas into account, this was a highly plausible development. However, even though such a code would remain very useful for such offices, it may have too much and too strong undesirable effects on scientists. In this respect it is interesting to compare it with 'ethical review' procedures for research proposals. Pettit (1992) argues that such procedures endanger valuable research on human beings. Without precautionary measures, 'it is likely to carry us along a degenerating trajectory', avoiding all kinds of important research which might lead to ethical blockades. Hence, the question is whether a general code is possible that is not the start of a degenerating trajectory but a useful new point of reference in the interest of science and society.

SUGGESTIONS FOR FURTHER READING

The suggestions for further reading mainly indicate the most important references in the text.

1 *Research Programs and Research Strategies*

The main references in Section 1.1 were to Lakatos (1970, 1978), Kuhn (1962/1969), Laudan (1977), and Zandvoort (1986, 1988) and in the second section to Nowak (1980), Nowakowa (1974, 1994), Krajewski (1977), and Hamminga (1983). An elaborated application of similar ideas to an interdisciplinary field of research, viz. cognitive science, is provided by Von Eckardt (1993).

2 *Observational Laws and Proper Theories*

For the distinction between observational laws and proper theories see Nagel (1961), Hempel (1966), Sneed (1971), Balzer, Moulines and Sneed (1987). There are numerous publications about one or more of the epistemological positions. The main ones for our survey are Van Fraassen (1980), Hacking (1983), and Niiniluoto (1987a). Leplin (1984) provides a useful collection of position papers. Rescher (1992) presents a kind of 'conceptual idealism', which has more similarities with 'constructive realism' than one might think at first sight. ICR gives a thorough analysis of the arguments needed for the transitions: instrumentalism → constructive empiricism → referential realism → constructive realism.

3 *Explanation and Reduction of Laws*

The main references were to Nagel (1961, Chapter 11), Schaffner (1967), Sklar (1967), Nickles (1973) and Causey (1977). In Chapter 12 we will indicate the nature of reduction of laws from the structuralist perspective on theories. For a detailed exposition, see Balzer, Moulines, and Sneed (1987, Chapter VI). Humphreys (1989) provides a systematic study of explanation, with emphasis on causal explanation in the social, medical and physical sciences. Schaffner (1993) concentrates on biology and medicine. P.S. Churchland (1986), Kim (1996, 2000) and Bickle (1998) focus on explanation and reduction in mind-body research.

4 *Explanation and Description by Specification*

For Section 4.1 the references to Hempel (1965, Chapter 12), Von Wright (1971) and Stegmüller (1975) were crucial, for Section 4.2 the ones to Nagel (1961, Chapter 12.I) and Hempel (1965, Chapter 11). There are many approaches to functional explanations, e.g. Schaffner (1993, Chapter 8) and Wouters (1999). An analysis of causal explanation in terms of abnormal factors is given by Hart and Honoré (1959/1985^2). For an analysis of explanation by specification from the point of view of the logic of questions, see Kuipers and Wisniewski (1994).

5 *Reduction and Correlation of Concepts*

For Section 5.1 the main references were the same as those in Chapter 3, with special emphasis on Causey (1977). Regarding the mind-body identity theories and reduction P.S. Churchland (1986), Macdonald (1989), and Kim (1996) are useful, the last one also for more general purposes, as was illustrated in Section 5.2. The indication of research strategies in Section 5.3 was highly inspired by Looijen (1998/2000). Kim (2000) and Bickle (1998) develop their own version of moderate reductionism, with emphasis on mind-body research, see the reading suggestions for the next chapter. An interesting evaluation of research strategies, with the emphasis on the prospects for the social sciences, leading to a preference for nonreductive strategies of unification, is presented by Kincaid (1997).

6 *Levels, Styles, and Mind-Body Research*

The main references were to Burton (1993), Darden and Maull (1977), Abrahamsen (1987), Bechtel (1988a), Millikan (1984, 1993), Guichard (1997), Mackor (1997), Engel (1977, 1980), Panhuysen and Tuiten (1993, 1995). Darden (1991) is an illuminating case study, with general lessons, in interfield research. For an extreme reductionist portrait of goals and means of mind-body research one may consult P.M. Churchland (1990). Some other studies in philosophy of science with emphasis on mind-body research are P.S. Churchland (1986), Dennett (1991) and Thompson (1995) (specifically on color vision). See also Bechtel (1986), Bechtel and Richardson (1993), and Kincaid (1997) for other examples and analyses of interfield research. Specifically on the status of connectionism, see Smolensky (1988), Rantala and Vadén (1995) and Bickle (1993, 1998), where the latter includes structuralist representations of folk psychology and connectionism. Some studies in the philosophy of mind with much attention to issues involving the philosophy of science are Bechtel (1988b), Searle (1992) and Kim (1996). Kim (2000) develops a 'functional model of reduction' of properties, according to which the property to be reduced is first construed as 'a second-order property defined by its causal role'. The actual reduction then amounts to the derivation of this causal role from the entities and properties of the reducing theory. In the case that the property to be reduced can be functionally explained in the biological sense, Kim's model amounts to such a functional explanation followed by the reduction in one of our senses of the proximate causal component of that explanation. However, Kim also wants to apply his model to purely physical cases, e.g. temperature, where it seems less convincing. Bickle (1998) takes up the challenge of P.S. Churchland (1986, p. 279) to elaborate an 'intertheoretic-reduction formulation' of the mind-body problem. He rightly claims that by doing so 'philosophy of mind becomes a specialized part of applied philosophy of science' (p. 45). His structuralist approach is more directly than ours in line with the 'bible' of structuralism, viz. Balzer et al. (1987). If one gets lost in the treatment of the set-theoretical details of the reductive links (in Sections 3.2–3.4), one is advised to jump to Section 3.5, because they are not essential for Bickle's main points and examples, which are very stimulating indeed. For some hints about his main points, see the end of our Section 3.3.1. The constitutive role of the body and the environment for the mind and for behavior is impressively elaborated by Clark (1997) and Keijzer (2001).

7 *Testing and Further Separate Evaluation of Theories*

The idea of separate (and comparative) HD-evaluation after falsification is primarily an instrumentalist or pragmatist idea, which was elaborated to some extent by Laudan (1977). For falsifying general hypotheses, see (Popper 1934/1959). For a survey of classical and Bayesian approaches to statistical testing, see Schaffner (1993) and Howson and Urbach (1989). In (Lakatos 1978) and (Wolpert 1992) one can find several indications and illustrations of complications in applying the HD-method.

8 *Empirical Progress and Pseudoscience*

Laudan (1977) deals with comparison of theories in a way which is globally similar to ours in Section 8.1 but not in the details. The main reference in Sections 8.2 and 8.3 is to (Lakatos, 1978). In particular, in the introduction one finds his view on science and pseudoscience, in the first chapter his view on falsificationism and the methodology of research programs, and in the second chapter his view on the comparison of methodologies. See Donovan et al. (1988) for the empirical evaluation of methodologies. For a sophisticated view on science and pseudoscience, with special attention to causal fallacies, see Thagard (1988).

9 *Progress in Nomological, Explicative and Design Research*

The book *From instrumentalism to constructive realism* (ICR) deals extensively with confirmation, empirical progress and truth approximation, and their mutual relations. For the quantitative approach to truthlikeness and truth approximation, see Niiniluoto (1987a). For further references, see ICR. For the (dis-)similarities between nomological and design research, see also Section 10.6.

10 *Design Research Programs*

The main references were to Bodewitz et al. (1988), and Vos (1991). Since the nature of design research is not very popular in the philosophy of science, there is not much literature about it. However, the philosophy of technology is popular, but addresses more social issues. As suggested in the concluding remarks, Bijker (1995) is not only relevant for our purposes, but could be improved a lot by using the above approach, more or less in the same way as Vos used it for drug research.

11 *Computational Philosophy of Science*

The main references in Section 11.1 were to Langley et al. (1987), Thagard (1988) and Shrager and Langley (1990). In Section 11.2 it was to Thagard (1992). For recent surveys, see Darden (1997), Langley (1998/2000) and Valdés-Pérez (1999).

12 *The Structuralist Approach to Theories*

The start of the structuralist approach of theories was made by Suppes (1957), and refined by Sneed (1971). The main textbooks on the structuralist approach are Diederich (1981), Balzer (1982), Stegmüller (1973/1986), Giere (1984), and Balzer et al. (1987). For examples, see (Balzer et al., 2000) and the references in Subsection 12.3.3, in particular at the end.

13 *'Default-Norms' in Research Ethics*

The basic references were to (Merton, 1942) and (Broad and Wade, 1982). There is a growing literature on research ethics and, more broadly, science ethics: some stimulating examples are (Lafollette, 1992; Bulger et al., 1993; Kitcher and Cartwright, 1996; Ziman, 1994).

EXERCISES

1 *Research Programs and Research Strategies*

(1) Describe in some detail a research program in your own field along the lines that have been suggested for the structure, development, and success of research programs.

(2) Describe in some detail an example of interaction between research programs, as far as possible, in the general terms of Subsections 1.2.2–1.2.5.

(3) Describe in some detail and along the lines suggested in Subsections 1.2.6 and 1.2.7 an example of research guided by idealization and concretization or by an interesting theorem.

(4) Discuss the advantages and disadvantages of program-bound education and research.

(5) *Paper-task*: Describe the research you are interested in along the lines of this chapter, assuming that at least something like a research program can be discerned in it. The following set-up usually works out well.

Start with an informal substantial exposition of the program. This is not only useful for other readers, it also serves to contextualize yourself. If there is some program description available, this might be useful for both purposes. Then one presents the main lines of the portrait of the program resulting from the attempt to characterize it in terms of the types, structure and development presented in this chapter.

Next one answers in separate sections at least the following questions in some detail:

- Is there interaction with other programs and of what type? Are there possibilities for interaction that have been overlooked?
- What are the internal strategies used, if any, for program development.
- What types of success does the program have? In what sense is it progressive?

Of course, when terms and distinctions presented in this chapter do not seem to be appropriate for the case, you should try to introduce more adequate ones for the occasion.

2 *Observational Laws and Proper Theories*

(1) Show how one pair of examples, mentioned in Section 2.1, of proper theories and observational laws explained by them (except (b) and (b*) dealt with in the

text) satisfies the definitions in Subsection 2.2.2 and is in accordance with the hierarchy presented in Subsection 2.2.3.

(2) Apply the analysis of theory-laden observation in Section 2.3 to one pair of examples.

(3) Analyze one of the examples of proper theories in terms of the structure of theories presented in Section 2.4.

(4) Determine in detail your epistemological position with respect to one pair of examples.

(5) *Paper-task*: Look for another pair of examples in your own field or interest and repeat tasks 1–4 in some detail. Start by giving extensive accounts of the proper theory and the observational law in question.

3 *Explanation and Reduction of Laws*

(1) Select one of the examples of Section 3.1 and look for two (university) textbook presentations. Make a comparison, using your own standards of clear exposition.

(2) Reconstruct the explanation of the law of combining volumes by Avogadro's version of Dalton's theory of the atom (see Subsection 1.1.7) in terms of the model.

(3) Reconstruct the explanation of Avogadro's hypothesis (see Subsection 1.1.7) by the kinetic theory in terms of the model. Hint: as far as possible use the decomposed explanation of the ideal gas law.

(4) *Paper-task*: Reconstruct the explanation of a law by a theory in your field of interest and discuss your reconstruction in terms of the presented models of explanation and reduction.

4 *Explanation and Description by Specification*

(1) Fill in the scheme for the intentional thought process for the search for an explanation of Jane's opening the window.

(2) Spell out the train of thought in a functional explanation of a particular example of a biological trait.

(3) Describe the thought process for an explanation of a particular event by causal specification.

(4) Spell out the strong version of explanation by causal specification.

(5) Make a set-up of the meaning analysis and the thought process for the general idea of an explanation by specification.

(6) Look for a 'clear and distinct' fourth subtype of explanation by specification, and argue why it does or does not fit the general scheme on second thought.

(7) *Paper-task*: Give an example of an intentional, functional, or (specific) causal explanation: first in informal terms, then try to reconstruct, along the lines of 'explanation by specification', both the meaning analysis as well as the thought process. In what respects are the suggested lines not suitable? What alternative would you prefer?

5 *Reduction and Correlation of Concepts*

(1) Show that TT-$IH^{=}$ implies the (Supervenience or) Realization Hypothesis.

(2) Show that the water example could have been analyzed as a combination of a thing-with-property: water in liquid form.

(3) Show that the reduction of the concept of temperature can be transformed into the reduction of the 'equal thermal state' relation. Use Appendix 1 of Chapter 2.

(4) Reformulate the survey of Subsection 5.2.3 for (genuine) binary relations.

(5) Look for cases of multiple reduction of concepts in your field of interest, and discuss its prospects for (global) singular reduction.

(6) *Paper-task*: Look for cases of concept reduction of one type or another in your field of interest and try to analyze it along similar lines to those in the text.

6 *Levels, Styles, and Mind-Body Research*

(1) Apply in some detail Burton's four types of interlevel methods, the matrix of levels and styles (LS-matrix), and the matrix taxonomy of interfield research (IF-matrix) to your primary field of interest, and discuss the mutual relations.

(2) Discuss the relation between the given decomposition of proper functions and the historical evolutionary component, $CE_s(x, y, z)$, of a specific functional statement.

(3) Spell out in some detail the causal explication of the intentional style resulting from the functional explication of the intentional style and the causal explication of the functional style.

(4) Look in the literature about mind-body research for an example of a more or less accepted correlation between MB-concepts and for a biophysicalist explanation of some MB-correlation. Discuss the prospects for some kind of reduction?

(5) Similarly, look for a case of functional decomposition, and discuss it in terms of reduction of concepts and laws.

(6) Evaluate the Principles I–VI for the relation between the macro- and micro-level of an individual and its environment.

(7) Explain on the basis of the causal explications of the functional and the intentional style in Section 6.2 that, *prima facie* interlevel, functional and intentional transitions can be reconstructed, by monolevel causal, functional or (only for the macro-level) intentional analysis and conceptual interlevel analysis.

(8) *Paper-task*: Look for a concrete example of interlevel and interstyle mind-body research. Give first an informal exposition, then an exposition in terms of the Principles I–VI and the IC-principle, and finally in terms of the IE-model and its principles.

7 *Testing and Further Separate Evaluation of Theories*

(1) Look in a recent issue of a leading journal (e.g. *Nature*, *Science*) in which the authors speak about confirmation (or verification, or corroboration) of some hypothesis or theory. Does it concern (conditional-) deductive or probabilistic

confirmation? If deductive, are the principles P1 and P2 used, or incompatible alternatives? If probabilistic, is it merely structural confirmation, or is there inductive confirmation involved? Present the case in some detail and evaluate it.

(2) On the basis of the exposition of the development of Dalton's theory in Chapter 1 and this chapter, reconstruct the evaluation report in the three phases distinguished in Subsection 1.1.7.

(3) Reconstruct on the basis of a textbook of the discipline of your interest a general and/or individual test implication of some theory.

(4) Identify and prove the main logical claims made in Section 7.1.

(5) Identify and prove the main logical claims made in Section 7.2.

(6) On the basis of a textbook exposition, give an example of a general and individual test implication of a (comparative or non-comparative) statistical nature.

(7) Identify two possible complicating factors in the evaluation of Dalton's theory in one or two of its phases (see Subsection 1.1.7).

(8) *Paper-task:* Give a schematic presentation of the macro- and the micro-argument for an application of the HD-method to a certain theory. That is, look for a (perhaps popular) scientific publication in the field of your interest on the basis of which you can specify a theory, and reconstrue a general and an individual test implication and all the corresponding arguments. Specify at least two possible points for contesting actual or possible falsification or confirmation.

8 *Empirical Progress and Pseudoscience*

(1) Make a comparison of the evaluation reports of Dalton's theory in the phases distinguished in Subsection 1.1.7.

(2) Look in the scientific literature for a clear application of the rule of success, and present the example in some detail, including the role of the relevant comparative success hypothesis.

(3) Reconstrue one of the examples of theory comparison in Thagard (1992), either along the asymmetric lines of Subsection 8.1.1 or the symmetric lines of Subsection 8.1.4.

(4) Search in Popper (1959), (1963) or (1983) for remarks on the elimination of theories after falsification and report the findings.

(5) Give the main lines of Lakatos' view (Lakatos, 1978) on the methodology of research programs (Chapter 1) or the evaluation of methodologies (Chapter 2).

(6) Give a detailed comparison of the views of pseudoscience in (Lakatos, 1978, Introduction), (Thagard, 1988, Chapter 9), and Section 8.3.

(7) *Paper-task*: Analyze a publication of a *prima facie* pseudoscientific nature in terms of Section 8.3 and/or (Thagard, 1988, Chapter 9).

9 *Progress in Nomological, Explicative and Design Research*

X, Y, Z, T indicate subsets of Mp.

(1) Prove that, if Y is a subset of X, then the weak claim of hypothesis Y, viz. T is a subset of Y, implies the weak claim of hypothesis X.

(2) Prove that the strong claim '$X = T$' is equivalent to:

(i) $X \subseteq T$ and $T \subseteq X$
(ii) $X - T = \emptyset$ and $T - X = \emptyset$
(iii) $X \triangle T (=_{\text{def}} (T - X) \cup (X - T)) = \emptyset$

(3) Prove: $\langle X, Y, T \rangle$ satisfies (DP) and (UP) iff $Y \triangle T \subseteq X \triangle T$.

(4) Draw an electric circuit with one bulb and three switches for which '$p_0 \leftrightarrow p_1 \& p_2 \& p_3$' is the true theory, where p_0 indicates that the bulb lights and p_i that switch i is on. Prove that '$p_0 \leftrightarrow p_1 \& p_2$' is closer to the truth than '$p_0 \leftrightarrow p_1$'.

(5) Prove that (DP) and (UP) applied to $\langle X, Y, T \rangle$, for fixed T, is a reflexive partial order (i.e. reflexive, anti-symmetric and transitive) relation on the powerset $P(X)$ of Mp, i.e. the set of all subsets of X.

(6) Recall that $Q(X)$ indicates the set of supersets of X, that is, the set of subsets of Mp which include X, called the co-powerset of X. Prove that $Q(X)$ represents the set of hypotheses following from theory X by showing that all weak claims associated with members of $Q(X)$ follow from the strong claim of X.

(7)(a) Prove the first half of the equivalence theorem according to which the clause (UP), $Y - T \subseteq X - T$, is equivalent to the condition (DF), $Q(X) \cap Q(T) \subseteq Q(Y) \cap Q(T)$. Start by first proving $Q(X) \cap Q(T) = Q(X \cup T)$.

(b) Prove also the second half, i.e. (DP), $X \cap T \subseteq Y \cap T$, is equivalent to (UF), $Q(Y) \cap Q(cT) \subseteq Q(X) \cap Q(cT)$.

(8) *Paper-task*: Select one of the following publications: Niiniluoto (1987b), Oddie (1987), Schurz and Weingartner (1987), Brink and Heidema (1987), Zwart (1998, Chapter 6). Describe the main similarities and differences, and compare the advantages and disadvantages between the selected approach and the one in this chapter.

10 *Design Research Programs*

(1) Explain why the R3-line of refinement is essentially compatible with the R4/5/6-lines, and indicate how they can be integrated (see Subsection 10.2.3).

(2) Give an informal description of the refinements R1 and R2 and discuss the possibility of combining them with the other refinements. Try to visualize R1.

(3) Starting from a *W*/*I*/*N*-division (Section 10.5) show that the 'new' types of problems, viz. $O(x_1) \cap N$ and $W - O(x_1)$, cannot interfere, assuming constant *W*/*I*/*N*. Show also that the 'new' definitions of 'improvement' and 'concession' reduce to the original one by assuming I to be empty.

(4) Give a critical evaluation of the comparison of product and truth approximation (Section 10.6).

(5) *Paper-task*: Describe in detail an example of a design research program in which, on the one hand, the distinction between factual and desired properties and, on the other, the distinction between structural and functional properties can be

made explicit. Explain what the heuristic principles for design programs amount to in the example. Give also a description of potential applications and potential realizations.

11 *Computational Philosophy of Science*

(1) Look in your own field for a quantitative or qualitative law that might be (re-) discovered with one of the programs discussed in Subsection 11.1.1. Try to indicate the main heuristic operations that could be used.

(2) Give examples in your own field of the four types of abduction indicated in Subsection 11.1.2.2. Give also an example of concept formation along the lines suggested there.

(3) Discuss the pro's and con's of theory comparison according to PI (see Subsection 11.1.2.3 and 11.2.1.1).

(4) Look in your own field for an example of theory comparison with a generally accepted outcome in favor of one of them, and check whether PES/EM would lead to the same conclusion.

(5) *Paper-task*: Check one of the mentioned examples of ECHO-comparisons 2.2.2 (Lavoisier, Darwin, Wegener), 2.3 (Copernicus, Newton, Huygens), that is, consult the reports of Thagard c.s. and check the reproducibility claim by PES/EM. Discuss the various possible aims of an evaluation computer program.

12 *The Structuralist Approach to Theories*

(1) Show that the statement approach's concept of a potential falsifier is made essentially redundant by the notion of a counter-example.

(2) (Subsection 12.3.3) (a) Show that as long as the identity constraint for mass is not taken into consideration *GCPM* has no empirical content, and hence neither does *CPM*. Show that with this constraint *CPM* still lacks empirical content, but *GCPM* gets it. (b) Show that the naive and the refined theory of the periodic table have empirical content. (c) Show that the psycho-analytic theory has (partial) empirical content and prove the theorem about neurosis generation.

(3) Indicate in one of the following examples that they could have been represented in the structuralist way. For example, the atomic theory (Chapter 1), the kinetic theory, the genetic theory, and Olson's theory of collective goods (Chapter 3).

(4) Show that the absolute empirical content ($Mpp - \pi M$) is the union of the partial empirical content ($Mpp - Mpart$) and the relative empirical content ($Mpart - \pi M$). Show also that, if a theory has relative (and/or partial) empirical content it has absolute empirical content. Finally, show that a theory may have absolute empirical content without having relative empirical content, and that in this case the absolute empirical content coincides with the partial empirical content.

(5) Show that the empirical claim becomes trivial in the third way of determining I, that is, by auto-determination. Look for an example of auto-determination in your own field.

(6) Derive the definitions of links that result if the stratification is assumed to disappear (i.e. $Mp = Mpp$ and π is the identity function).

(7) Prove that all singleton sets $\{x\}$, for x in S, belong to C, and hence that a constraint does not exclude any individual potential model.

(8) *Paper-task*: Make a start with a structuralist reconstruction of an empirical or non-empirical theory. Try to answer questions like: does the theory have empirical content, is it epistemologically stratified, do constraints play a role, is there a theory-net, etc.?

13 *'Default-Norms' in Research Ethics*

(1) Discuss one of the forms of strategic behavior displayed by scientists which conflicts in some way with Merton's norm of disinterestedness. Try to indicate by some examples where your limits lie, for yourself and for others.

(2) *Paper-task*: Look for a recent published debate about a moral dilemma dealing with the integrity of scientific researchers. Analyze the standpoints of the opponents in detail. Formulate your own opinion. Show the preliminary version to a colleague and ask for her/his detailed opinion. Include that opinion in your paper as neutrally as possible. Finally, give your detailed opinion about her/his opinion.

General paper-task

Look for an example of a substantial or methodological concept or intuition in your sphere of interest for which an explication is according to you desirable. Explain why the explication is desirable. How would you start the explication? Consider the possibility of using the guidelines for explication in Chapter 1.

NOTES

FOREWORD

[1] ICR starts with a more extended survey of positions and ends with a sketch of the favorite epistemological position, resulting from the different analyses in that book, viz. constructive realism, which may be seen as a background pattern of thinking guiding many, if not most, scientists.

CHAPTER 1

[1] Of course, not every aspect of this true description may have our interest. It may or may not be possible to restrict the observational vocabulary, and hence the program, to some sub-vocabulary of what we find really interesting.

[2] For example, the intended concept is usually not uniquely determined, such that there are degrees of freedom in explication, leaving room for diverging explications. For this reason one may even prefer to speak of concept modelling instead of concept explication (Brink, 1989).

[3] Lakatos's concept of negative heuristic coincides with the intention of keeping fixed the hard core as long as possible. Hence, this notion does not add something to the notion of the hard core (in the sense of Lakatos).

[4] Unfortunately, the term 'idea program' is linguistically not very attractive.

[5] The branched portrait of evaluation of a research program is mostly a product of the structuralists; the idea of a sequence of improving theories comes from Lakatos, following Popper; the pattern of idealization and concretization originates with Nowak (1974, 1980), and is further developed by others, e.g. Krajewski (1977).

[6] The notion of 'explanatory success' should be taken here in the liberal sense of derivational success, that is, facts that can be derived from the theory.

[7] For this purpose, however, it is in addition (almost) necessary that there are no new (types of) counter-examples, a condition which Popper and Lakatos just presuppose.

[8] For a brief history of the successive and competing research programs in high-energy physics, see (Cushing, 1982); for the history of the NMR-(nuclear magnetic resonance) program, engaged in asymmetric reductive co-operation with chemical and biological research programs, see (Zandvoort, 1986).

[9] ICR grew out of research and teaching in at least two rather different, if not opposing, research programs, the Carnap–Hintikka program of inductive (confirmation) logic and Popper's program of truth approximation.

CHAPTER 2

[1] Though this terminology has a systematic background, it is certainly not very attractive; one might prefer to speak of an experimental observation theory.

[2] The complications are mainly due to the fact that the social and cultural world is constructed by humans in a sense not applicable to the natural world. It is a topic of increasing interest, e.g. Bhaskar

(1979), Giddens (1984), Searle (1995), Tuomela (1995), Balzer and Tuomela (1997), to mention a few.

[3] The important restriction to asymptotic behavior is neglected in all references to this example in the main text of the book. Contrary to Bickle's suggestion (1998, p. 26), the asymptotic nature of the equivalence between temperature and mean molecular kinetic energy is unproblematic for the reduction of the ideal gas law. As we will see in this appendix, although temperature is asymptotically defined, the relevant identification is unrestrictedly that of equal thermal states with equal mean kinetic energy. The abovementioned equivalence (KTR, see the appendix) results from this identity and some empirical laws and is usually, somewhat misleadingly, used in expositions of the reduction of the gas law, as we will do in the next chapter.

[4] Recall that descriptive programs are always in observation terms, whereas explanatory programs may or may not introduce theoretical terms. If they do, they are called theoretical programs.

CHAPTER 3

[1] Nagel (1961), Hempel (1965, Chapters 10 and 12), Tuomela (1973) are good examples of advanced textbooks paying substantial attention to subsumption explanation of laws by theories. Mahner and Bunge (1997, pp. 107–109) stress the point of the relative superficiality of subsumption explanations of individual events by making a distinction between merely subsumption explanations and 'mechanismic' explanations. As a matter of fact, the latter amount to theoretical explanations of the law enabling the former.

[2] Special thanks to Francisco Flores for preventing this mistake.

[3] See Note 2 of Chapter 2 for some general references about the constitution and construction of the social world.

[4] From the examples above, it will already be clear that a 'causal correlation' may, but need not, be of a statistical nature. It may well be of a deterministic nature, as in the case of genotypes determining phenotypes. When 'correlation' is meant in the statistical sense it will be explicitly called a 'statistical correlation'.

[5] The unavoidable idealizations involved in modelling auxiliary hypotheses in the application step seem to be the main reason for Cartwright (1983) to relativize the deductive character of explanation and to speak of the 'simulacrum account'.

[6] Starting from Einstein's non-exhaustive division of theories consisting of theoretical laws, i.e. proper theories in our sense, in principle and constructive theories, Flores (1999) has introduced the distinction between framework and interaction (or causal mechanism) theories. Whereas the former roughly correspond to principle theories, such as Newtonian mechanics, thermodynamics, and the special theory of relativity, the latter covers a larger set of theories than Einstein's constructive ones. For example, it not only includes the kinetic theory of gases, but also Newton's theory of universal gravitation, which had no place in Einstein's division. According to Flores, interaction theories presuppose and use framework theories, but cannot be derived from them. Flores relates his distinction to 'top-down explanations' (TDEs), advocated by Kitcher (1989), and bottom-up explanations (BUEs), advocated by Salmon (1989). He further claims 'Only laws that belong to an interaction theory, and particularly those that belong to constructive theories, receive BCEs. Laws belonging to framework theories, on the other hand, always receive TDEs.' He illustrates his claim with Einstein's replacement of his earlier BUE-like explanation of '$E = mc^2$', using Maxwell's theory, by his TDE starting from the special theory of relativity (constancy of light velocity, principle of relativity) and certain conservation principles, such as the conservation of momentum. The latter is generally considered to be more satisfactory and to do more justice to the theoretical and general nature of the target law.

In our terminology, we would say that theoretical laws get TDEs from framework proper theories, whereas observational laws get BUEs from interaction or causal mechanism proper theories.

Moreover, whereas we would certainly claim that the five-step model applies to BUEs, it is a matter of decomposing a number of TDEs, whether the model also applies in these cases or whether some substantially new steps have to be added.

[7] The distinction between micro-reduction and ontological identification roughly corresponds to what Moulines (1984) calls heterogeneous and, strangely enough, homogeneous ontological reduction, leaving room for mixed versions as well. However, following Nagel, we prefer to reserve the terms 'heterogenous' and 'homogeneous' for the epistemological distinction given in the text.

CHAPTER 4

[1] Instead of 'useful' the term 'rational' might have been used. However, 'rational' is laden with several connotations, see e.g. below. The term 'goal-rational' is essentially adequate, but sounds rather technical.

[2] Whereas IP was already used as a three-place predicate, it now also is used as a two-place predicate. This and similar ambiguities will nowhere be a problem, for we will always indicate the relevant arguments.

[3] This decomposition should not be confused with the terminologically similar decomposition of the internal phase of a research program (Chapter 1); as a rule it will be at most something like a micro-version.

[4] The causal effectiveness component should not be read as the (quasi-)causal law: 'for all x, y, and z, $D(x, z)$ and $BU(x, y, z)$ cause (as a rule) $P(x, y)$'. It just claims that in this particular case, the mental states played a significant causal role. In this connection it is also interesting to note, in anticipation of Section 6.2, that an explanation by intentional specification, assuming the causal component, can also be reconstrued as an explanation by causal specification, where the mental states function as the 'abnormal' or at least special causal factors, viz. mental factors.

[5] See Pettit (1993) for a (realistic) 'resilience' interpretation of such behavior: as soon as there are threats to straightforward 'as-if-utilistic' behavior, people tend to make utilistic calculations.

[6] Although claiming that the appendix is at present non-functional may not be totally uncontroversial, the Penguin Medical Encyclopedia (of 1985) even leaves room for the possibility that it never was: 'Any purpose that it may have served in some remote ancestor has vanished in the course of evolution.' However, in contrast to a mere causal effect, as in the case of heart sounds, it is difficult to see how part of an organ could evolutionary evolve as a by-product of one kind or another.

[7] Although it will be generally more easy to test the 'presentist' versions of the relevant test implications, Wouters (1999) makes it clear, using the example of the thymus, that it may well be misleading to study adult individuals in their present environment. After many efforts, it became clear that the thymus only persists (and probably originated) due to its functionality in young individuals.

[8] In the case of the causal style, this might be a reason to speak instead of the structural style.

CHAPTER 5

[1] The structuralist representation of theories, to be presented in Chapter 12, conceives theories as sets of such set-theoretical structures.

[2] To avoid confusion, we will systematically use the prefix 'base', e.g. base-objects, base-terms, when we want to refer to individuals of the lower ontological level and to their properties and relations. Moreover, we will speak of base-sets in the sense of, usually large, sets of base-objects. However, as soon as we want to refer to aggregates of such base-objects and to structures built upon such base-

sets, as well as classes and types of such structures, we will use the prefix 'micro', leading to micro-aggregates, micro-structures, micro-classes and micro-types.

3 Note that it is a *partial* representation, at least relative to a more encompassing vocabulary, including not only substance terms, but also, for instance, the volume, pressure and temperature of the macro-object.

4 It should be stressed that the example in the present context has not yet anything to do with the aggregation phase of the substance water. That is, a_o may well be water in solid form (ice) or in gaseous form (vapor). The three phases are typical attributes (in particular properties) of the substance water. The example of substance terms nevertheless makes it already clear, in particular by including mixture-terms, that macro-types may be rather complex predicates, i.e. they may involve a family of predicates that are themselves conjunctions of more elementary predicates, all belonging to different, more elementary families.

5 The set is indicated by '*Mp*' since, from the structuralistic perspective, it forms the set of *p*otential *m*odels of theories about aggregates.

6 In the next subsection, the chemical kind partition might be deleted, but functions for mass and position in a time-interval, hence velocity and kinetic energy, will be added. For the case of mind-body reduction we will see in Chapter 6 that a (neuro-)structure at least specifies a base-set representing neurons, and a relation on it representing the asymmetric connection relation, and a function representing the time-related activation function of neurons.

7 Note that we do not use the term 'base-type', for that is the plausible term for types of base-objects. As far as properties are concerned, Kim (2000, p. 84) calls our micro-types more correctly 'micro-based properties', but we prefer the shortest name.

8 It should be noted that aggregate concepts do not presuppose aggregation in the sense of the aggregation step in the micro-reduction of laws, whereas, conversely, such a step presupposes such concepts, viz. the terms used in the resulting aggregated law.

9 We distinguish 'ontological realization' from 'causal realization'; only in the latter case are there causal processes involved which are not covered by the structures.

10 The notion of a realization-class essentially corresponds to the set of what Kim (1996, p. 223) calls 'supervenience bases' of a macro-type. In line with this terminology, the indicated set might also be called the 'subvenience-class' of the macro-type. Kim (p. 224) has stressed that the ideas of realizability and supervenience are more or less the same. In supervenience (base) terms, our version of the Realization Hypothesis amounts to three claims: (1) the structure of an aggregate is a sufficient condition for the macro-type, called a supervenience base for that macro-type; (2) a macro-type (usually) has multiple supervenience bases; (3) a macro-type is conceptually distinct from its set of supervenience bases. However, we prefer to use the term 'base' in another sense and use the term 'the realization-class of structures' instead of 'the set of supervenience bases' of a macro-type. Hence, our 'base'-notions should not be confused with Kim's notion of 'supervenience base'.

11 Note that we could also define 'micro-classes' as the classes of structures which are classified as having the same micro-type. However, these classes are less interesting, than the 'macro-induced' realization-classes, giving rise to 'macro-induced' realization-types.

12 The determination of the frequency intervals of the colors in the spectrum of colors is a nice example, despite its problematic nature in other respects.

13 The example of the temperature function is, strictly speaking, not an example of a function between macro-objects, but a real-valued function on macro-objects, which can be conceived as a family of monadic predicates. A genuine function between macro-objects is just a special type of relation between macro-objects. So we need an example of a genuine macro-relation, e.g. 'being in the same thermal state'. Since this is an equivalence relation (the 0-th law of thermodynamics), it gives rise to equivalence-classes and hence to 'equivalence-properties', viz. 'being in a certain thermal state', which brings us back to a case of a monadic attribute-predicate.

14 Our convention is to call the predicates *P* and *Q* *ontologically identical* wrt *A* if and only if the following holds: the corresponding subsets of *A* are *ontologically equal* iff the simultaneous satisfaction claims of *P* and *Q* and hence the simultaneous membership claims of the corresponding

subsets of A are *ontologically equivalent*. The adjective 'ontological' here means 'on ontological grounds'.

[15] Macro-structures are intended to represent macro-aggregates, i.e. aggregates of macro-objects, in terms of macro-types. Hence, macro-structures do not represent the structure of macro-objects, but of aggregates of macro-objects. Micro-structures represent macro-objects.

[16] Note that the concept of a rigid body can only be defined by conceiving a rigid body not only as a macro-object but also as an aggregate of its constituting particles, viz. by requiring that their mutual distances remain constant (Balzer, Moulines and Sneed, 1987). Hence, one does not speak of the reduction of the concept of a rigid body. Note, however, that this does not prevent the possibility of (micro-)reduction of rigid body mechanics to particle mechanics.

[17] Our current convention is to call the predicates P and Q *causally correlated* wrt A if and only if the following holds: the corresponding subsets of A are *equal due to causal correlation* iff the simultaneous satisfaction claims of P and Q and hence the simultaneous membership claims of the corresponding subsets of A are *causally equivalent*.

[18] Two isotopic atoms of one chemical element are atoms in which the nuclei have the same number of protons, hence similar chemical behavior, but different numbers of neutrons, leading generally to different nuclear behavior (e.g. radioactivity).

[19] For a stimulating discussion of genuine and artificial disjunctive properties, see Kim (1996, 2000).

[20] The analysis in Section 5.2 was particularly stimulated by critical interaction with Rick Looijen.

CHAPTER 6

[1] See (Stiekema, to appear) for a completely different account of the relation between symbolism and connectionism. According to Stiekema, they represent two different cognitive modules co-operating in a specific way. The corresponding research programs are invited to co-operate from this perspective.

[2] Instead of talking about levels of aggregation one might also talk about levels of organization, but this term typically reflects the functional style.

[3] These terms are a variant of those proposed by Abrahamsen (1987).

[4] If one field itself already integrates two styles, e.g. functional and causal, the other should concern the third style, i.e. the intentional, in order to speak of interstyle interfield research. The situation is similar for interlevel research.

[5] Note that, although the second question itself is primarily of an (interstyle but) monolevel nature, research along biophysicalist lines looks for answers of an interlevel nature. However, this does not alter the fact that there is an important difference between interstyle explications and the reduction of concepts and laws in Chapter 3 and 5, which were primarily of a monostyle and interlevel nature.

[6] The magnetosome example was introduced by Dretske (1988), the bee dance example by Millikan (1984) and the soap avoidance example by Millikan (1993). As mentioned in the main text, Guichard (1997) presented a detailed reconstruction of them.

[7] Of course, not every desire is causally positive for reproduction and survival, think of suicidal wishes.

[8] One particular form is a 'kind law' stating that all tokens of a certain thing-predicate, i.e. tokens of a certain kind of entity, are also tokens of a certain attribute-predicate, i.e. share a certain property.

[9] Of course, an approximative step in a later phase of the reduction of a law is not excluded.

[10] This example and the ones about standardized therapies below were brought to my attention in a paper by Dineke Hubbeling (1997).

[11] Although Chalmers (1996) is rather pessimistic about the possibility of reducing phenomenal states, notably consciousness, he is rather optimistic about the perspectives for the reduction of (functionalistically conceived) psychological states. To be sure, if this optimism comes true, the

science of man has made real progress, even if the riddle of the nature and causation of consciousness has then not yet been solved.

[12] This may be compared with the way in which the ideal gas behavior might be called the causal-ontological consequence of the kinetic behavior of molecules.

[13] A monist ontology does not of course prevent the recognition of different ontological levels of aggregation or organization. The point is that all aggregates result essentially from interacting lower-level biophysical entities.

[14] We will not pay attention to this claim as such. It seems that the six principles are compatible with, but do not necessarily follow from, general systems theory.

[15] In the case of multiple type-type identities, it seems plausible to assume that the two relevant direction criteria point in the same direction. However, in general it is not evident that the construction of the levels as such and the principles do not conflict with one another. In particular, the direction of levels excludes the possibility of 'downward supervenience' and 'upward multiplicity', whereas Principle VI seems to claim one or both of these opposite possibilities. Hence, the intuition behind VI will have to be explicated in another way.

[16] Bechtel describes extensively the development of the so-called prototype theory of concepts, originating with Wittgenstein and Rosch, also mentioned by Burton (1993) for similar purposes. It is an interfield research program linking philosophy, linguistics, psychology and cultural anthropology.

CHAPTER 7

[1] Recall from Subsection 2.4.3 that we use in this book the term 'epistemological position' in a broad sense, as a view concerning the nature of (claims to) knowledge, including relevant ontological and semantic presuppositions. 'Methodology' is conceived as the way in which claims to knowledge are evaluated in terms of evidence.

[2] When we speak of individual and general facts we assume that these facts are formulated in observation terms and that the hypotheses describing them have been sufficiently confirmed to be accepted. Of course, we may nevertheless be mistaken, we just assume for the time being that we are not.

[3] If one finds the term 'counter-example' to have a realist, or falsificationist, flavor, one may replace it systematically by 'problem' or 'failure'.

[4] The distinction between testing and evaluation somewhat corresponds to Schaffner's distinction between local and global evaluation (Schaffner, 1993, Chapter V), but the precise relation is difficult to indicate.

[5] McAllister (1996) convincingly argues for three main claims about plausibility considerations, at least as far as they are of an, in a broad sense, aesthetic nature. First, in normal science they govern the preferences between equally successful theories, and only between such theories. Second, they originate from 'aesthetic induction' applied to the accepted background knowledge. Three, the opposing parties in scientific revolutions arise from the strong differences in the willingness to sacrifice dominant aesthetic considerations to empirical success.

[6] This follows directly from the general definition of conditional probability, viz. $p(A/B) =_{\text{def}} p(A\&B)/p(B)$, according to which $p(E/H)/p(E) = p(E\&H)/(p(H)p(E)) = p(H/E)/\, p(H)$.

[7] The P2-related arguments concern the first and the second argument in Fitelson's Table 1, and the second in Table 2. Of the other two, the example of 'unintuitive' confirmation is rebutted in ICR (Chapter 3) with a similar case against the difference measure. The other one is related to the 'grue-paradox', for which Chapter 2 and 3 of ICR claim to present an illuminating analysis in agreement with P2.

[8] Harré (1993, Chapter 3) distinguishes three dimensions of generality: substance, experimental, spatio-temporal, and their possible restrictions. We take Harré's first and third dimension into account in the domain, and the second dimension in the initial conditions (see below).

[9] Here we neglect the famous asymmetry problems about causal explanations, in particular concerning individual facts. The standard example is that the length of a flag pole cannot be explained by the length of its shadow. Note, however, that this problem does not arise when general facts are involved. For example, the proportionality between the length of both can be explained by the classic theory of light propagation.

[10] Note that the above treatments of 'C_0 and *not*-F_0' and 'C_0 and F_0' and the ones to come about '*not*-C_0 and *not*-F_0' and '*not*-C_0 and F_0' are essentially in agreement with the way of dealing with (non-)black (non-)ravens, in the context of testing the raven hypothesis in Chapter 7.

[11] For a survey of classical and Bayesian approaches to statistical testing, see Schaffner (1993, Chapter V). For a more detailed presentation, see Howson and Urbach (1989).

[12] Informally: for all sufficiently large random samples s of individuals from domain D satisfying condition C, it holds, with high probability, that the ratio of individuals satisfying F is in a certain region.

[13] Informally: for all sufficiently large samples $s1$ and $s2$ of individuals of domains $D1$ and $D2$, satisfying C, it holds, with high probability, that the ratio in $s1$ satisfying F is significantly larger than the percentage in $s2$ satisfying F.

CHAPTER 8

[1] This definition may be conceived as being too strict. In Subsection 8.1.4, dealing with symmetric forms of theory comparison, we will define the notion of 'almost more successful'.

[2] In ICR it is argued that this definition of 'more successful', and hence what will be based on it in this chapter, is naive in at least two senses. In ICR it is called more specifically the 'basic' definition, to distinguish it from the 'refined' definition.

[3] Note that,when applied to not yet falsified theories, it concerns one aspect of Popper's idea of severe testing: severe testing as comparative testing.

[4] However, in ICR (Chapter 9) it is argued that the theory realist may even have good reasons to prefer a theory which is explanatory more successful, but instantially somewhat less successful. The reason is that the preferred theory may still be closer to the theoretical truth, in which case the other theory misses the extra counter-examples due to a lack of observational sensitivity.

[5] Note that crucial experiments are a kind of symmetric comparative evaluation, designed to generate good reasons for asymmetric comparative evaluation. In a later subsection we will deal with them and show, in some detail, how they stimulate the application of RS.

[6] Note that in the asymmetric approach an individual problem of one theory can only become a non-problem for the other. The possibilities of becoming an individual success or a neutral instance are not differentiated. The situation is, though technically somewhat different, essentially similar for the way general successes are treated in the asymmetric approach. In sum, the asymmetric approach is less differentiated than the two symmetric ones as far as neutral results are concerned.

[7] It is clear what the differences are between the three resulting definitions of more success, the asymmetric one of Subsection 8.1.1 and the symmetric micro and macro ones of the present subsection. Which one will be preferred in a specific case will depend largely on pragmatic considerations.

[8] See Note 5 of Chapter 7 dealing with the role of aesthetic criteria as presented by McAllister (1996). See also Subsection 9.2.1.

[9] If one uses the quantitative evaluation matrix, the consequence is that simplicity only comes into play in the case of quantitatively equal success, where the latter possibility is just a matter of sheer accident.

[10] It is plausible to distinguish some grades of testability. One might call a general testable conditional (GTC) *directly testable*, and a theory, merely, *testable* if it does not coincide with a GTC, but has GTCs as (general) test implications. Note that this definition of testable theories comes close to

imposing the FIT-ness condition of Simon and Groen (1973): finite (i.e. by a finite number of observations) and irrevocable testability.

11 A similar difference applies to Lakatos' construal of 'naive falsificationism' and our explication of it.

12 This even suggests a mere ordering of a theory in terms of the shells, the outer ones will be given up c.q. changed earlier than the inner ones (see Darden, 1991).

13 Besides irresponsible dogmatism, pseudoscience frequently suffers from lack of reproducibility and from causal and statistical fallacies. As is clear from our exposition, testing a GTI of a theory will lead to a success of that theory only if the corresponding ITIs can be reproduced. Causal fallacies arise from 'similarity thinking', i.e. taking similarities as proofs of causal relations, and from irresponsible 'correlation thinking', i.e. taking correlations as proofs of causal relations. Since similarities are neither necessary nor sufficient for causal relations, similarity-thinking can, at most, be of some heuristic value. Since correlations are not sufficient for causal relations, they can at most play the role of (quasi-)necessary conditions (not strict, in view of the possibility of counteracting forces). See Thagard (1988) for causal fallacies deriving from similarity thinking. Moreover, as is well-known, there are lots of types of statistical fallacies.

CHAPTER 9

1 In informal formulations with 'more' and 'fewer', these phrases are used in the set-theoretic sense. For example, 'Y has more true consequences than X' means that the set of true consequences of X is a proper subset of the set of true consequences of Y.

2 See (Kuipers, 1997 or ICR, Section 8.1) for a detailed analysis. See also (Zwart, 1998) for that purpose, and a creative recycling of a part of Popper's intuitions.

3 In this terminology, 'inductive' is supposed to exclusively refer to observational induction and not to theoretical or referential induction, distinguished in Subsection 2.2.2.

4 Vos (1991) nicely illustrates this possibility for drug research: given a certain interesting substance, the subsequent aim of research may become the search for an appropriate disease to cure with it.

CHAPTER 10

1 Hence, RP should *not* be read as 'research program'.

2 There may, of course, be properties of which we cannot seriously wish that they are present or absent. They can be left out of the picture. Hence, the set of relevant properties RP is the union of W and $(RP - W)$; the latter respectively correspond to $Q(T)$ and $Q(Mp - T)$ in the previous chapter. From these correspondences it becomes also clear that the properties in this chapter are 'distributed' features in the sense of the previous chapter. For example, a property is (un-)desired if and only if it is (un-)desired for all realizations of the intended profile, or, in the terms of the previous chapter, if and only if all structures realizing the desired profile (do not) have the relevant property. In the context of design research, the distributed character of the relevant properties is more or less self-evident. This is a happy fact, for it functions as a safeguard against similar logical problems for the comparative definitions to be given in the next subsection as arose with Popper's original definition of 'more truthlikeness' (for an indication, see Subsection 9.1.2).

3 We leave it as an exercise to verify that the corresponding figure resembles Figure 9.3 of the previous chapter.

CHAPTER 11

[1] This chapter profited a lot from Alexander van den Bosch's detailed criticism. For a survey study of 'rationality in discovery', see Bosch (2001).

[2] Davis' book is certainly not restricted to common sense knowledge, as the title of the book claims.

[3] Chapter 3 of (Shrager and Langley, 1990) is, however, devoted to such problems.

[4] As far as we know, there are only a few logical approaches to standard abduction, viz. the already mentioned belief revision developed by Peter Gärdenfors and others (Gärdenfors, 1988), non-monotonic logic approaches by Konolige (1990, 1996), logic programming (see Kakas et al., 1998), and the semantic tableau approach (Meyer and Pirri, 1993; Aliseda, 1997).

[5] See Langley (2000) for an extended version. For some other recent surveys of approaches and results, see Darden (1997) and Valdés-Pérez (1999).

CHAPTER 12

[1] This chapter profited a lot from Wolfgang Balzer's detailed criticism.

[2] This consideration is rather similar to Giere's (1999) point of view. Enlarging the scope from set-theoretic structures to models of one kind or another, he argues that a 'representational' view of models 'is much more adequate to the needs of empirical science' than the standard or 'instantial' view of models. The latter corresponds to the statement view in the sense that models may or may not be instances making a (set of) statement(s) true.

[3] Assuming some idealizations for the 'completeness claim' '$SB \subseteq E$', e.g. that the objects can be conceived as point-masses.

[4] For a critical discussion of the main historical and reductive claims in (Hettema and Kuipers, 1988) see (Scerri, 1997).

CHAPTER 13

[1] See, e.g. Mulkay (1977) and Ziman (1994). Ziman not only shows that present-day research cannot be characterized by Merton's CUDOS norms, but even argues that its conditions can better be characterized by PLACE: Proprietary, Local, Authoritarian, Commissional, Expert work.

[2] From a personal communication with Matthias Kaiser, chair of the ICSU Standing Committee of Ethics and Responsibility in Science.

REFERENCES

Abrahamsen, A. (1987): Bridging boundaries versus breaking boundaries: psycholinguistics in perspective, *Synthese*, 72, 355–388.

Achinstein, P. (1990): Hypotheses, probability, and waves, *British Journal for the Philosophy of Science*, 41, 117–147.

Adams, E. (1959): The foundation of rigid body mechanics and the derivation of its laws from those of particle mechanics, *The axiomatic method*, L Henkin et al. (eds.), North-Holland, Amsterdam, pp. 250–265.

Aliseda, A. (1997): *Seeking Explanations: Abduction in Logic, Philosophy of Science and Artificial Intelligence*, Dissertation Stanford, ILLC Dissertations Series (1997–04), Amsterdam.

Andel, P. van (1994): Anatomy of the unsought finding, *British Journal for the Philosophy of Science*, 45, 631–648.

Anscombe, E. (1957): *Intention*, Basil Blackwell, Oxford.

Balzer, W. (1982): *Empirische Theorien: Modelle, 1Strukturen, Beispiele*, Vieweg, Braunschweig/ Wiesbaden.

Balzer, W. (1996): Theoretical terms: recent developments, in: *Structuralist Theory of Science*, Balzer, W. and Moulines, C.U. (eds.), Berlin, De Gruyter Verlag, pp. 139–166.

Balzer, W. and Dawe, C. (1986): Structure and comparison of genetic theories, *The British Journal for the Philosophy of Science*, (1): 37.1, 55–69; (2): 37.2, 177–191.

Balzer, W. and Dawe, C. (1997): *Models for genetics*, Peter Lang, Frankfurt/M.

Balzer, W. and Marcou, Ph. (1989): A reconstruction of Sigmund Freud's early theory of the unconscious, in: Westmeyer, 1989, pp. 13–31.

Balzer, W. Moulines, C.U. and Sneed, J.D. (1987): *An architectonic for science*, Reidel, Dordrecht.

Balzer, W., Sneed J.D. and Ulises Moulines, C. (2000): *Structuralist Knowledge Representation. Paradigmatic examples*, Poznan Studies, Rodopi, Amsterdam.

Balzer, W. and Tuomela, R. (1997): A fixed point approach to collective actions, in: *Contemporary Action Theory*, Holmström-Hintikka, G. (eds.), Kluwer AP, Dordrecht, pp. 115–142.

Bechtel, W. (ed.) (1986): *Integrating scientific disciplines*, Dordrecht, Nijhoff.

Bechtel, W. (1988a): *Philosophy of Science. An overview for cognitive science*, Erlbaum, Hillsdale.

Bechtel, W. (1988b): *Philosophy of Mind. An overview for cognitive science*, Erlbaum, Hillsdale.

Bechtel, W. and Richardson, R. (1993): *Discovering complexity. Decomposition and localization as strategies in scientific research*, Princeton University Press, Princeton.

Bickle, J. (1993): Connectionism, eliminativism, and the semantic view of theories, *Erkenntnis*, 39, 359–382.

Bickle, J. (1998): *Psychoneural Reduction. The New Wave*, MIT Press, Cambridge.

Bhaskar, R. (1979): *The possibility of naturalism. A philosophical critique of the contemporary human sciences*, Harvester Press, Brighton.

Bijker, W. (1995): *Of bicycles, bakelites, and bulbs*, MIT press, Cambridge Ma.

Blaug, M. (1980): *A methodological appraisal of Marxian economics*, Cambridge University Press, Cambridge Ma.

Blumenthal, A. (1987): The emergence of psycholinguistics, *Synthese*, 72, 313–324.

Bodewitz, H., De Vries, G. and Weeder, P. (1988): Towards a cognitive model for technology-oriented R&D processes, *Research Policy*, 17, 213–224.

Boehm, T. Folkman, J., Browder. T. and O'Reilly, M.S. (1979): Anti-angiogenic therapy of experimental cancer does not induce acquired drug resistance, *Nature*, 390, 404–407.

Bosch, A. van den (2001): *Rationality in Discovery. A study of logic, cognition, computation and neuropharmacology*, PhD-dissertation, University of Groningen, ILLC Dissertation Series 2001–02, Amsterdam.

Boyd, R.N. (1984): The current status of scientific realism, in: *Scientific realism*, Leplin, J. (ed.) University of California Press, Berkeley, pp. 41–82.

Brink, C. (1989): Verisimilitude: views and reviews, *History and philosophy of logic*, 10, 181–201.

Brink, C. and Heidema, J. (1987): A verisimilar ordering of theories phrased in a propositional language, *The British Journal for the Philosophy of Science*, 38, 533–549.

Broad, W. and Wade, N. (1982): *Betrayers of the Truth*, Simon and Schuster, New York

Bulger, R., Heitman, E. and Reiser, S. (1993): *The ethical dimensions of the biological sciences*, Cambridge University Press, Cambridge MA.

Bunge, M. (1977a): Levels and reduction, *American Journal of Physiology*, 233(3), R75–82.

Bunge, M. (1977b): The GST challange to classical philosophies of science, *International Journal of General Systems*, 4, 29–37.

Burton, R. (1993): Interdependent methodologies of cognitive science, in: *Philosophy and the cognitive sciences*, Casati, R. and White, G. (eds.), Austrian Ludwig Wittgenstein Society, Vienna/Kirchberg, pp. 77–80.

Carnap, R. (1950/1963^2): *Logical Foundations of Probability*, University of Chicago Press, Chicago, (1963^2 with a new foreword).

Cartwright, N. (1983): *How the laws of physics lie*, Clarendon Press, Oxford.

Causey, R. (1977): *Unity of science*, Reidel, Dordrecht.

Chalmers, D. (1996): *The conscious mind*, Oxford University Press, Oxford.

Churchland, P.M. (1990): *A neurocomputational perspective: the nature of mind and the structure of science*, MIT Press, Cambridge, MA.

Churchland, P.S. (1986): *Neurophilosophy*, MIT Press, Cambridge MA.

Clark, A. (1997): *Being There. Putting brain, body, and world together again*, MIT Press, Cambridge MA.

Compton, K. (1988): 0–1 laws in logic and combinatorics, in: *Proceedings 1987 NATO Adv. Study Inst. on Algorithms and Order*, Rival, I. (ed.) Reidel, Dordrecht, pp. 353–383.

Cools, K., Hamminga, B. and Kuipers, T. (1994): Truth approximation by concretization in capital structure theory, in: *Idealization VI: Idealization in economics*, Hamminga, B. and de Marchi, N.B. (eds.), *Poznan Studies*, Vol. 38, Amsterdam, Rodopi, 205–228.

Crisler, Ch. (1994): An evolutionist looks at the origin of religion, *Truth-seeker*, 121.1, 2 pages, http:/ / truth-seeker.com/1994archive/121_1/ts211q.html

Cushing, J. (1982): Models and methodologies in current theoretical high-energy physics, *Synthese*, 50.1, 5–101.

Darden, L. (1991): *Theory change in science: strategies from Mendelian genetics*, Oxford University Press, Oxford.

Darden, L. (1997): Recent work in computational scientific discovery, in: *Proceedings of the Nineteenth Annual Conference of the Cognitive Science Society*, Shafto, M. and Langley, P. (eds.), Lawrence Erlbaum, Hillsdale, pp. 161–166.

Darden, L. and Maull, N. (1977): Interfield theories, *Philosophy of science*, 43, 44–64.

Davidson, D. (1963): Actions, reasons, and causes, *Journal of Philosophy*, 60, 685–700.

Davis, E. (1990): *Representations of common sense knowledge*, Kaufmann, San Mateo.

Dennett, D. (1987): *The intentional stance*, Cambridge, MA. MIT Press.

Dennett, D. (1991): *Consciousness explained*, Allen Lane The Penguin Press, London.

Dennett, D. (1996): *Kinds of minds*, Basic Books.

Diederich, W. (1981): *Strukturalistische Rekonstruktionen*, Vieweg Verlag, Braunschweig.

Dolman, H. and Gramsbergen, J.B. (1981): *Kontroverses in de populatiegenetika, een wetenschapsfilosofische analyse*, masters thesis theoretical biology, University of Groningen.

Donovan, A. et al. (eds.), (1988): *Scrutinizing science: empirical studies of scientific change*, Reidel, Dordrecht.

Dray, W. (1957): *Laws and explanation in history*, Oxford University Press, Oxford.

Dretske, F. (1988): *Explaining behavior*, Cambridge, MIT Press.

Duschl, R. (1990): *Restructuring science teaching. The importance of theories and their development*, Teachers College Press, New York.

Earman, J. (1992): *Bayes or bust. A critical examination of Bayesian confirmation theory*, MIT Press, Cambridge, MA.

Eckardt, B. von (1993): *What is cognitive science?*, MIT Press, Cambridge, MA.

Eliasmith, C. and Thagard, P. (1997): Waves, particles, and explanatory coherence, *British Journal for the Philosophy of Science*, 48, 1–19.

Emmelkamp, P., Bouman, T. and Blaauw, E. (1994): Individualized versus standardized therapy: a comparative evaluation with obsessive-compulsive patients, *Clinical psychology and psychotherapy*, Vol. 1 (2), 95–100.

Engel, G. (1977): The need for a new medical model; a challenge for biomedicine, *Science*, 196, 129–136.

Engel, G. (1980): The clinical application of the biopsychosocial model, *American Journal of Psychiatry*, 137, 535–544.

Festa, R. (1986): A measure for the distance between an interval hypothesis and the truth, *Synthese*, 67, 273–320.

Festa, R. (1993): *Optimum inductive methods. A study in inductive probability theory, Bayesian statistics and verisimilitude*, Kluwer, Dordrecht.

Festa, R. (1995): Verisimilitude, disorder, and optimum prior probabilities, in: Kuipers and Mackor, 1995, pp. 299–320.

Festa, R. (1999): Bayesian confirmation, in: *Experience, Reality, and Scientific Explanation*, Galavotti, M.C. and Pagnini, A. (eds.), Kluwer, Dordrecht, pp. 55–87.

Feyerabend, P. (1962): Explanation, reduction, and empiricism, *Minnesota Studies in the Philosophy of Science*, Vol. III, 28–97.

Feyerabend, P. (1975): *Against Method*, NLB, London.

Fischbach, G.D. (1992): Mind and brain, *Scientific American*, 267.3, 24–33.

Fitelson, B. (1999): The plurality of Bayesian measures of confirmation and the problem of measure sensitivity, *Philosophy of Science*, Supplement to Volume 66.3, S362-S378.

Fleck, L. (1935/1979): *Entstehung und Entwicklung einer wissenschaftlichen Tatsache*, Franfurt am Main, 1935; translated as *Genesis and development of a scientific fact*, University of Chicago Press, Chicago, 1979.

Flores, F. (1999): Einstein's theory of theories and types of theoretical explanation, *International Studies in Philosophy of Science*, 13.2, 123–134.

Fodor, J. (1975): *The language of thought*, Harvard University Press, Cambridge MA.

Fraassen, B. van, (1980): *The scientific image*, Clarendon, Oxford.

Fraassen, B. van, (1989): *Laws and symmetry*, Clarendon Press, Oxford.

Franklin, A. (1986): *The neglect of experiment*, Cambridge University Press, Cambridge.

Franssen, M. (1997): *Some contributions to methodological individualism in the social sciences*, Ph.D. Thesis, University of Amsterdam.

Franssen, M. (1998): The not-so-trivial truth of methodological individualism, paper 20th World Congress of Philosophy, 1998, http:/ /www.bu.edu/wcp/Papers/Scie/ScieFran.htm.

Gähde, U. (1983): *T-Theoretizität und Holismus*, Peter Lang, Frankfurt.

Gärdenfors, P. (1988): *Knowledge in flux: modeling the dynamics of epistemic states*, MIT Press, Cambridge, MA.

Gardiner, P. (1952): *The nature of historical explanation*, Oxford University Press, Oxford.

Gemes, K. (1990): Horwich and Hempel on hypothetico-deductivism, *Philosophy of Science*, 56, 609–702.

Giddens, A. (1984): *The constitution of society*, Polity Press, Cambridge.

Giere, R. (1984): *Understanding scientific reasoning*, Holt, Rinehart and Winston, New York.

Giere, R. (1985): Constructive realism, in: *Images of science*, Churchland, P. and Clifford, C. (eds.), The University of Chicago Press, Chicago, pp. 75–98.

Giere, R. (1996): The feminism question in the philosophy of science, in: *Feminism, science, and the philosophy of science*, Hankinson Nelson, L. and Nelson, J. (eds.), Kluwer, Dordrecht, 1996, pp. 3–15.

Giere, R. (1999): Using models to represent reality, in: *Model-based reasoning in scientific discovery*, Magnani, L., Nersessian, N. and Thagard, P. (eds.), Kluwer, Dordrecht, pp. 41–57.

Girill, T. (1976): Evaluating micro-explanations, *Erkenntnis*, 10, 387–405.

Gispen, W. (ed.), (1993): *Neurobiology of learning and memory*, Wiley, New York.

Gomperts, W. (1996): Herinneringen aan het hiervoormaals. Een veldstudie van reïncarnatietherapie, *Maandblad Geestelijke Volksgezondheid*, 7/8, 777–800.

Groot, A.D. de (1961/1969): *Methodologie*, Mouton, Den Haag, 1961; translated as *Methodology*, Mouton, New York, 1969.

Grove, A., Halpern, J. and Koller, D. (1996): Asymptotic conditional probabilities: the non-unary case, *Journal of Symbolic Logic*, 61, 250–276.

Grünbaum, A. (1984): *The foundations of psychoanalysis: a philosophical critique*, University of California Press, Berkeley.

Guichard, L. (1995): The causal efficacy of propositional attitudes, in: Kuipers and Mackor, 1995, pp. 373–392.

Guichard, L. (1997): *Stress research and the mind-body problem*, PhD-manuscript, Groningen.

Hacking, I. (1983): *Representing and intervening*, Cambridge University Press, Cambridge.

Hamminga, B. (1983): *Neoclassical theory structure and theory development*, Springer, Berlin.

Harré, R. (1986): *Varieties of realism*, Blackwell, Oxford.

Harré, R. (1993): *Laws of nature*, Duckworth, London.

Hart, H. and Honoré, T. (1959/1985^2): *Causation in the law*, Clarendon Press, Oxford.

Hempel, C. (1965): *Aspects of scientific explanation*, The Free Press, New York.

Hempel, C. (1966): *Philosophy of natural science*, Prentice-Hall, Englewood Cliffs.

Hempel, C. (1967): Reduction: ontological and linguistic facets, *Philosophy, science and method*, Morgenbesser, S. et al. (ed.) St. Martin's, New York, pp. 179–199.

Hempel, C.G. (1970): On the 'standard conception' of scientific theories, *Minnesota studies in the philosophy of science*, Vol. IV, Minneapolis.

Hettema, H. and Kuipers, T. (1988/2000): The periodic table: its formalization, status, and relation to atomic theory, *Erkenntnis*, 28, 387–408. Revised version, entitled 'The formalisation of the Periodic Table', in: Balzer et al., 2000, pp. 285–305.

Hintikka, J. (1988): On the incommensurability of theories, *Philosophy of Science*, 55.1, 25–38.

Holland, J. et al. (1986): *Induction. Processes of inference, learning and discovery*, MIT Press, Cambridge, MA.

Holmström-Hintikka, G. (1991): *Action, Purpose and Will. A formal theory*, Acta Philosophica Fennica, Vol. 50.

Holton, G. (1973): *Introduction to concepts and theories in physical science*, 2 Edn, Addison-Wesley, Reading, MA.

Hooker, C.A. (1981): Towards a general theory of reduction. Part I, II, III, *Dialogue*, 20, 38–59, 201–236, 496–529.

Howson, C. and Urbach, P. (1989): *Scientific reasoning: the Bayesian approach*, Open Court, La Salle.

Hubbeling, D. (1997): Is philosophy of mind important for psychiatry? paper, University of Warwick.

Humphreys, P. (1989): *The chances of explanation*, Princeton University Press, Princeton.

Janssen, M. (1987): Utilistic reduction of the macroeconomic consumption function, *LMPS-VIII Abstracts*, Moscow, Vol. 5.2, 386–390.

Janssen, M. (1993): *Micro-foundations*, Routledge, London.

Janssen, M. and Kuipers, T. (1989): Stratification of general equilibrium theory: a synthesis of reconstructions, *Erkenntnis*, 30, 183–205.

Kaiser, M. (1996): Towards more Secrecy in Science? – Comments on Some Structural Changes in Science and their Implications for an Ethics of Science, *Perspectives on Science*, 4.2, 207–230.

Kakas, A., Kowalski, R. and Toni, F. (1998): The role of abduction in logic programming, *Handbook of Logic in Artificial Intelligence and Logic Programming*, Vol. 5, Gabbay, D., Hogger, C. and Robinson, J. (eds.), Oxford University Press, Oxford, pp. 235–324.

Kandel, E. (2000): Research in the Kandel Lab, http:/ /fido.cpmc.columbia.edu/kandel/research/.

Keijzer, F. (2001): *Representation and behavior*, MIT Press, Cambridge, MA.

Kemeny, J. (1953): A logical measure function, *The Journal of Symbolic Logic*, 18.4, 289–308.
Kemeny, J. and Oppenheim, P. (1956): On reduction, *Philosophical Studies*, 7, 6–19.
Kim, J. (1996): *Philosophy of mind*, Westview Press, Boulder.
Kim, J. (2000): *Mind in a Physical World*, MIT Press, Cambridge.
Kincaid, H. (1997): *Individualism and the Unity of Science*, Rowman and Littlefield, Totowa.
Kitcher, P. (1989): Explanatory unification and the causal structure of the world, in: *Minnesota Studies in the Philosophy of Science*, Vol. XIII, Kitcher, P. and Salmon, W. (eds.) University of Minnesota Press, Minnesota, pp. 410–503.
Kitcher, P. and Cartwright, N. (eds.) (1996): *Science and Ethics*, special issue of *Perspectives on Science*, 4.2.
KNAW (2000): *Wetenschappelijk onderzoek: dilemma's en verleidingen*, Koninklijke Nederlandse Akademie voor Wetenschappen, Amsterdam.
Köbben, A. and Tromp, H. (1999): *De onwelkome boodschap*, Mets, Amsterdam.
Kolfschoten, F. van (1993): *Valse Vooruitgang*, Veen, Amsterdam.
Konolige, K. (1990): A General Theory of Abduction, in: *Automated Abduction, Working Notes*. Spring Symposium Series of the AAA. Stanford University, Stanford, pp. 62–66.
Konolige, K. (1996): Abductive Theories in Artificial Intelligence, in: *Principles of Knowledge Representations*, Brewka G. (ed.), CSLI Publications. Stanford University, Stanford.
Krajewski, W. (1977): *Correspondence principle and growth of science*, Reidel, Dordrecht.
Kuhn, T. (1962/1969): *The structure of scientific revolutions*, University of Chicago Press, Chicago.
Kuhn, T. (1963): The function of dogma in scientific research, in: *Scientific change*, Ch. 11. Crombie, A. (ed.) Basic Books, New York.
Kuhn, T. (1990): The road since structure, *PSA 1990*, Vol. 2, pp. 3–13.
Kuipers, T. (1978): *Studies in inductive probability and rational expectation*, Reidel, Dordrecht.
Kuipers, T. (1982): The reduction of phenomenological to kinetic thermostatics, *Philosophy of Science*, 49.1, 107–119.
Kuipers, T. (1984), Utilistic reduction in sociology: the case of collective goods, in: *Reduction in science*, Balzer W. et al. (eds.), Reidel Dordrecht, pp. 239–267.
Kuipers, T. (1985): The paradigm of concretization: the law of Van der Waals, *Poznan studies in the philosophy of the sciences and the humanities*, Vol. 8, Rodopi, Amsterdam, pp. 185–199.
Kuipers, T. (1987): *What is closer-to-the-truth? A parade of approaches to truthlikeness*, *Poznan Studies*, Vol. 10, Rodopi, Amsterdam.
Kuipers, T. (1997): The dual foundation of qualitative truth approximation, *Erkenntnis*, 47.2, 145–179.
Kuipers, T. (1999a): Kan schoonheid de weg wijzen naar de waarheid? *Algemeen Nederlands Tijdschrift voor Wijsbegeerte*, 91.3, 174–193. Revised and extended version to appear in *Synthese*.
Kuipers, T. (1999b): Abduction aiming at empirical progress or even at truth approximation, *Scientific Discovery and Creativity*, Meheus, J. and Nickles, T. (eds.) special issue of *Foundations of Science*, Vol. 4.3. 307–323.
Kuipers, T. (2000/ICR): *From instrumentalism to constructive realism*, Synthese Library 287, Kluwer AP, Dordrecht.
Kuipers, T. and Mackor, A.R. (eds.) (1995): *Cognitive patterns in science and common sense*, Poznan Studies, Vol. 45, Rodopi, Amsterdam.
Kuipers, T., Vos, R. and Sie, H. (1992): Design research programs and the logic of their development, *Erkenntnis*, 37.1, 37–63.
Kuipers, T. and Wisniewski, A. (1994): An erotetic approach to explanation by specification, *Erkenntnis*, 40.3, 377–402.
Kyburg, H. (1968), *Philosophy of science: a formal approach*, New York, McMillan.
Lafollette, M. (1992): *Stealing into print. Fraud, Plagiarism, and Misconduct in Scientific Publishing*, University of California Press.
Lakatos, I. (1970): Falsification and the methodology of scientific research programmes, *Criticism and the growth of knowledge*, Lakatos I. and Musgrave I. (eds.) Cambridge Univercity Press, pp. 91–196; reprinted in Lakatos (1978), pp. 8–101.

Lakatos, I. (1976): *Proofs and refutations: the logic of mathematical discovery*, Cambridge University Press, Cambridge.

Lakatos, I. (1978): *The methodology of scientific research programmes*, Worrall, J. and Currie G. (eds.) Cambridge University Press, Cambridge.

Lakoff, G. (1987): *Women, fire, and dangerous things: what categories reveal about the mind*, University of Chicago Press, Chicago.

Langley, P. (1998): The computer-aided discovery of scientific knowledge, in: *Discovery Science. First International Conference* (1998, Fukuoka, Japan), Arikawa, S. and Motoda, H. (eds.) Springer, Berlin, 1998, pp. 25–39.

Langley, P. (2000): The computational support of scientific discovery, *International Journal of Human-Computer Studies*, 53, 393–410.

Langley, P., Simon, H., Bradshaw, G. and Zytkow, J. (1987): *Scientific discovery. Computational explorations of the creative mind*, MIT Press, Cambridge.

Lastowski, K. (1977): The method of idealization in population genetics, *Poznan Studies*, Vol. 3.1–4, Grüner, Amsterdam, pp. 199–212.

Latour, B. (1987): *Science in action*, Open University Press, Milton Keynes.

Laudan, L. (1977): *Progress and its problems*, University of California Press, Berkeley.

Lawson, E. and McCauley, R. (1990): *Rethinking religion: connecting cognition and culture*, Cambridge University Press, Cambridge.

Leplin, J. (ed.) (1984): *Scientific realism*, University of California Press, Berkeley.

Levitt, S. and Donohue III, J. (1999): Legalized abortion and crime (as yet unpublished paper, reported in Scientific American), Dec. 1999, 13–15.

Looijen, R. (1987): Emergence and reduction in biology, *LMPS-VIII Abstracts*, Moscow, Vol. 2, 265–268.

Looijen, R. (1995): On the distinction between habitat and niche, and some implications for species' differentiation, in Kuipers and Mackor, 1995, pp. 87–108.

Looijen, R. (1998/2000): *Holism and reductionism in biology and ecology. The mutual dependence of higher and lower level research programmes*, PhD-dissertation. Revised version *Episteme*, Vol. 23, 2000, Kluwer AP, Dordrecht.

Macdonald, C. (1989): *Mind-body identity theories*, Routledge, London.

Mackie, J.L. (1965): Causes and conditions, *American Philosophical Quarterly*, 2.4, 245–264.

Mackor, A.R. (1995): Intentional psychology is a biological discipline, in Kuipers and Mackor, 1995, pp. 329–348.

Mackor, A.R. (1997): *Meaningful and rule-guided behaviour: a naturalistic approach. A teleofunctional argument against the alleged gap between the natural and the social sciences*, PhD-dissertation, University of Groningen, Groningen.

Mahner, M. and Bunge, M. (1997): *Foundations of biophilosophy*, Springer, Berlin.

Marek, W. and Truszczynski, M. (1993): *Nonmonotonic logic: context-dependent reasoning*, Springer, Berlin.

Matthews, M. (1994): *Science teaching. The role of history and philosophy of science*, Routledge, London.

Maull, N. (1977): Unifying science without reduction, *Studies in History and Philosophy of Science*, 8.2, 143–162.

Mayer, M. and Pirri, F. (1993): First Order Abduction via Tableau and Sequent Calculi, Bulletin of the IGPL 1, pp. 99–117.

Mayr, D. (1976/1981): Investigations of the concept of reduction, I: *Erkenntnis*, 10, 1976, 275–294, II: *Erkenntnis*, 16, 1981, 109–129.

McAllister, J. (1996): *Beauty and revolution in science*, Cornell University Press, Ithaca.

Merton, R.K. (1942): Science and Technology in a Democratic Order, *Journal of Legal and Political Sociology*, 1, 115–126. Reprinted under several titles, e.g. The Normative Structure of Science, in: *The Sociology of Science*, Merton, R.K. (ed.) The University of Chicago Press, 1973, pp. 267–278.

Millikan, R.G. (1984): *Language, thought and other biological categories*, MIT Press, Cambridge MA.

Millikan, R.G. (1993): *White queen psychology and other essays for Alice*, MIT Press, Cambridge MA.

Milne, P. (1996): *Log[P(h/eb)/ P(h/b)]* is the one true measure of confirmation, *Philosophy of Science*, 63, 21–26.

Minsky, M. (1985): *The society of mind*, Simon and Schuster, New York.

Moffitt, T.E. (1993): Adolescence-limited and life-course-persistent antisocial behavior: a developmental taxonomy, *Psychological Review*, 100, 674–701.

Moulines, C. (1984): Ontological reduction in the natural sciences, *Reduction in Science*, Balzer W. et al. (ed.), Reidel, Dordrecht, pp. 51–70.

Mulkay, M. (1977): Some connections between the quantitative history of science, the social history of science, and the sociology of science, in: *Proceedings of the International Seminar on Science Studies*, Löppönen P. (ed.), Academy of Finland, Helsinki, pp. 54–76.

Mura, A. (1990): When probabilistic support is inductive, *Philosophy of Science*, 57, 278–289.

Nagel, E. (1961): *The structure of science*, Routledge and Kegan Paul, London.

Nagel, E. (1979): Issues in the logic of reductive explanations, *Teleology revisited*, Columbia University Press. New York, 95–117.

NAS (1989): *On Being a Scientist: Responsible Conduct in Research*, National Academy of Sciences, Washington, USA.

Nelson-Gray, R., Herbert, J., Herbert, D., Sigmon, S. and Brannon, S. (1989): Effectiveness of matched, mismatched and package treatments of depression, *Journal of Behavior Therapy and Experimental Psychiatry*, 20.4, 281–294.

Newell, A. and Simon, H.A. (1972): *Human problem solving*, Prentice-Hall, Englewood Cliffs, N.J.

Nickles, Th. (1973): Two concepts of intertheoretic reduction, *The Journal of Philosophy*, 70.12, 181–201.

Nierop, M. van (1989): *Denken in tweespalt*, Eburon, Delft.

Niiniluoto, I. (1987a): *Truthlikeness*, Reidel, Dordrecht.

Niiniluoto, I. (1987b): How to define verisimilitude, in: Kuipers, 1987, pp. 11–24.

Niiniluoto, I. (1999): *Critical Scientific Realism*, Oxford University Press, Oxford.

Nowak, G. and Thagard, P. (1992a): Newton, Descartes, and explanatory coherence, in: *Philosophy of science, cognitive psychology and educational theory and practice*, Duschl, R. and Hamilton, R. (eds.) Albany: SUNY Press, pp. 99–115.

Nowak, G. and Thagard, P. (1992b): Copernicus, Ptolemy, and explanatory coherence, in: *Cognitive models of science*, Giere, R. (ed.) Minneapolis: University of Minnesota Press, pp. 274–309.

Nowak, L. (1974): Galileo of the social sciences, *Revolutionary World*, Vol. 8, pp. 5–11.

Nowak, L. (1977): On the structure of marxist dialectics, *Erkenntnis*, 11, 341–363.

Nowak, L. (1980): *The structure of idealization*, Reidel, Dordrecht.

Nowakowa, I. (1974): The concept of dialectical correspondence, *Dialectics and Humanism*, 3, 51–55.

Nowakowa, I. (1994): *The dynamics of idealizations*, Poznan Studies Vol. 34, Rodopi, Amsterdam.

Oddie, G. (1987): The picture theory of truthlikeness, in: Kuipers, 1987, pp. 25–46.

O'Keefe, J. and Nadel, L. (1978): *The hippocampus as a cognitive map*, Clarendon Press, Oxford.

Olson, M. (1965): *The logic of collective action*, Harvard University Press, Cambridge, MA.

Oppenheim, P. and Putnam, H. (1958): Unity of science as a working hypothesis, *Minnesota studies in the philosophy of science*, Vol. II, University of Minnesota Press, pp. 3–36.

Panhuysen, G. and Tuiten, A. (1993): How mental and physiological mechanisms gear into one other, in: *Recent trends in theoretical psychology*, Stam, H., Mos, L., Thorngate W. and Kaplan, B. (eds.) Springer, New York, pp. 283–293.

Panhuysen, G. and Tuiten, A. (1995): On the importance of being interactionist, in: *Trends and issues in theoretical psychology*, Lubek, I., van Hezewijk, R., Pheterson, G. and Tolman, Ch. Springer, New York, pp. 334–340.

Panofsky, W. and Phillips, M. (1955, 1962^2): *Classical electricity and magnetism*, Addison-Wesley, London.

Pardo, J., Pardo, P. and Raichle, M. (1993): Neural correlates of self-induced dysphoria, *American Journal of Psychiatry*, 150.5, 713–719.

Pearce, D. and Rantala, V. (1985): Approximative explanation is deductive-nomological, *Philosophy of Science*, 52, 126–140.

Peirce, C.S. (1934): *Collected Papers. Vol. 5*, Cambridge University Press, Cambridge, MA.

Pettit, Ph. (1992): Instituting a research ethics. Chilling and cautionary tales, *Bioethics*, 6.2, 89–112.

Pettit, Ph. (1993): *The common mind. An essay on psychology, society and politics*, Oxford University Press, New York. Paperback edition of 1996, with new postscript.

Pettit, Ph. (1996): Functional explanation and virtual selection, *The British Journal for the Philosophy of Science*, 47, 291–302.

Pirri, F. (1995): *Abduction*, PhD-dissertation, Université de Paris VI, Pierre et Marie Curie, Paris.

Pluhar, E. (1978): Emergence and Reduction, *Studies in History and Philosophy of Science*, 9.4, 279–289.

Popper, K.R. (1934/1959): *Logik der Forschung*, Vienna, 1934, translated as *The logic of scientific discovery*, Hutchinson, London, 1959.

Popper, K.R. (1963): *Conjectures and refutations*, Routledge and Kegan Paul, London.

Popper, K.R. (1982): Scientific reduction and the essential incompleteness of all science, *The open universe*, Rowman and Littlefield, Totowa, pp. 131–174.

Popper, K.R. (1983): *Realism and the aim of science*, Routledge, London.

Popper, K.R. and Miller, D. (1983): A proof of the impossibility of inductive probability, *Nature*, 302, 687–688.

Radder, H. (1988): *The material realization of science*, Van Gorcum Assen.

Radder, H. (1996): *In and about the world. Philosophical studies of science and technology*, State University of New York Press, New York.

Raine, A., Stoddard, J., Bihrle, S. and Buchsbaum, M. (1998): Prefrontal glucose deficits in murderers lacking psychosocial deprivation, *Neuropsychiatry, neuropsychology, and behavioral neurology*, 11.1, 1–7.

Rantala, V. (1997): Explanatory translation: beyond functionalism and reductionism, in: *Logic and scientific method*, Vol. 1, 10th LMPS-International Congress, Dalla Chiara, M.L., Doets, K. and van Benthem, J. (eds.) Kluwer, Dordrecht.

Rantala, V. and Vaden, T. (1994): Idealization in cognitive science, in: *Idealization VII: Structuralism, Idealization and Approximation*, Kuokkanen, M. (ed.) *Poznan Studies, Vol. 42*, Amsterdam, Rodopi, pp. 179–200.

Rescher, N. (1992): *A system of pragmatic idealism. Vol. I: Human knowledge in idealistic perspective*, Princeton University Press, Princeton.

Rosch, E. (1978): Principles of categorization, *Cognition and categorization*, Lawrence Erlbaum, Hillsdale, pp. 28–48.

Ruse, M. (1973): *The philosophy of biology*, Hutchinson, London.

Salmon, W. (1969): Partial entailment as a basis for inductive logic, in: *Essays in honor of Carl G. Hempel*, Rescher, N. (ed.) Reidel, Dordrecht, pp. 47–82.

Salmon, W. (1989): Four decades of scientific explanation, in: *Minnesota Studies in the Philosophy of Science*, Vol. XIII, Kitcher, P. and Salmon, W. (eds.) University of Minnesota Press, Minnesota, pp.3–219.

Saren, M. (1984): A classification and review of models of the intra-firm innovation process, *R and D Management*, 14.1, 11–24.

Saviotti, P. (1988): A characteristics approach to technological evolution and competition, Presented at the conference 'Recent trends in the economics of technological change', Manchester, 21–22 March.

Saviotti, P. (1996): *Technological evolution, variety and competition*, Edward Elgar, Cheltenham.

Scerri, E. (1997): Has the periodic table been successfully axiomatized? *Erkenntnis*, 47.2, 229–243.

Schäfer, W. et al. (eds.): (1983): *Finalization in science: the social orientation of scientific progress*, Reidel, Dordrecht.

Schaffner, K. (1967): Approaches to reduction, *Philosophy of Science*, 34, 137–147.

Schaffner, K. (1993): *Discovery and explanation in biology and medicine*, University of Chicago Press, Chicago.

Schlesinger, G. (1995): Measuring degrees of confirmation, *Analysis*, 55.3, 208–212.

Schlick, M. (1938): *Gesammelte Aufsätze*, Gerold, Wien.

Schulte, D., Künzel, R., Pepping, G. and Sculte-Bahrenberg, T. (1992): Tailor-made versus standardized therapy of phobic patients, *Advances in Behaviour Research and Therapy*, 14, 67–92.

Schurz, G. (1990): Paradoxical consequences of Balzer's and Gähde's criteria of theoriticity, *Erkenntnis*, 32, 161–214.

Schurz, G. and Weingartner, P. (1987): Verisimilitude defined by relevant consequence-elements, in Kuipers, 1987, pp. 47–77.

Schwartz, J. (1991): Reduction, elimination, and the mental, *Philosophy of Science*, 58.2, 203–220.

Searle, J. (1992): *The rediscovery of the mind*, MIT Press, Cambridge, MA.

Searle, J. (1995): *The construction of social reality*, Allan Lane, London.

Selz, O. (1924): *Die Gesetze der produktiven und reproduktiven Geistestätigkeit. Kurzfgefasste Darstellung*, Cohen, Bonn.

Shapere, D. (1982): The concept of observation in science and philosophy, *Philosophy of Science*, 49.4, 485–525.

Shrager, J. and Langley, P. (eds.) (1990): *Computational models of scientific discovery and theory formation*, Kaufmann, San Mateo.

Sie, H. (1989): *Industrieel onderzoek en haar relatie tot academisch onderzoek*, masters-thesis philosophy of science, University of Groningen.

Simon, H. (1969): The architecture of complexity, in: *The sciences of the artificial*, MIT Press, Cambridge MA. pp. 84–118.

Simon, H. (1980): *The sciences of the artificial*, 2nd Edn, MIT Press, Cambridge, MA.

Simon, H.A. and Groen, G.J. (1973): Ramsey eliminability and the testability of scientific theories, *The British Journal for the Philosophy of Science*, 24, 367–380.

Sklar, L. (1967): Types of inter-theoretic reduction, *The British Journal for the Philosophy of Science*, 18, 109–124.

Sklar, L. (1993): *Physics and Chance*, Cambridge University Press, Cambridge, MA.

Slezak, P. (1989): Scientific discovery by computer as empirical refutation of the strong program, *Social Studies of Science*, 19.4, 563–695, continued discussion in 21.1, 1991, 143–156.

Smolensky, P. (1988): On the proper treatment of connectionism, *Behavioral and brain sciences*, 11, 1–74.

Smullyan, R. (1968): *First order logic*, Springer Verlag, Berlin.

Sneed, J. (1971): *The logical structure of mathematical physics*, Reidel, Dordrecht.

Sober, E. (1998): Simplicity (in scientific theories), *Routledge Encyclopedia of Philosophy*, Vol. 8, pp. 780–783.

Sober, E. (2000a): *Philosophy of Biology*, second edition, Westview Press, Boulder/Oxford.

Sober, E. (2000b): Quine's two dogmas, *Supplementary Volume LXXIV*, Aristotelian Society, London, pp. 237–280.

Solomon, M. (1996): Information and the Ethics of Information Control in Science, *Perspectives on Science*, 4.2, 195–206.

Spiegel, D. et al. (1989): Effects of psychosocial treatment on survival of patients with metastatic breast cancer, *Lancet*, 2, 888–891.

Spiegel, D. (1993): *Living beyond limits*, Time Books, New York.

Steen, W. van der (1982): *Algemene methodologie voor biologen*, Bohn, Scheltema and Holkema, Utrecht-Antwerpen.

Stegmüller, W. (1973): *Theorie und Erfahrung*, Band II, Teil D, Springer, Berlin.

Stegmüller, W. (1973) *Carnap II: Normative Theorie des Induktiven Räsonierens*, Berlin: Springer.

Stegmüller, W. (1975): Hermeneutik und Wissenschaftstheorie: Erklären und Verstehen nach Georg Hendrik von Wright, in: *Hauptströmungen der Gegenwartsphilosophie. Band II*, Kröner, Stuttgart, pp. 103–146.

Stegmüller, W. (1986): *Realismus und Strukturalismus, Theorie und Erfahrung*, Band II, Teil H, Springer, Berlin.

Stiekema, E. (to appear) *A developmental theory of meaning*, PhD-dissertation, University of Groningen.

Stillings, N. et al. (1987): *Cognitive science. An introduction*, MIT Press, Cambridge.

Suppes, P. (1957): *Introduction to logic*, Van Nostrand, New York.

Tan, Y.-H. (1992): *Non-monotonic reasoning: logical architecture and philosophical applications*, Free University, Amsterdam.

Thagard, P. (1988): *Computational philosophy of science*, MIT Press, Cambridge.

Thagard, P. (1992): *Conceptual revolutions*, Princeton University Press.

Thompson, E. (1995): *Colour vision. A study in cognitive science and the philosophy of perception*, Routledge, London.

Tiger, L. (1980): *Optimism, the Biology of Hope*, Simon and Schuster, New York.

Toulmin, S. (1953): *The philosophy of science*, Hutchinson, London.

Tuomela (1973): *Theoretical concepts*, Springer, Berlin.

Tuomela (1995): *The importance of us: a philosophical study of basic social notions*, Stanford University Press, Stanford.

Valdés-Pérez, R. (1999): Principles of human computer collaboration for knowledge discovery in science, *Artificial Intelligence*, 107(2), 335–346.

Vos, R. (1991): *Drugs looking for diseases*, Kluwer, Dordrecht.

Vos, R. (1995): The logic and epistemology of the concept of drug and disease profile, in Kuipers and Mackor, 1995, pp. 69–86.

Weeder, P. and Kester, D. (1982): Variatie en selectie: de constructie van een industrieel produkt. Het geval Tenax, *Kennis en Methode*, VI.3, 221–251.

Wenegrat, B. (1990): *The divine archetype. The sociobiology and psychology of religion*, D.C. Heath and Company/Lexington, Massachusetts/Toronto.

Westmeyer, W. (ed.) (1989): *Psychological theories from a structuralist point of view*, Springer, Berlin.

Westmeyer, W. (ed.) (1992): *The structuralist program in psychology: foundations and applications*, Hogrefe and Huber, Seattle.

Withaar, H. (1979): De Context of Persuasion–een voorstel tot ruilverkaveling, *Kennis en Methode*, 3.1, 41–50.

Wolpert, L. (1992): *The unnatural nature of science*, Faber and Faber, London.

Woolgar, S. (1988): *Science: the very idea*, Tavistock, London.

Wouters, A. (1999): *Explanation without a cause*, dissertation University Utrecht, Quaestiones Infinitae. Vol. XXIX, ZENO Institute of Philosophy, Utrecht.

Wright, G. von (1971): *Explanation and understanding*, Routledge, London.

Zandvoort, H. (1982) An extension of Sneed's reconstruction of classical particle mechanics to complex applications, and an alternative approach to special force laws, *Erkenntnis*, 18, 39–63

Zandvoort, H. (1984): Lakatos and Nagel: a fruitful confrontation, *Zeitschrift fur allgemeine Wissenschaftstheorie*, XV-2, 299–307.

Zandvoort, H. (1986): *Models of scientific development and the case of NMR*, Synthese Library 184, Reidel, Dordrecht.

Zandvoort, H. (1988): Nuclear magnetic resonance and the acceptability of guiding assumptions, in: *Scrutinizing science: empirical studies of scientific change*, Donovan, A. et al. (eds.) Reidel, Dordrecht, pp. 337–358.

Zandvoort, H. (1989): De struktuur van theorieën en hun rol in verklaringen, De ontwikkeling van theorieën: onderzoeksprogramma's, in: *Wetenschapsleer*, Korthals, M. (ed.) Open Universiteit, Heerlen, pp. 24–47, 48–69.

Zandvoort, H. (1995): Concepts of interdisciplinarity and environmental science, in: *Cognitive patterns of science and common sense*, in Kuipers and Mackor, 1995, pp. 45–68.

Ziman, J. (1994): *Prometheus Bound: Science in a Dynamic Steady State*, Cambridge University Press, Cambridge.

Zwart, S.D. (1998): *Approach to the truth. Verisimilitude and truthlikeness*, dissertation Groningen, ILLC-Dissertation-Series-1998–02. Revised version to appear in the Synthese Library of Kluwer AP.

INDEX OF NAMES

INDEX OF SUBJECTS

SYNTHESE LIBRARY

1. J. M. Bochénski, *A Precis of Mathematical Logic*. Translated from French and German by O. Bird. 1959 ISBN 90-277-0073-7
2. P. Guiraud, *Problèmes et méthodes de la statistique linguistique*. 1959 ISBN 90-277-0025-7
3. H. Freudenthal (ed.), *The Concept and the Role of the Model in Mathematics and Natural and Social Sciences*. 1961 ISBN 90-277-0017-6
4. E. W. Beth, *Formal Methods*. An Introduction to Symbolic Logic and to the Study of Effective Operations in Arithmetic and Logic. 1962 ISBN 90-277-0069-9
5. B. H. Kazemier and D. Vuysje (eds.), *Logic and Language*. Studies dedicated to Professor Rudolf Carnap on the Occasion of His 70th Birthday. 1962 ISBN 90-277-0019-2
6. M. W. Wartofsky (ed.), *Proceedings of the Boston Colloquium for the Philosophy of Science, 1961–1962*. [Boston Studies in the Philosophy of Science, Vol. I] 1963 ISBN 90-277-0021-4
7. A. A. Zinov'ev, *Philosophical Problems of Many-valued Logic*. A revised edition, edited and translated (from Russian) by G. Küng and D.D. Comey. 1963 ISBN 90-277-0091-5
8. G. Gurvitch, *The Spectrum of Social Time*. Translated from French and edited by M. Korenbaum and P. Bosserman. 1964 ISBN 90-277-0006-0
9. P. Lorenzen, *Formal Logic*. Translated from German by F.J. Crosson. 1965 ISBN 90-277-0080-X
10. R. S. Cohen and M. W. Wartofsky (eds.), *Proceedings of the Boston Colloquium for the Philosophy of Science, 1962–1964*. In Honor of Philipp Frank. [Boston Studies in the Philosophy of Science, Vol. II] 1965 ISBN 90-277-9004-0
11. E. W. Beth, *Mathematical Thought*. An Introduction to the Philosophy of Mathematics. 1965 ISBN 90-277-0070-2
12. E. W. Beth and J. Piaget, *Mathematical Epistemology and Psychology*. Translated from French by W. Mays. 1966 ISBN 90-277-0071-0
13. G. Küng, *Ontology and the Logistic Analysis of Language*. An Enquiry into the Contemporary Views on Universals. Revised ed., translated from German. 1967 ISBN 90-277-0028-1
14. R. S. Cohen and M. W. Wartofsky (eds.), *Proceedings of the Boston Colloquium for the Philosophy of Sciences, 1964–1966*. In Memory of Norwood Russell Hanson. [Boston Studies in the Philosophy of Science, Vol. III] 1967 ISBN 90-277-0013-3
15. C. D. Broad, *Induction, Probability, and Causation*. Selected Papers. 1968 ISBN 90-277-0012-5
16. G. Patzig, *Aristotle's Theory of the Syllogism*. A Logical-philosophical Study of *Book A* of the *Prior Analytics*. Translated from German by J. Barnes. 1968 ISBN 90-277-0030-3
17. N. Rescher, *Topics in Philosophical Logic*. 1968 ISBN 90-277-0084-2
18. R. S. Cohen and M. W. Wartofsky (eds.), *Proceedings of the Boston Colloquium for the Philosophy of Science, 1966–1968, Part I*. [Boston Studies in the Philosophy of Science, Vol. IV] 1969 ISBN 90-277-0014-1
19. R. S. Cohen and M. W. Wartofsky (eds.), *Proceedings of the Boston Colloquium for the Philosophy of Science, 1966–1968, Part II*. [Boston Studies in the Philosophy of Science, Vol. V] 1969 ISBN 90-277-0015-X
20. J. W. Davis, D. J. Hockney and W. K. Wilson (eds.), *Philosophical Logic*. 1969 ISBN 90-277-0075-3
21. D. Davidson and J. Hintikka (eds.), *Words and Objections*. Essays on the Work of W. V. Quine. 1969, rev. ed. 1975 ISBN 90-277-0074-5; Pb 90-277-0602-6
22. P. Suppes, *Studies in the Methodology and Foundations of Science. Selected Papers from 1951 to 1969*. 1969 ISBN 90-277-0020-6
23. J. Hintikka, *Models for Modalities*. Selected Essays. 1969 ISBN 90-277-0078-8; Pb 90-277-0598-4

24. N. Rescher *et al.* (eds.), *Essays in Honor of Carl G. Hempel.* A Tribute on the Occasion of His 65th Birthday. 1969 ISBN 90-277-0085-0
25. P. V. Tavanec (ed.), *Problems of the Logic of Scientific Knowledge.* Translated from Russian. 1970 ISBN 90-277-0087-7
26. M. Swain (ed.), *Induction, Acceptance, and Rational Belief.* 1970 ISBN 90-277-0086-9
27. R. S. Cohen and R. J. Seeger (eds.), *Ernst Mach: Physicist and Philosopher.* [Boston Studies in the Philosophy of Science, Vol. VI]. 1970 ISBN 90-277-0016-8
28. J. Hintikka and P. Suppes, *Information and Inference.* 1970 ISBN 90-277-0155-5
29. K. Lambert, *Philosophical Problems in Logic.* Some Recent Developments. 1970 ISBN 90-277-0079-6
30. R. A. Eberle, *Nominalistic Systems.* 1970 ISBN 90-277-0161-X
31. P. Weingartner and G. Zecha (eds.), *Induction, Physics, and Ethics.* 1970 ISBN 90-277-0158-X
32. E. W. Beth, *Aspects of Modern Logic.* Translated from Dutch. 1970 ISBN 90-277-0173-3
33. R. Hilpinen (ed.), *Deontic Logic.* Introductory and Systematic Readings. 1971 *See also* No. 152. ISBN Pb (1981 rev.) 90-277-1302-2
34. J.-L. Krivine, *Introduction to Axiomatic Set Theory.* Translated from French. 1971 ISBN 90-277-0169-5; Pb 90-277-0411-2
35. J. D. Sneed, *The Logical Structure of Mathematical Physics.* 2nd rev. ed., 1979 ISBN 90-277-1056-2; Pb 90-277-1059-7
36. C. R. Kordig, *The Justification of Scientific Change.* 1971 ISBN 90-277-0181-4; Pb 90-277-0475-9
37. M. Čapek, *Bergson and Modern Physics.* A Reinterpretation and Re-evaluation. [Boston Studies in the Philosophy of Science, Vol. VII] 1971 ISBN 90-277-0186-5
38. N. R. Hanson, *What I Do Not Believe, and Other Essays.* Ed. by S. Toulmin and H. Woolf. 1971 ISBN 90-277-0191-1
39. R. C. Buck and R. S. Cohen (eds.), *PSA 1970.* Proceedings of the Second Biennial Meeting of the Philosophy of Science Association, Boston, Fall 1970. In Memory of Rudolf Carnap. [Boston Studies in the Philosophy of Science, Vol. VIII] 1971 ISBN 90-277-0187-3; Pb 90-277-0309-4
40. D. Davidson and G. Harman (eds.), *Semantics of Natural Language.* 1972 ISBN 90-277-0304-3; Pb 90-277-0310-8
41. Y. Bar-Hillel (ed.), *Pragmatics of Natural Languages.* 1971 ISBN 90-277-0194-6; Pb 90-277-0599-2
42. S. Stenlund, *Combinators, γ Terms and Proof Theory.* 1972 ISBN 90-277-0305-1
43. M. Strauss, *Modern Physics and Its Philosophy.* Selected Paper in the Logic, History, and Philosophy of Science. 1972 ISBN 90-277-0230-6
44. M. Bunge, *Method, Model and Matter.* 1973 ISBN 90-277-0252-7
45. M. Bunge, *Philosophy of Physics.* 1973 ISBN 90-277-0253-5
46. A. A. Zinov'ev, *Foundations of the Logical Theory of Scientific Knowledge (Complex Logic).* Revised and enlarged English edition with an appendix by G. A. Smirnov, E. A. Sidorenka, A. M. Fedina and L. A. Bobrova. [Boston Studies in the Philosophy of Science, Vol. IX] 1973 ISBN 90-277-0193-8; Pb 90-277-0324-8
47. L. Tondl, *Scientific Procedures.* A Contribution concerning the Methodological Problems of Scientific Concepts and Scientific Explanation. Translated from Czech by D. Short. Edited by R.S. Cohen and M.W. Wartofsky. [Boston Studies in the Philosophy of Science, Vol. X] 1973 ISBN 90-277-0147-4; Pb 90-277-0323-X
48. N. R. Hanson, *Constellations and Conjectures.* 1973 ISBN 90-277-0192-X

49. K. J. J. Hintikka, J. M. E. Moravcsik and P. Suppes (eds.), *Approaches to Natural Language.* 1973 ISBN 90-277-0220-9; Pb 90-277-0233-0
50. M. Bunge (ed.), *Exact Philosophy.* Problems, Tools and Goals. 1973 ISBN 90-277-0251-9
51. R. J. Bogdan and I. Niiniluoto (eds.), *Logic, Language and Probability.* 1973 ISBN 90-277-0312-4
52. G. Pearce and P. Maynard (eds.), *Conceptual Change.* 1973 ISBN 90-277-0287-X; Pb 90-277-0339-6
53. I. Niiniluoto and R. Tuomela, *Theoretical Concepts and Hypothetico-inductive Inference.* 1973 ISBN 90-277-0343-4
54. R. Fraissé, *Course of Mathematical Logic* – Volume 1: *Relation and Logical Formula.* Translated from French. 1973 ISBN 90-277-0268-3; Pb 90-277-0403-1 (For *Volume 2* see under No. 69).
55. A. Grünbaum, *Philosophical Problems of Space and Time.* Edited by R.S. Cohen and M.W. Wartofsky. 2nd enlarged ed. [Boston Studies in the Philosophy of Science, Vol. XII] 1973 ISBN 90-277-0357-4; Pb 90-277-0358-2
56. P. Suppes (ed.), *Space, Time and Geometry.* 1973 ISBN 90-277-0386-8; Pb 90-277-0442-2
57. H. Kelsen, *Essays in Legal and Moral Philosophy.* Selected and introduced by O. Weinberger. Translated from German by P. Heath. 1973 ISBN 90-277-0388-4
58. R. J. Seeger and R. S. Cohen (eds.), *Philosophical Foundations of Science.* [Boston Studies in the Philosophy of Science, Vol. XI] 1974 ISBN 90-277-0390-6; Pb 90-277-0376-0
59. R. S. Cohen and M. W. Wartofsky (eds.), *Logical and Epistemological Studies in Contemporary Physics.* [Boston Studies in the Philosophy of Science, Vol. XIII] 1973 ISBN 90-277-0391-4; Pb 90-277-0377-9
60. R. S. Cohen and M. W. Wartofsky (eds.), *Methodological and Historical Essays in the Natural and Social Sciences. Proceedings of the Boston Colloquium for the Philosophy of Science, 1969–1972.* [Boston Studies in the Philosophy of Science, Vol. XIV] 1974 ISBN 90-277-0392-2; Pb 90-277-0378-7
61. R. S. Cohen, J. J. Stachel and M. W. Wartofsky (eds.), *For Dirk Struik. Scientific, Historical and Political Essays.* [Boston Studies in the Philosophy of Science, Vol. XV] 1974 ISBN 90-277-0393-0; Pb 90-277-0379-5
62. K. Ajdukiewicz, *Pragmatic Logic.* Translated from Polish by O. Wojtasiewicz. 1974 ISBN 90-277-0326-4
63. S. Stenlund (ed.), *Logical Theory and Semantic Analysis.* Essays dedicated to Stig Kanger on His 50th Birthday. 1974 ISBN 90-277-0438-4
64. K. F. Schaffner and R. S. Cohen (eds.), *PSA 1972. Proceedings of the Third Biennial Meeting of the Philosophy of Science Association.* [Boston Studies in the Philosophy of Science, Vol. XX] 1974 ISBN 90-277-0408-2; Pb 90-277-0409-0
65. H. E. Kyburg, Jr., *The Logical Foundations of Statistical Inference.* 1974 ISBN 90-277-0330-2; Pb 90-277-0430-9
66. M. Grene, *The Understanding of Nature.* Essays in the Philosophy of Biology. [Boston Studies in the Philosophy of Science, Vol. XXIII] 1974 ISBN 90-277-0462-7; Pb 90-277-0463-5
67. J. M. Broekman, *Structuralism: Moscow, Prague, Paris.* Translated from German. 1974 ISBN 90-277-0478-3
68. N. Geschwind, *Selected Papers on Language and the Brain.* [Boston Studies in the Philosophy of Science, Vol. XVI] 1974 ISBN 90-277-0262-4; Pb 90-277-0263-2
69. R. Fraissé, *Course of Mathematical Logic* – Volume 2: *Model Theory.* Translated from French. 1974 ISBN 90-277-0269-1; Pb 90-277-0510-0 (For *Volume 1* see under No. 54)

70. A. Grzegorczyk, *An Outline of Mathematical Logic.* Fundamental Results and Notions explained with all Details. Translated from Polish. 1974 ISBN 90-277-0359-0; Pb 90-277-0447-3
71. F. von Kutschera, *Philosophy of Language.* 1975 ISBN 90-277-0591-7
72. J. Manninen and R. Tuomela (eds.), *Essays on Explanation and Understanding.* Studies in the Foundations of Humanities and Social Sciences. 1976 ISBN 90-277-0592-5
73. J. Hintikka (ed.), *Rudolf Carnap, Logical Empiricist.* Materials and Perspectives. 1975 ISBN 90-277-0583-6
74. M. Čapek (ed.), *The Concepts of Space and Time.* Their Structure and Their Development. [Boston Studies in the Philosophy of Science, Vol. XXII] 1976 ISBN 90-277-0355-8; Pb 90-277-0375-2
75. J. Hintikka and U. Remes, *The Method of Analysis.* Its Geometrical Origin and Its General Significance. [Boston Studies in the Philosophy of Science, Vol. XXV] 1974 ISBN 90-277-0532-1; Pb 90-277-0543-7
76. J. E. Murdoch and E. D. Sylla (eds.), *The Cultural Context of Medieval Learning.* [Boston Studies in the Philosophy of Science, Vol. XXVI] 1975 ISBN 90-277-0560-7; Pb 90-277-0587-9
77. S. Amsterdamski, *Between Experience and Metaphysics.* Philosophical Problems of the Evolution of Science. [Boston Studies in the Philosophy of Science, Vol. XXXV] 1975 ISBN 90-277-0568-2; Pb 90-277-0580-1
78. P. Suppes (ed.), *Logic and Probability in Quantum Mechanics.* 1976 ISBN 90-277-0570-4; Pb 90-277-1200-X
79. H. von Helmholtz: *Epistemological Writings. The Paul Hertz / Moritz Schlick Centenary Edition of 1921 with Notes and Commentary by the Editors.* Newly translated from German by M. F. Lowe. Edited, with an Introduction and Bibliography, by R. S. Cohen and Y. Elkana. [Boston Studies in the Philosophy of Science, Vol. XXXVII] 1975 ISBN 90-277-0290-X; Pb 90-277-0582-8
80. J. Agassi, *Science in Flux.* [Boston Studies in the Philosophy of Science, Vol. XXVIII] 1975 ISBN 90-277-0584-4; Pb 90-277-0612-2
81. S. G. Harding (ed.), *Can Theories Be Refuted?* Essays on the Duhem-Quine Thesis. 1976 ISBN 90-277-0629-8; Pb 90-277-0630-1
82. S. Nowak, *Methodology of Sociological Research.* General Problems. 1977 ISBN 90-277-0486-4
83. J. Piaget, J.-B. Grize, A. Szeminśska and V. Bang, *Epistemology and Psychology of Functions.* Translated from French. 1977 ISBN 90-277-0804-5
84. M. Grene and E. Mendelsohn (eds.), *Topics in the Philosophy of Biology.* [Boston Studies in the Philosophy of Science, Vol. XXVII] 1976 ISBN 90-277-0595-X; Pb 90-277-0596-8
85. E. Fischbein, *The Intuitive Sources of Probabilistic Thinking in Children.* 1975 ISBN 90-277-0626-3; Pb 90-277-1190-9
86. E. W. Adams, *The Logic of Conditionals.* An Application of Probability to Deductive Logic. 1975 ISBN 90-277-0631-X
87. M. Przełęcki and R. Wójcicki (eds.), *Twenty-Five Years of Logical Methodology in Poland.* Translated from Polish. 1976 ISBN 90-277-0601-8
88. J. Topolski, *The Methodology of History.* Translated from Polish by O. Wojtasiewicz. 1976 ISBN 90-277-0550-X
89. A. Kasher (ed.), *Language in Focus: Foundations, Methods and Systems.* Essays dedicated to Yehoshua Bar-Hillel. [Boston Studies in the Philosophy of Science, Vol. XLIII] 1976 ISBN 90-277-0644-1; Pb 90-277-0645-X

90. J. Hintikka, *The Intentions of Intentionality and Other New Models for Modalities.* 1975 ISBN 90-277-0633-6; Pb 90-277-0634-4
91. W. Stegmüller, *Collected Papers on Epistemology, Philosophy of Science and History of Philosophy.* 2 Volumes. 1977 Set ISBN 90-277-0767-7
92. D. M. Gabbay, *Investigations in Modal and Tense Logics with Applications to Problems in Philosophy and Linguistics.* 1976 ISBN 90-277-0656-5
93. R. J. Bogdan, *Local Induction.* 1976 ISBN 90-277-0649-2
94. S. Nowak, *Understanding and Prediction.* Essays in the Methodology of Social and Behavioral Theories. 1976 ISBN 90-277-0558-5; Pb 90-277-1199-2
95. P. Mittelstaedt, *Philosophical Problems of Modern Physics.* [Boston Studies in the Philosophy of Science, Vol. XVIII] 1976 ISBN 90-277-0285-3; Pb 90-277-0506-2
96. G. Holton and W. A. Blanpied (eds.), *Science and Its Public: The Changing Relationship.* [Boston Studies in the Philosophy of Science, Vol. XXXIII] 1976 ISBN 90-277-0657-3; Pb 90-277-0658-1
97. M. Brand and D. Walton (eds.), *Action Theory.* 1976 ISBN 90-277-0671-9
98. P. Gochet, *Outline of a Nominalist Theory of Propositions.* An Essay in the Theory of Meaning and in the Philosophy of Logic. 1980 ISBN 90-277-1031-7
99. R. S. Cohen, P. K. Feyerabend, and M. W. Wartofsky (eds.), *Essays in Memory of Imre Lakatos.* [Boston Studies in the Philosophy of Science, Vol. XXXIX] 1976 ISBN 90-277-0654-9; Pb 90-277-0655-7
100. R. S. Cohen and J. J. Stachel (eds.), *Selected Papers of Léon Rosenfield.* [Boston Studies in the Philosophy of Science, Vol. XXI] 1979 ISBN 90-277-0651-4; Pb 90-277-0652-2
101. R. S. Cohen, C. A. Hooker, A. C. Michalos and J. W. van Evra (eds.), *PSA 1974. Proceedings of the 1974 Biennial Meeting of the Philosophy of Science Association.* [Boston Studies in the Philosophy of Science, Vol. XXXII] 1976 ISBN 90-277-0647-6; Pb 90-277-0648-4
102. Y. Fried and J. Agassi, *Paranoia.* A Study in Diagnosis. [Boston Studies in the Philosophy of Science, Vol. L] 1976 ISBN 90-277-0704-9; Pb 90-277-0705-7
103. M. Przełęcki, K. Szaniawski and R. Wójcicki (eds.), *Formal Methods in the Methodology of Empirical Sciences.* 1976 ISBN 90-277-0698-0
104. J. M. Vickers, *Belief and Probability.* 1976 ISBN 90-277-0744-8
105. K. H. Wolff, *Surrender and Catch.* Experience and Inquiry Today. [Boston Studies in the Philosophy of Science, Vol. LI] 1976 ISBN 90-277-0758-8; Pb 90-277-0765-0
106. K. Kosík, *Dialectics of the Concrete.* A Study on Problems of Man and World. [Boston Studies in the Philosophy of Science, Vol. LII] 1976 ISBN 90-277-0761-8; Pb 90-277-0764-2
107. N. Goodman, *The Structure of Appearance.* 3rd ed. with an Introduction by G. Hellman. [Boston Studies in the Philosophy of Science, Vol. LIII] 1977 ISBN 90-277-0773-1; Pb 90-277-0774-X
108. K. Ajdukiewicz, *The Scientific World-Perspective and Other Essays, 1931-1963.* Translated from Polish. Edited and with an Introduction by J. Giedymin. 1978 ISBN 90-277-0527-5
109. R. L. Causey, *Unity of Science.* 1977 ISBN 90-277-0779-0
110. R. E. Grandy, *Advanced Logic for Applications.* 1977 ISBN 90-277-0781-2
111. R. P. McArthur, *Tense Logic.* 1976 ISBN 90-277-0697-2
112. L. Lindahl, *Position and Change.* A Study in Law and Logic. Translated from Swedish by P. Needham. 1977 ISBN 90-277-0787-1
113. R. Tuomela, *Dispositions.* 1978 ISBN 90-277-0810-X
114. H. A. Simon, *Models of Discovery and Other Topics in the Methods of Science.* [Boston Studies in the Philosophy of Science, Vol. LIV] 1977 ISBN 90-277-0812-6; Pb 90-277-0858-4

115. R. D. Rosenkrantz, *Inference, Method and Decision.* Towards a Bayesian Philosophy of Science. 1977 ISBN 90-277-0817-7; Pb 90-277-0818-5
116. R. Tuomela, *Human Action and Its Explanation.* A Study on the Philosophical Foundations of Psychology. 1977 ISBN 90-277-0824-X
117. M. Lazerowitz, *The Language of Philosophy.* Freud and Wittgenstein. [Boston Studies in the Philosophy of Science, Vol. LV] 1977 ISBN 90-277-0826-6; Pb 90-277-0862-2
118. Not published 119. J. Pelc (ed.), *Semiotics in Poland, 1894–1969.* Translated from Polish. 1979 ISBN 90-277-0811-8
120. I. Pörn, *Action Theory and Social Science.* Some Formal Models. 1977 ISBN 90-277-0846-0
121. J. Margolis, *Persons and Mind.* The Prospects of Nonreductive Materialism. [Boston Studies in the Philosophy of Science, Vol. LVII] 1977 ISBN 90-277-0854-1; Pb 90-277-0863-0
122. J. Hintikka, I. Niiniluoto, and E. Saarinen (eds.), *Essays on Mathematical and Philosophical Logic.* 1979 ISBN 90-277-0879-7
123. T. A. F. Kuipers, *Studies in Inductive Probability and Rational Expectation.* 1978 ISBN 90-277-0882-7
124. E. Saarinen, R. Hilpinen, I. Niiniluoto and M. P. Hintikka (eds.), *Essays in Honour of Jaakko Hintikka on the Occasion of His 50th Birthday.* 1979 ISBN 90-277-0916-5
125. G. Radnitzky and G. Andersson (eds.), *Progress and Rationality in Science.* [Boston Studies in the Philosophy of Science, Vol. LVIII] 1978 ISBN 90-277-0921-1; Pb 90-277-0922-X
126. P. Mittelstaedt, *Quantum Logic.* 1978 ISBN 90-277-0925-4
127. K. A. Bowen, *Model Theory for Modal Logic.* Kripke Models for Modal Predicate Calculi. 1979 ISBN 90-277-0929-7
128. H. A. Bursen, *Dismantling the Memory Machine.* A Philosophical Investigation of Machine Theories of Memory. 1978 ISBN 90-277-0933-5
129. M. W. Wartofsky, *Models.* Representation and the Scientific Understanding. [Boston Studies in the Philosophy of Science, Vol. XLVIII] 1979 ISBN 90-277-0736-7; Pb 90-277-0947-5
130. D. Ihde, *Technics and Praxis.* A Philosophy of Technology. [Boston Studies in the Philosophy of Science, Vol. XXIV] 1979 ISBN 90-277-0953-X; Pb 90-277-0954-8
131. J. J. Wiatr (ed.), *Polish Essays in the Methodology of the Social Sciences.* [Boston Studies in the Philosophy of Science, Vol. XXIX] 1979 ISBN 90-277-0723-5; Pb 90-277-0956-4
132. W. C. Salmon (ed.), *Hans Reichenbach: Logical Empiricist.* 1979 ISBN 90-277-0958-0
133. P. Bieri, R.-P. Horstmann and L. Krüger (eds.), *Transcendental Arguments in Science.* Essays in Epistemology. 1979 ISBN 90-277-0963-7; Pb 90-277-0964-5
134. M. Marković and G. Petrović (eds.), *Praxis.* Yugoslav Essays in the Philosophy and Methodology of the Social Sciences. [Boston Studies in the Philosophy of Science, Vol. XXXVI] 1979 ISBN 90-277-0727-8; Pb 90-277-0968-8
135. R. Wójcicki, *Topics in the Formal Methodology of Empirical Sciences.* Translated from Polish. 1979 ISBN 90-277-1004-X
136. G. Radnitzky and G. Andersson (eds.), *The Structure and Development of Science.* [Boston Studies in the Philosophy of Science, Vol. LIX] 1979 ISBN 90-277-0994-7; Pb 90-277-0995-5
137. J. C. Webb, *Mechanism, Mentalism and Metamathematics.* An Essay on Finitism. 1980 ISBN 90-277-1046-5
138. D. F. Gustafson and B. L. Tapscott (eds.), *Body, Mind and Method.* Essays in Honor of Virgil C. Aldrich. 1979 ISBN 90-277-1013-9
139. L. Nowak, *The Structure of Idealization.* Towards a Systematic Interpretation of the Marxian Idea of Science. 1980 ISBN 90-277-1014-7

140. C. Perelman, *The New Rhetoric and the Humanities.* Essays on Rhetoric and Its Applications. Translated from French and German. With an Introduction by H. Zyskind. 1979
ISBN 90-277-1018-X; Pb 90-277-1019-8
141. W. Rabinowicz, *Universalizability.* A Study in Morals and Metaphysics. 1979
ISBN 90-277-1020-2
142. C. Perelman, *Justice, Law and Argument.* Essays on Moral and Legal Reasoning. Translated from French and German. With an Introduction by H.J. Berman. 1980
ISBN 90-277-1089-9; Pb 90-277-1090-2
143. S. Kanger and S. Öhman (eds.), *Philosophy and Grammar.* Papers on the Occasion of the Quincentennial of Uppsala University. 1981 ISBN 90-277-1091-0
144. T. Pawlowski, *Concept Formation in the Humanities and the Social Sciences.* 1980
ISBN 90-277-1096-1
145. J. Hintikka, D. Gruender and E. Agazzi (eds.), *Theory Change, Ancient Axiomatics and Galileo's Methodology.* Proceedings of the 1978 Pisa Conference on the History and Philosophy of Science, Volume I. 1981 ISBN 90-277-1126-7
146. J. Hintikka, D. Gruender and E. Agazzi (eds.), *Probabilistic Thinking, Thermodynamics, and the Interaction of the History and Philosophy of Science.* Proceedings of the 1978 Pisa Conference on the History and Philosophy of Science, Volume II. 1981 ISBN 90-277-1127-5
147. U. Mönnich (ed.), *Aspects of Philosophical Logic.* Some Logical Forays into Central Notions of Linguistics and Philosophy. 1981 ISBN 90-277-1201-8
148. D. M. Gabbay, *Semantical Investigations in Heyting's Intuitionistic Logic.* 1981
ISBN 90-277-1202-6
149. E. Agazzi (ed.), *Modern Logic – A Survey.* Historical, Philosophical, and Mathematical Aspects of Modern Logic and Its Applications. 1981 ISBN 90-277-1137-2
150. A. F. Parker-Rhodes, *The Theory of Indistinguishables.* A Search for Explanatory Principles below the Level of Physics. 1981 ISBN 90-277-1214-X
151. J. C. Pitt, *Pictures, Images, and Conceptual Change.* An Analysis of Wilfrid Sellars' Philosophy of Science. 1981 ISBN 90-277-1276-X; Pb 90-277-1277-8
152. R. Hilpinen (ed.), *New Studies in Deontic Logic.* Norms, Actions, and the Foundations of Ethics. 1981 ISBN 90-277-1278-6; Pb 90-277-1346-4
153. C. Dilworth, *Scientific Progress.* A Study Concerning the Nature of the Relation between Successive Scientific Theories. 3rd rev. ed., 1994 ISBN 0-7923-2487-0; Pb 0-7923-2488-9
154. D. Woodruff Smith and R. McIntyre, *Husserl and Intentionality.* A Study of Mind, Meaning, and Language. 1982 ISBN 90-277-1392-8; Pb 90-277-1730-3
155. R. J. Nelson, *The Logic of Mind.* 2nd. ed., 1989 ISBN 90-277-2819-4; Pb 90-277-2822-4
156. J. F. A. K. van Benthem, *The Logic of Time.* A Model-Theoretic Investigation into the Varieties of Temporal Ontology, and Temporal Discourse. 1983; 2nd ed., 1991 ISBN 0-7923-1081-0
157. R. Swinburne (ed.), *Space, Time and Causality.* 1983 ISBN 90-277-1437-1
158. E. T. Jaynes, *Papers on Probability, Statistics and Statistical Physics.* Ed. by R. D. Rozenkrantz. 1983 ISBN 90-277-1448-7; Pb (1989) 0-7923-0213-3
159. T. Chapman, *Time: A Philosophical Analysis.* 1982 ISBN 90-277-1465-7
160. E. N. Zalta, *Abstract Objects.* An Introduction to Axiomatic Metaphysics. 1983
ISBN 90-277-1474-6
161. S. Harding and M. B. Hintikka (eds.), *Discovering Reality.* Feminist Perspectives on Epistemology, Metaphysics, Methodology, and Philosophy of Science. 1983
ISBN 90-277-1496-7; Pb 90-277-1538-6
162. M. A. Stewart (ed.), *Law, Morality and Rights.* 1983 ISBN 90-277-1519-X

163. D. Mayr and G. Süssmann (eds.), *Space, Time, and Mechanics.* Basic Structures of a Physical Theory. 1983 ISBN 90-277-1525-4
164. D. Gabbay and F. Guenthner (eds.), *Handbook of Philosophical Logic.* Vol. I: Elements of Classical Logic. 1983 ISBN 90-277-1542-4
165. D. Gabbay and F. Guenthner (eds.), *Handbook of Philosophical Logic.* Vol. II: Extensions of Classical Logic. 1984 ISBN 90-277-1604-8
166. D. Gabbay and F. Guenthner (eds.), *Handbook of Philosophical Logic.* Vol. III: Alternative to Classical Logic. 1986 ISBN 90-277-1605-6
167. D. Gabbay and F. Guenthner (eds.), *Handbook of Philosophical Logic.* Vol. IV: Topics in the Philosophy of Language. 1989 ISBN 90-277-1606-4
168. A. J. I. Jones, *Communication and Meaning.* An Essay in Applied Modal Logic. 1983 ISBN 90-277-1543-2
169. M. Fitting, *Proof Methods for Modal and Intuitionistic Logics.* 1983 ISBN 90-277-1573-4
170. J. Margolis, *Culture and Cultural Entities.* Toward a New Unity of Science. 1984 ISBN 90-277-1574-2
171. R. Tuomela, *A Theory of Social Action.* 1984 ISBN 90-277-1703-6
172. J. J. E. Gracia, E. Rabossi, E. Villanueva and M. Dascal (eds.), *Philosophical Analysis in Latin America.* 1984 ISBN 90-277-1749-4
173. P. Ziff, *Epistemic Analysis.* A Coherence Theory of Knowledge. 1984 ISBN 90-277-1751-7
174. P. Ziff, *Antiaesthetics.* An Appreciation of the Cow with the Subtile Nose. 1984 ISBN 90-277-1773-7
175. W. Balzer, D. A. Pearce, and H.-J. Schmidt (eds.), *Reduction in Science.* Structure, Examples, Philosophical Problems. 1984 ISBN 90-277-1811-3
176. A. Peczenik, L. Lindahl and B. van Roermund (eds.), *Theory of Legal Science.* Proceedings of the Conference on Legal Theory and Philosophy of Science (Lund, Sweden, December 1983). 1984 ISBN 90-277-1834-2
177. I. Niiniluoto, *Is Science Progressive?* 1984 ISBN 90-277-1835-0
178. B. K. Matilal and J. L. Shaw (eds.), *Analytical Philosophy in Comparative Perspective.* Exploratory Essays in Current Theories and Classical Indian Theories of Meaning and Reference. 1985 ISBN 90-277-1870-9
179. P. Kroes, *Time: Its Structure and Role in Physical Theories.* 1985 ISBN 90-277-1894-6
180. J. H. Fetzer, *Sociobiology and Epistemology.* 1985 ISBN 90-277-2005-3; Pb 90-277-2006-1
181. L. Haaparanta and J. Hintikka (eds.), *Frege Synthesized.* Essays on the Philosophical and Foundational Work of Gottlob Frege. 1986 ISBN 90-277-2126-2
182. M. Detlefsen, *Hilbert's Program.* An Essay on Mathematical Instrumentalism. 1986 ISBN 90-277-2151-3
183. J. L. Golden and J. J. Pilotta (eds.), *Practical Reasoning in Human Affairs.* Studies in Honor of Chaim Perelman. 1986 ISBN 90-277-2255-2
184. H. Zandvoort, *Models of Scientific Development and the Case of Nuclear Magnetic Resonance.* 1986 ISBN 90-277-2351-6
185. I. Niiniluoto, *Truthlikeness.* 1987 ISBN 90-277-2354-0
186. W. Balzer, C. U. Moulines and J. D. Sneed, *An Architectonic for Science.* The Structuralist Program. 1987 ISBN 90-277-2403-2
187. D. Pearce, *Roads to Commensurability.* 1987 ISBN 90-277-2414-8
188. L. M. Vaina (ed.), *Matters of Intelligence.* Conceptual Structures in Cognitive Neuroscience. 1987 ISBN 90-277-2460-1

189. H. Siegel, *Relativism Refuted*. A Critique of Contemporary Epistemological Relativism. 1987 ISBN 90-277-2469-5
190. W. Callebaut and R. Pinxten, *Evolutionary Epistemology*. A Multiparadigm Program, with a Complete Evolutionary Epistemology Bibliograph. 1987 ISBN 90-277-2582-9
191. J. Kmita, *Problems in Historical Epistemology*. 1988 ISBN 90-277-2199-8
192. J. H. Fetzer (ed.), *Probability and Causality*. Essays in Honor of Wesley C. Salmon, with an Annotated Bibliography. 1988 ISBN 90-277-2607-8; Pb 1-5560-8052-2
193. A. Donovan, L. Laudan and R. Laudan (eds.), *Scrutinizing Science*. Empirical Studies of Scientific Change. 1988 ISBN 90-277-2608-6
194. H.R. Otto and J.A. Tuedio (eds.), *Perspectives on Mind*. 1988 ISBN 90-277-2640-X
195. D. Batens and J.P. van Bendegem (eds.), *Theory and Experiment*. Recent Insights and New Perspectives on Their Relation. 1988 ISBN 90-277-2645-0
196. J. Österberg, *Self and Others*. A Study of Ethical Egoism. 1988 ISBN 90-277-2648-5
197. D.H. Helman (ed.), *Analogical Reasoning*. Perspectives of Artificial Intelligence, Cognitive Science, and Philosophy. 1988 ISBN 90-277-2711-2
198. J. Woleński, *Logic and Philosophy in the Lvov-Warsaw School*. 1989 ISBN 90-277-2749-X
199. R. Wójcicki, *Theory of Logical Calculi*. Basic Theory of Consequence Operations. 1988 ISBN 90-277-2785-6
200. J. Hintikka and M.B. Hintikka, *The Logic of Epistemology and the Epistemology of Logic*. Selected Essays. 1989 ISBN 0-7923-0040-8; Pb 0-7923-0041-6
201. E. Agazzi (ed.), *Probability in the Sciences*. 1988 ISBN 90-277-2808-9
202. M. Meyer (ed.), *From Metaphysics to Rhetoric*. 1989 ISBN 90-277-2814-3
203. R.L. Tieszen, *Mathematical Intuition*. Phenomenology and Mathematical Knowledge. 1989 ISBN 0-7923-0131-5
204. A. Melnick, *Space, Time, and Thought in Kant*. 1989 ISBN 0-7923-0135-8
205. D.W. Smith, *The Circle of Acquaintance*. Perception, Consciousness, and Empathy. 1989 ISBN 0-7923-0252-4
206. M.H. Salmon (ed.), *The Philosophy of Logical Mechanism*. Essays in Honor of Arthur W. Burks. With his Responses, and with a Bibliography of Burk's Work. 1990 ISBN 0-7923-0325-3
207. M. Kusch, *Language as Calculus vs. Language as Universal Medium*. A Study in Husserl, Heidegger, and Gadamer. 1989 ISBN 0-7923-0333-4
208. T.C. Meyering, *Historical Roots of Cognitive Science*. The Rise of a Cognitive Theory of Perception from Antiquity to the Nineteenth Century. 1989 ISBN 0-7923-0349-0
209. P. Kosso, *Observability and Observation in Physical Science*. 1989 ISBN 0-7923-0389-X
210. J. Kmita, *Essays on the Theory of Scientific Cognition*. 1990 ISBN 0-7923-0441-1
211. W. Sieg (ed.), *Acting and Reflecting*. The Interdisciplinary Turn in Philosophy. 1990 ISBN 0-7923-0512-4
212. J. Karpiński, *Causality in Sociological Research*. 1990 ISBN 0-7923-0546-9
213. H.A. Lewis (ed.), *Peter Geach: Philosophical Encounters*. 1991 ISBN 0-7923-0823-9
214. M. Ter Hark, *Beyond the Inner and the Outer*. Wittgenstein's Philosophy of Psychology. 1990 ISBN 0-7923-0850-6
215. M. Gosselin, *Nominalism and Contemporary Nominalism*. Ontological and Epistemological Implications of the Work of W.V.O. Quine and of N. Goodman. 1990 ISBN 0-7923-0904-9
216. J.H. Fetzer, D. Shatz and G. Schlesinger (eds.), *Definitions and Definability*. Philosophical Perspectives. 1991 ISBN 0-7923-1046-2
217. E. Agazzi and A. Cordero (eds.), *Philosophy and the Origin and Evolution of the Universe*. 1991 ISBN 0-7923-1322-4

218. M. Kusch, *Foucault's Strata and Fields.* An Investigation into Archaeological and Genealogical Science Studies. 1991 ISBN 0-7923-1462-X
219. C.J. Posy, *Kant's Philosophy of Mathematics.* Modern Essays. 1992 ISBN 0-7923-1495-6
220. G. Van de Vijver, *New Perspectives on Cybernetics.* Self-Organization, Autonomy and Connectionism. 1992 ISBN 0-7923-1519-7
221. J.C. Nyíri, *Tradition and Individuality.* Essays. 1992 ISBN 0-7923-1566-9
222. R. Howell, *Kant's Transcendental Deduction.* An Analysis of Main Themes in His Critical Philosophy. 1992 ISBN 0-7923-1571-5
223. A. García de la Sienra, *The Logical Foundations of the Marxian Theory of Value.* 1992 ISBN 0-7923-1778-5
224. D.S. Shwayder, *Statement and Referent.* An Inquiry into the Foundations of Our Conceptual Order. 1992 ISBN 0-7923-1803-X
225. M. Rosen, *Problems of the Hegelian Dialectic.* Dialectic Reconstructed as a Logic of Human Reality. 1993 ISBN 0-7923-2047-6
226. P. Suppes, *Models and Methods in the Philosophy of Science: Selected Essays.* 1993 ISBN 0-7923-2211-8
227. R. M. Dancy (ed.), *Kant and Critique: New Essays in Honor of W. H. Werkmeister.* 1993 ISBN 0-7923-2244-4
228. J. Woleński (ed.), *Philosophical Logic in Poland.* 1993 ISBN 0-7923-2293-2
229. M. De Rijke (ed.), *Diamonds and Defaults.* Studies in Pure and Applied Intensional Logic. 1993 ISBN 0-7923-2342-4
230. B.K. Matilal and A. Chakrabarti (eds.), *Knowing from Words.* Western and Indian Philosophical Analysis of Understanding and Testimony. 1994 ISBN 0-7923-2345-9
231. S.A. Kleiner, *The Logic of Discovery.* A Theory of the Rationality of Scientific Research. 1993 ISBN 0-7923-2371-8
232. R. Festa, *Optimum Inductive Methods.* A Study in Inductive Probability, Bayesian Statistics, and Verisimilitude. 1993 ISBN 0-7923-2460-9
233. P. Humphreys (ed.), *Patrick Suppes: Scientific Philosopher.* Vol. 1: Probability and Probabilistic Causality. 1994 ISBN 0-7923-2552-4
234. P. Humphreys (ed.), *Patrick Suppes: Scientific Philosopher.* Vol. 2: Philosophy of Physics, Theory Structure, and Measurement Theory. 1994 ISBN 0-7923-2553-2
235. P. Humphreys (ed.), *Patrick Suppes: Scientific Philosopher.* Vol. 3: Language, Logic, and Psychology. 1994 ISBN 0-7923-2862-0
Set ISBN (Vols 233–235) 0-7923-2554-0
236. D. Prawitz and D. Westerståhl (eds.), *Logic and Philosophy of Science in Uppsala.* Papers from the 9th International Congress of Logic, Methodology, and Philosophy of Science. 1994 ISBN 0-7923-2702-0
237. L. Haaparanta (ed.), *Mind, Meaning and Mathematics.* Essays on the Philosophical Views of Husserl and Frege. 1994 ISBN 0-7923-2703-9
238. J. Hintikka (ed.), *Aspects of Metaphor.* 1994 ISBN 0-7923-2786-1
239. B. McGuinness and G. Oliveri (eds.), *The Philosophy of Michael Dummett.* With Replies from Michael Dummett. 1994 ISBN 0-7923-2804-3
240. D. Jamieson (ed.), *Language, Mind, and Art.* Essays in Appreciation and Analysis, In Honor of Paul Ziff. 1994 ISBN 0-7923-2810-8
241. G. Preyer, F. Siebelt and A. Ulfig (eds.), *Language, Mind and Epistemology.* On Donald Davidson's Philosophy. 1994 ISBN 0-7923-2811-6
242. P. Ehrlich (ed.), *Real Numbers, Generalizations of the Reals, and Theories of Continua.* 1994 ISBN 0-7923-2689-X

243. G. Debrock and M. Hulswit (eds.), *Living Doubt*. Essays concerning the epistemology of Charles Sanders Peirce. 1994 ISBN 0-7923-2898-1
244. J. Srzednicki, *To Know or Not to Know*. Beyond Realism and Anti-Realism. 1994 ISBN 0-7923-2909-0
245. R. Egidi (ed.), *Wittgenstein: Mind and Language*. 1995 ISBN 0-7923-3171-0
246. A. Hyslop, *Other Minds*. 1995 ISBN 0-7923-3245-8
247. L. Pólos and M. Masuch (eds.), *Applied Logic: How, What and Why*. Logical Approaches to Natural Language. 1995 ISBN 0-7923-3432-9
248. M. Krynicki, M. Mostowski and L.M. Szczerba (eds.), *Quantifiers: Logics, Models and Computation*. Volume One: Surveys. 1995 ISBN 0-7923-3448-5
249. M. Krynicki, M. Mostowski and L.M. Szczerba (eds.), *Quantifiers: Logics, Models and Computation*. Volume Two: Contributions. 1995 ISBN 0-7923-3449-3
Set ISBN (Vols 248 + 249) 0-7923-3450-7
250. R.A. Watson, *Representational Ideas from Plato to Patricia Churchland*. 1995 ISBN 0-7923-3453-1
251. J. Hintikka (ed.), *From Dedekind to Gödel*. Essays on the Development of the Foundations of Mathematics. 1995 ISBN 0-7923-3484-1
252. A. Wiśniewski, *The Posing of Questions*. Logical Foundations of Erotetic Inferences. 1995 ISBN 0-7923-3637-2
253. J. Peregrin, *Doing Worlds with Words*. Formal Semantics without Formal Metaphysics. 1995 ISBN 0-7923-3742-5
254. I.A. Kieseppä, *Truthlikeness for Multidimensional, Quantitative Cognitive Problems*. 1996 ISBN 0-7923-4005-1
255. P. Hugly and C. Sayward: *Intensionality and Truth*. An Essay on the Philosophy of A.N. Prior. 1996 ISBN 0-7923-4119-8
256. L. Hankinson Nelson and J. Nelson (eds.): *Feminism, Science, and the Philosophy of Science*. 1997 ISBN 0-7923-4162-7
257. P.I. Bystrov and V.N. Sadovsky (eds.): *Philosophical Logic and Logical Philosophy*. Essays in Honour of Vladimir A. Smirnov. 1996 ISBN 0-7923-4270-4
258. Å.E. Andersson and N-E. Sahlin (eds.): *The Complexity of Creativity*. 1996 ISBN 0-7923-4346-8
259. M.L. Dalla Chiara, K. Doets, D. Mundici and J. van Benthem (eds.): *Logic and Scientific Methods*. Volume One of the Tenth International Congress of Logic, Methodology and Philosophy of Science, Florence, August 1995. 1997 ISBN 0-7923-4383-2
260. M.L. Dalla Chiara, K. Doets, D. Mundici and J. van Benthem (eds.): *Structures and Norms in Science*. Volume Two of the Tenth International Congress of Logic, Methodology and Philosophy of Science, Florence, August 1995. 1997 ISBN 0-7923-4384-0
Set ISBN (Vols 259 + 260) 0-7923-4385-9
261. A. Chakrabarti: *Denying Existence*. The Logic, Epistemology and Pragmatics of Negative Existentials and Fictional Discourse. 1997 ISBN 0-7923-4388-3
262. A. Biletzki: *Talking Wolves*. Thomas Hobbes on the Language of Politics and the Politics of Language. 1997 ISBN 0-7923-4425-1
263. D. Nute (ed.): *Defeasible Deontic Logic*. 1997 ISBN 0-7923-4630-0
264. U. Meixner: *Axiomatic Formal Ontology*. 1997 ISBN 0-7923-4747-X
265. I. Brinck: *The Indexical 'I'*. The First Person in Thought and Language. 1997 ISBN 0-7923-4741-2
266. G. Hölmström-Hintikka and R. Tuomela (eds.): *Contemporary Action Theory*. Volume 1: Individual Action. 1997 ISBN 0-7923-4753-6; Set: 0-7923-4754-4

267. G. Hölmström-Hintikka and R. Tuomela (eds.): *Contemporary Action Theory.* Volume 2: Social Action. 1997 ISBN 0-7923-4752-8; Set: 0-7923-4754-4
268. B.-C. Park: *Phenomenological Aspects of Wittgenstein's Philosophy.* 1998 ISBN 0-7923-4813-3
269. J. Paśniczek: *The Logic of Intentional Objects.* A Meinongian Version of Classical Logic. 1998 Hb ISBN 0-7923-4880-X; Pb ISBN 0-7923-5578-4
270. P.W. Humphreys and J.H. Fetzer (eds.): *The New Theory of Reference.* Kripke, Marcus, and Its Origins. 1998 ISBN 0-7923-4898-2
271. K. Szaniawski, A. Chmielewski and J. Woleński (eds.): *On Science, Inference, Information and Decision Making.* Selected Essays in the Philosophy of Science. 1998 ISBN 0-7923-4922-9
272. G.H. von Wright: *In the Shadow of Descartes.* Essays in the Philosophy of Mind. 1998 ISBN 0-7923-4992-X
273. K. Kijania-Placek and J. Woleński (eds.): *The Lvov–Warsaw School and Contemporary Philosophy.* 1998 ISBN 0-7923-5105-3
274. D. Dedrick: *Naming the Rainbow.* Colour Language, Colour Science, and Culture. 1998 ISBN 0-7923-5239-4
275. L. Albertazzi (ed.): *Shapes of Forms.* From Gestalt Psychology and Phenomenology to Ontology and Mathematics. 1999 ISBN 0-7923-5246-7
276. P. Fletcher: *Truth, Proof and Infinity.* A Theory of Constructions and Constructive Reasoning. 1998 ISBN 0-7923-5262-9
277. M. Fitting and R.L. Mendelsohn (eds.): *First-Order Modal Logic.* 1998 Hb ISBN 0-7923-5334-X; Pb ISBN 0-7923-5335-8
278. J.N. Mohanty: *Logic, Truth and the Modalities from a Phenomenological Perspective.* 1999 ISBN 0-7923-5550-4
279. T. Placek: *Mathematical Intiutionism and Intersubjectivity.* A Critical Exposition of Arguments for Intuitionism. 1999 ISBN 0-7923-5630-6
280. A. Cantini, E. Casari and P. Minari (eds.): *Logic and Foundations of Mathematics.* 1999 ISBN 0-7923-5659-4 set ISBN 0-7923-5867-8
281. M.L. Dalla Chiara, R. Giuntini and F. Laudisa (eds.): *Language, Quantum, Music.* 1999 ISBN 0-7923-5727-2; set ISBN 0-7923-5867-8
282. R. Egidi (ed.): *In Search of a New Humanism.* The Philosophy of Georg Hendrik von Wright. 1999 ISBN 0-7923-5810-4
283. F. Vollmer: *Agent Causality.* 1999 ISBN 0-7923-5848-1
284. J. Peregrin (ed.): *Truth and Its Nature (if Any).* 1999 ISBN 0-7923-5865-1
285. M. De Caro (ed.): *Interpretations and Causes.* New Perspectives on Donald Davidson's Philosophy. 1999 ISBN 0-7923-5869-4
286. R. Murawski: *Recursive Functions and Metamathematics.* Problems of Completeness and Decidability, Gödel's Theorems. 1999 ISBN 0-7923-5904-6
287. T.A.F. Kuipers: *From Instrumentalism to Constructive Realism.* On Some Relations between Confirmation, Empirical Progress, and Truth Approximation. 2000 ISBN 0-7923-6086-9
288. G. Holmström-Hintikka (ed.): *Medieval Philosophy and Modern Times.* 2000 ISBN 0-7923-6102-4
289. E. Grosholz and H. Breger (eds.): *The Growth of Mathematical Knowledge.* 2000 ISBN 0-7923-6151-2

290. G. Sommaruga: *History and Philosophy of Constructive Type Theory*. 2000 ISBN 0-7923-6180-6
291. J. Gasser (ed.): *A Boole Anthology*. Recent and Classical Studies in the Logic of George Boole. 2000 ISBN 0-7923-6380-9
292. V.F. Hendricks, S.A. Pedersen and K.F. Jørgensen (eds.): *Proof Theory*. History and Philosophical Significance. 2000 ISBN 0-7923-6544-5
293. W.L. Craig: *The Tensed Theory of Time*. A Critical Examination. 2000 ISBN 0-7923-6634-4
294. W.L. Craig: *The Tenseless Theory of Time*. A Critical Examination. 2000 ISBN 0-7923-6635-2
295. L. Albertazzi (ed.): *The Dawn of Cognitive Science*. Early European Contributors. 2001 ISBN 0-7923-6799-5
296. G. Forrai: *Reference, Truth and Conceptual Schemes*. A Defense of Internal Realism. 2001 ISBN 0-7923-6885-1
297. V.F. Hendricks, S.A. Pedersen and K.F. Jørgensen (eds.): *Probability Theory*. Philosophy, Recent History and Relations to Science. 2001 ISBN 0-7923-6952-1
298. M. Esfeld: *Holism in Philosophy of Mind and Philosophy of Physics*. 2001 ISBN 0-7923-7003-1
299. E.C. Steinhart: *The Logic of Metaphor*. Analogous Parts of Possible Worlds. 2001 ISBN 0-7923-7004-X

Previous volumes are still available.

KLUWER ACADEMIC PUBLISHERS – DORDRECHT / BOSTON / LONDON